OXFORD SERIES ON MATERIALS MODELLING

Oxford Series on Materials Modelling

Materials modelling is one of the fastest growing areas in the science and engineering of materials, both in academe and in industry. It is a very wide field covering materials phenomena and processes that span ten orders of magnitude in length and more than twenty in time. A broad range of models and computational techniques has been developed to model separately atomistic, microstructural and continuum processes. A new field of multi-scale modelling has also emerged in which two or more length scales are modelled sequentially or concurrently. The aim of this series is to provide a pedagogical set of texts spanning the atomistic and microstructural scales of materials modelling, written by acknowledged experts. Each book will assume at most a rudimentary knowledge of the field it covers and it will bring the reader to the frontiers of current research. It is hoped that the series will be useful for teaching materials modelling at the postgraduate level.

APS, London
RER, Livermore, California

1. M.W. Finnis: *Interatomic forces in condensed matter*
2. K. Bhattacharya: *Microstructure of martensite—Why it forms and how it gives rise to the shape-memory effects*
3. V.V. Bulatov, W. Cai: *Computer simulations of dislocations*
4. A.S. Argon: *Strengthening mechanisms in crystal plasticity*
5. L.P. Kubin: *Dislocations, mesoscale simulations and plastic flow*

Forthcoming:
T.N. Todorov: *Electrical conduction in nanoscale systems*
D.N. Theodorou, V. Mavrantzas: *Multiscale modelling of polymers*

Dislocations, Mesoscale Simulations and Plastic Flow

Ladislas P. Kubin
Laboratoire d'Étude des Microstructures, CNRS-ONERA, Châtillon, France

OXFORD
UNIVERSITY PRESS

OXFORD
UNIVERSITY PRESS

Great Clarendon Street, Oxford, OX2 6DP,
United Kingdom

Oxford University Press is a department of the University of Oxford.
It furthers the University's objective of excellence in research, scholarship,
and education by publishing worldwide. Oxford is a registered trade mark of
Oxford University Press in the UK and in certain other countries

First Edition published in 2013

Impression: 1

British Library Cataloguing in Publication Data
Data available

ISBN 978–0–19–852501–1

Printed and bound by
CPI Group (UK) Ltd, Croydon, CR0 4YY

PREFACE

Predicting the mechanical response of materials is the ultimate objective of dislocation theory. Some may even believe that this goal will never be reached. The challenge is, indeed, formidable, for several reasons that are stated in the introductory chapter of this book. After the period of euphoria of the years 1950–70, dislocation theory was confronted with several unsolvable problems. The question of bridging the huge range of time and length scales between electronic structure calculations of dislocation cores and the mechanical response of bulk materials reappeared by the end of the twentieth century under the name of *multiscale modelling*. The body of knowledge accumulated by physical metallurgy was renovated by the importation of new concepts from non-linear dynamics and statistical mechanics, the expansion of *ab initio* and atomistic simulations and the development of sophisticated experimental methods for investigating materials at all scales. Mesoscale simulations emerged in the same period with the objective of filling a strategic gap between atomic-scale studies and the continuum mechanical framework.

Like others in this series on materials modelling, the present book on *Dislocations, Mesoscale Simulations and Plastic Flow* is intended for graduate students and researchers in materials science and mechanical engineering. It discusses mesoscopic aspects of dislocations and crystal plasticity in elemental model materials, as well as the necessary background for establishing connections with other scales. Since this is a huge domain, the emphasis is on topics that require an update or a synthesis—areas of potential expansion and early seminal studies that are still of primordial importance. Abundant references are provided for further reading. Two books in this series are complementary to the present one and can also be consulted. The computational aspects of mesoscale simulations are treated in a chapter of *Computer Simulations of Dislocations* by Vassily V. Bulatov and Wei Cai. Alloying effects, on which little is said in the present book, are discussed in two chapters of *Strengthening Mechanisms in Crystal Plasticity* by Ali S. Argon.

The content of this book includes the individual and collective properties of dislocations in crystals and their mechanical response in the domain of temperatures where dislocation glide and do not climb. The approach adopted consists in going from early experimental and modelling studies to the present state of the art. General aspects are recalled in Chapter 1, whereas Chapters 2 and 3 discuss dislocations and plasticity in materials where plastic flow is governed, respectively, by interactions with obstacles and the lattice resistance.

There are few three-dimensional codes for dislocation dynamics simulations, but most of them are now public. For this reason, an objective of the present book is to present a comprehensive treatment of such simulations, their basic principles, limitations, achievements and potential developments. A guide to dislocation dynamics simulations and to the available existing codes is presented in Chapter 4. Chapter 5 illustrates the power of the method through a synthesis of representative applications, which range from nanomaterials to structural materials and from model to large-scale simulations.

Synthetizing the state of the art in a moving field is always a perilous exercise. Nevertheless, some rules for conducting simulations are not bound to change. It is important to draw a clear demarcation line between questions that we believe we understand, others that we do not understand well, or not at all, and others for which the answers are speculative. Performing simulations requires having in mind a specific problem to solve, a critical view of the relevant published literature and clear ideas about the input needed for this purpose. Mesoscale simulations reproduce to some extent the real behaviour of dislocations and provide a dynamic vision of plasticity, which is often absent from models, as well as numerical values that cannot always be accessed by other means. However, reproducing does not necessarily mean understanding. To become valuable, simulated outputs have to be connected to the world of real materials through comparison with experiment and, whenever possible, modelling.

Acknowledgements

I am grateful to many colleagues for clarifying discussions and critical comments on some parts of the book and for providing useful material. I would like to thank in particular Joel Bonneville, Daniel Caillard, Marc Fivel, Volker Mohles, Satish Rao, Klaus W. Schwarz, the group of Clain's Silicon Valley, Sandrine Brochard, Julien Godet, Laurent Pizzagalli and Jacques Rabier, as well as authors who kindly supplied me with original material and gave me their authorization for reproducing their published figures. I also would like to thank the series editors Robert E. Rudd and Adrian P. Sutton for their corrections and suggestions on several chapters of this book. Exchanges with Sönke Adlung, Jessica White and her predecessors at Oxford University Press were not only a source of stimulation, but also of answers to innumerable questions. My special gratitude goes to Benoit Devincre for a long-standing and fruitful collaboration and his constant help during the preparation of this book. I am also greatly indebted to Ronan Madec for providing me with many original figures. By the end of the 1980s, Fred Kocks suggested to Gilles Canova and myself that we should do some work together. This is how the first three-dimensional dislocation dynamics simulation was initiated. This book is dedicated to the memory of Gilles Canova.

CONTENTS

FIGURE PERMISSIONS

- Reproduced with permission of Elsevier

  Fig. 2.1a from Alshits and Indenbom 1986, ©1986, with permission of V. I. Alshits.
  Fig. 2.1b from Fusenig and Nembach 1993, ©1993.
  Fig. 2.11a from Bonneville and Escaig 1979, ©1979.
  Fig. 2.12 from Püschl 2002, ©2002.
  Fig. 2.14a,b from Rao *et al.* 2011, ©2011.
  Fig. 2.16 from Basinski and Basinski 1986, ©1986, with permission of S. J. Basinski.
  Fig. 2.19a from Chiu and Veyssière 2008, ©2008.
  Fig. 2.20a from Basinski 1974, ©1974.
  Fig. 2.21a from Liu *et al.* 1998, ©1998.
  Fig. 2.22 from Neubert and Rudolph 2005, ©2005.
  Fig. 2.27b from Kiener *et al.* 2008, ©2008.
  Fig. 3.1a from Smialek *et al.* 1970, ©1970.
  Fig. 3.3 from Louchet and Kubin 1975.
  Fig. 3.10a,b from Caillard 2010b, ©2010.
  Fig. 3.11 from Caillard 2010b, ©2010.
  Fig. 3.13 from Lachenmann and Schultz 1970, ©1970.
  Fig. 3.14 from Aono *et al.* 1980, ©1980.
  Fig. 3.18 from Farenc *et al.* 1995, ©1995.
  Fig. 3.21 from George 1997, ©1997.
  Fig. 3.26a from Rabier *et al.* 2004, ©2004.
  Fig. 5.8b from Nogaret *et al.* 2008, ©2008.
  Fig. 5.9a from Mohles 2004a, ©2008.
  Fig. 5.9b from Takahashi and Ghoniem 2008, ©2008.
  Fig. 5.16a from Motz *et al.* 2008, ©2008.
  Fig. 5.19 from Greer and De Hosson 2011, ©2011.
  Fig. 5.20a adapted from Rao *et al.* 2008, ©2008.
  Fig. 5.20b from Norfleet *et al.* 2008, ©2008.
  Fig. 5.22a,b from Motz *et al.* 2009, ©2009.
  Fig. 5.23 from Zhou *et al.* 2011, ©2011.
  Fig. 5.24 from Lee *et al.* 2009, ©2009.
  Fig. 5.25 from Mompiou *et al.* 2012, ©2012.
  Fig. C4 from Reid 1973, ©Pergamon Press 1973.

- Reproduced with permission of Taylor and Francis Ltd
  http://www.tandfonline.com

  Fig. 1.7a from Duesbery *et al.* 1992b.
  Fig. 2.21b from Huang and Winther 2007.
  Fig. 3.5a,b from Woodward and Rao 2001.
  Fig. 3.5c,d from Ito and Vitek 2001.
  Fig. 3.20a from Wessel and Alexander 1977.
  Fig. 3.23 from Omri *et al.* 1987.
  Fig. 3.24a,b from Imai and Sumino 1983.
  Fig. 3.25 from Yamashita *et al.* 1993.
  Fig. 5.8a from Rong *et al.* 2005.
  Fig. 5.14a,b from Déprés *et al.* 2004.

- Reproduced by courtesy of the American Physical Society

Fig. 2.13 from Rasmussen *et al.* 1997b, ©1997, with permission of K.W. Rasmussen.
Fig. 3.9 from Edagawa *et al.* 1997, with permission of K. Edagawa.
Fig. 3.22 from Kolar *et al.* 1996, with permission of H. R. Kolar.
Fig. 3.27 from Godet *et al.* 2006, with permission of J. Godet.
Fig. 4.2 from Clouet 2011b, ©2011, with permission of E. Clouet.
Fig. 5.1a from Wickham *et al.* 1999, with permission of K. W. Schwarz.
Fig. 5.13 from Gómez-García *et al.* 2006.
Fig. 5.21 from Tang *et al.* 2008, with permission of K. W. Schwarz.

- Reproduced with permission of John Wiley and Sons, Inc.

Fig. 2.22 (inset) from Rudolph 2005.
Fig. 3.1b from Werner 1987.
Fig. 3.2a from Richter 1970a.
Fig. 3.2b from Keh and Weissmann 1963.
Fig. 3.26b from Rabier *et al.* 2007.
Fig. 5.16b adapted from Fivel *et al.* 2011, with permission of M. Fivel.

- Reproduced with permission of the American Institute of Physics

Fig. 5.15a,b from Shehadeh *et al.* 2006, ©2006.
Fig. 5.17a,b from Schwarz *et al.* 2004, ©2004.
Fig. 5.18a from Schwarz and Tu 2009, ©2009.
Fig. 5.18b from Grydlik *et al.* 2012, ©2012.

- Reproduced with permission of Springer Science+Business Media

Fig. 2.23b from Mughabi 2009.
Fig. 2.24 from Hansen *et al.* 2010.
Fig. 3.6 from Moriarty *et al.* 2002b.

- Reproduced with permission of the publishers and authors cited

Fig. 1.1 from Read 1953, courtesy McGraw-Hill.
Fig. 1.6 from Haasen 1996, courtesy Cambridge University Press.
Fig. 1.7b from Hesselbarth and Steck 1992, courtesy Trans Tech Publications (Zurich-Durten).
Fig. 2.10 from Escaig 1968c, courtesy EDP Sciences.
http://publications.edpsciences.org/
Fig. 2.18 from Takeuchi 1975, courtesy Japan Institute of Metals.
Fig. 2.26 from Miguel *et al.* 2001, ©2001, courtesy Macmillan Publishers Ltd, *Nature* and M.C. Miguel.
Fig. 3.8 from Fig. 9, p. 121 of Moriarty *et al.* 2002a, courtesy Institute of Physics Publishing Ltd and J. Moriarty.
Fig. 4.5a from Fig. 9b, p. 10 of Aubry *et al.* 2011, courtesy Institute of Physics Publishing Ltd and W. Cai.
Fig. 4.5b from Fig. 3b, p. 7 of Aubry *et al.* 2011, courtesy Institute of Physics Publishing Ltd and W. Cai.
Fig. 4.9 adapted from Fig. 2, p. 691 of Van der Giessen and Needleman 1995, courtesy Institute of Physics Publishing Ltd and E. van der Giessen.
Fig. 5.10a from Csikor *et al.* 2007, courtesy of the American Association for the Advancement of Science and D. Weygand.
Fig. 5.10b from Devincre *et al.* 2008, courtesy of the American Association for the Advancement of Science.

All reasonable effort has been made to contact the holders of copyright in materials reproduced in this book. Any omissions will be rectified in future printings if notice is given to the publishers.

1

BACKGROUND AND DEFINITIONS

1.1 Introduction

Dislocations are linear crystal defects that carry plastic flow at the microscopic scale. Their history is recalled by Hirth (1985) in a brief but well-documented article. From the very beginning, the objective of dislocation theory was to establish in physical terms a predictive theory of plastic flow in crystalline solids. In spite of decades of efforts, such a theory does not exist yet. The reasons why it is so are well understood and can be listed as follows.

- As dislocations are metastable defects with high energy, their density cannot be drawn from equilibrium thermodynamics, as is the case for point defects; it is governed by the thermomechanical history of the considered material. When a sample is plastically deformed and unloaded, the dislocation microstructure that was generated under load relaxes but does not vanish, as plastic strains are not reversible. Most of the mechanical work done during plastic flow (about 90%) is dissipated into heat and the rest is principally stored in the form of elastic dislocation energy. Hence, dislocation-based plasticity is a highly dissipative, irreversible process driven far from equilibrium. In such conditions the Onsager-type treatment of linear return to equilibrium does not apply. In the non-linear regime, the thermodynamics of irreversible processes turns out to be of little help (Hillert and Agren, 2006).
- Driven systems brought far from equilibrium frequently exhibit spontaneous forms of organization. Such self-organization was extensively investigated in chemistry, biology, hydrodynamics, solid-state physics and many other domains (Haken, 2004). In deforming crystals, self-organization manifests itself by the emergence of various periodic, pseudo-periodic or more complex dislocation patterns and by the occurrence of an intermittent, scale-free behaviour of the mobile dislocations. The experimental aspects and the present understanding of these collective phenomena are discussed in Sections 2.7 and 5.4.
- The classical theory of dislocations, which is based largely on linear elasticity, stands at the mesoscale. The latter is intermediate between the atomic or electronic structure scales and the scale at which plasticity can be

described within a continuum framework. Establishing connections between the various approaches to dislocations and plasticity led in the 1990s to the emergence of modelling across scales (Phillips, 2001). *Multiscale modelling* benefited from a parallel expansion at all scales of methods for characterizing materials (Robertson *et al.*, 2011). Since then, continuous progress was recorded on this grand challenge problem. In particular, the range of dimensions on which dislocation-based plasticity is now developed is stretching from micron and sub-micron dimensions (Section 5.5) to the typical laboratory sample and the huge dimensions of interest to geophysicists (Cordier *et al.*, 2012).

In what follows it is assumed that the bases of dislocation theory are known at an elementary level. The literature on dislocations comprises several introductory textbooks (Read, 1953; Weertman and Weertman, 1992; Hull and Bacon, 2001). The classical monograph by Hirth and Lothe (1992) provides an exhaustive exposition of the elastic theory of dislocations. This theory can also be treated by means of a method developed by Mura (1991) in his book on the micromechanics of defects in solids, the eigenstrain method. Basic notions on the connection between dislocations and the mechanical response are found in other classical textbooks (Cottrell, 1953; Friedel, 1967; Nabarro, 1967) and in more recent ones (Mughrabi, 1993; Caillard and Martin, 2003). The series *Dislocations in Solids*, to which the name of Frank Nabarro is attached, contains 96 chapters devoted to all aspects of dislocation properties in a wide range of material classes (a table of contents of the sixteen volumes is found in Hirth and Kubin, 2010). Continuum approaches of plasticity that combine dislocation micromechanics and computing can be found in several books (Hosford, 1993; Raabe *et al.*, 2004; Dunne and Petrinic, 2005). Two books in this series are complementary to the present one. The first one (Argon, 2008) presents a broad overview of strengthening mechanisms in plasticity, and the second one presents models and methods for computer simulations of dislocations at all scales (Bulatov and Cai, 2006). Finally, a dynamic vision of dislocations is essential for understanding plasticity. This vision is provided in a book by Messerschmidt (2010*a*), which discusses dislocation dynamic processes and their relation to plastic deformation in metals, alloys, compounds, semiconductors, ceramics and quasicrystals (see Messerschmidt 2010*b* for video clips of *in situ* straining in a high-voltage electron microscope).

This introductory chapter is devoted to a short summary of the elementary properties of dislocations. It provides an introduction to Chapters 2 and 3, which discuss the body of knowledge that further serves as input to dislocation dynamics (DD) simulations (Chapter 4) and as a base for carrying out a variety of applications (Chapter 5). The topics covered are restricted to what constitutes the bulk of existing simulation studies—specifically, dislocation glide in model single crystals at low and medium temperatures. The materials that have drawn the most experimental and theoretical attention are face-centred cubic (fcc) metals, body-centred cubic (bcc) transition metals, hexagonal closed-packed (hcp) transition metals and diamond-cubic (dc) silicon.[1]

The forthcoming sections recall basic dislocation core properties and the application of linear elasticity to self-energies and self-stresses, dislocation interactions, Peach–Koehler forces and the line tension approximation. The definition of the effective stress on a dislocation leads to dislocation velocities, which can be obstacle-controlled or lattice-controlled, and to the Orowan law. The last two sections introduce multiscale modelling and DD simulations.

1.2 Dislocation core properties

This section discusses the properties of dislocation cores and of two thermally activated core mechanisms that significantly influence the mechanical response: cross-slip and the lattice resistance to dislocation motion. Heterogeneous dislocation nucleation at free surfaces is another important atomic-scale mechanism, as will be seen in Sections 3.4.9.2 and 5.5.5.4.

1.2.1 Core energy and structure

Figure 1.1 shows a portion of perfect dislocation loop in a crystallographic slip plane bounded by free surfaces. This dislocation line may have been created by a Volterra process (Volterra, 1907) that involved cutting the crystal along a bounded surface in a crystallographic plane, translating the upper part of the crystal with respect to the lower part by an elementary lattice vector **b**, the Burgers vector, and restoring the continuity of the crystal by pasting together the two lips of the cut.

In the continuum, the dislocation is a singularity line performing the accommodation between the sheared and non-sheared areas in the crystal. An atomic view reveals a highly distorted core region of typical radius $r_o = 1{-}2b$ (Fig. 1.1). One can notice in this figure how the core structure evolves when the dislocation character moves from screw to edge. In the large external region surrounding the core, the atomic displacements are small enough to be treated by linear elasticity.

Dislocation glide involves significant atomic rearrangements in the core region. This conservative motion propagates the elementary shear **b** in the slip plane. At the macroscopic scale, plastic flow is the composition of all the elementary shears produced by dislocations gliding on all active slip systems.

The core energy of a dislocation, E_{core}, is its formation energy in a perfect crystal; it depends on its line direction and on the core radius. The core radii and

[1] Throughout this book numerical estimates are performed using materials constants given in Appendix B for the most common fcc metals, simple hcp metals, Si and Ge, and in Table 3.1 for the most common bcc metals.

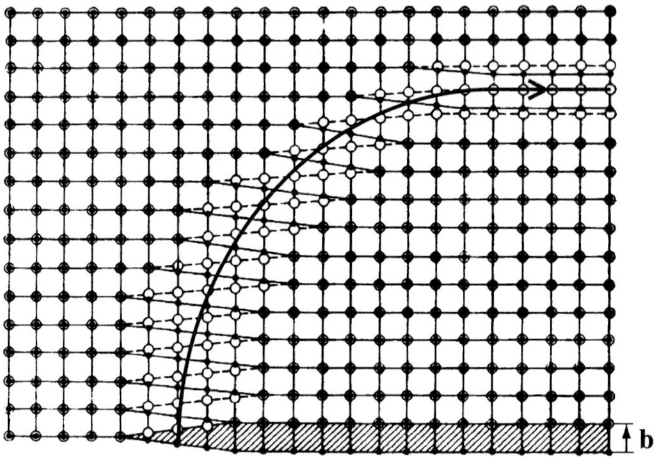

FIG. 1.1. A portion of perfect dislocation loop in a cubic crystal encloses an area sheared by an elementary translation **b**, the Burgers vector. Empty circles represent the atoms located immediately above the sheared area and full circles the atoms immediately below. The Burgers vector is defined according to a convention described in Section C.1. After Read 1953.

energies are in principle obtained from *ab initio* or atomistic calculations using potentials fitted to *ab initio* data. A few recent determinations are concerned with screw dislocations in bcc metals, which have compact cores. Specifically, we consider Ta and Mo (see Table 3, p. 26 of Yang *et al.* 2010) and Fe (Clouet, 2009, 2011*b*). The core radii and energies are $r_o = 1.75b$ and $E_{core} \approx 0.63$ eV/b for Ta, $r_o = 1.75b$ and $E_{core} \approx 0.82$ eV/b for Mo and $r_o = 1.2b$ and $E_{core} \approx 0.5$ eV/b for Fe. These values are compared to the corresponding elastic strain energies in Section 1.3.1.

In many cases the core energy of a perfect dislocation is reduced by a relaxation leading to more or less extended and anisotropic structures. For large enough planar spreadings the core is dissociated into partial dislocations with Burgers vectors that are no longer lattice translations, bounded by planar defects. In compact crystallographic structures, dissociation preferentially occurs in close-packed planes. The latter have the lowest stacking fault energy (SFE), γ, because the fault preserves the first and second atomic neighbours. The paradigmatic example is that of the splitting of perfect dislocations in the close-packed planes of fcc, dc and some hcp crystals. For a $\frac{1}{2}[\bar{1}10]$ perfect dislocation in a (111) plane, the reaction between Burgers vectors is written

$$\frac{1}{2}[\bar{1}10] \rightarrow \frac{1}{6}[\bar{2}11] + \frac{1}{6}[\bar{1}2\bar{1}] \tag{1.1}$$

A glissile configuration is formed, which consists of pairs of Shockley partials of mixed character with Burgers vectors of magnitude b_s. The dissociation width, d_o, results from an equilibrium between the repulsion between the Shockley partials, approximately $\mu b_s^2/4\pi d_o$, and the restoring force per unit surface γ exerted by the stacking fault. Hence, in dimensionless form, $d_o/b_s \approx \mu b_s/4\pi\gamma$. As discussed in Section B.1, discrepancies may be found between elastic predictions in linear anisotropic elasticity and transmission electron microscopy (TEM) measurements. They arise when the dissociation widths are so small, typically 1.5 nm and less, that the fields of the partials significantly overlap. A compilation of material constants related to this dissociation mode is given in Table B.1. Two examples of glissile, planar dissociations are shown in Fig. 1.2. In these high-resolution TEM micrographs, the dislocation lines are perpendicular to the view plane.

In bcc metals, the screw dislocations of Burgers vector $\mathbf{b} = \frac{1}{2}\langle 111\rangle$ are at the intersection of three $\{110\}$ and three $\{112\}$ potential slip planes. There is no dissociation and the screw dislocation cores exhibit a rather compact and almost isotropic structure, as mentioned above, with ternary symmetry (Section 3.2.2). In hcp metals, the relaxation preferentially occurs in the basal or prismatic slip planes, depending upon the electronic structure of the material considered (Section 3.3.1).

1.2.2 Cross-slip and the lattice resistance

1.2.2.1 Thermal activation At finite temperatures, thermal fluctuations allow dislocations to surmount localized energy barriers during their motion. The

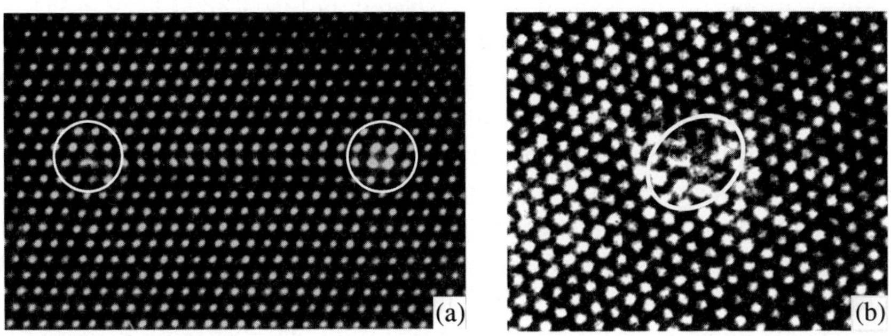

FIG. 1.2. High-resolution TEM images of dislocation cores (white circles). (a) Silicon. Dissociation of a mixed $\frac{1}{2}[1\bar{1}0]$ dislocation into two Shockley partials separated by a stacking fault in a (111) plane. The (111) planes are horizontal and viewed end-on; their spacing is 0.314 nm. Courtesy J. Thibault-Penisson. (b) Glissile core of an edge dislocation in a prismatic slip plane of hcp titanium. The view plane is the basal plane and the prismatic planes are seen end-on; their spacing is 0.255 nm. The core extension is about 1.2 nm, but the partials cannot be resolved. Courtesy S. Naka.

mesoscopic framework used to describe thermally activated dislocation motion is summarized in Appendix A. It is shown that thermal fluctuations provide dislocations with an energy of about 23–25k_BT, where k_B is the Boltzmann constant and T the absolute temperature. In practice, this means that a localized energy barrier of 1 eV can be overcome with the help of thermal fluctuations at a temperature of about 450 K. Above this temperature, which we call the *athermal transition temperature*, the obstacle no longer opposes any resistance to dislocation motion. Below this temperature, both mechanical work and thermal fluctuations contribute to the probability of activation. One can find in the literature short introductions to this framework (Schoeck, 2000; Hull and Bacon, 2001; Argon, 2008), as well as more complete descriptions (Kocks *et al.*, 1975; Schoeck, 1980; Taylor, 1992; Caillard and Martin, 2003).

The cross-slip mechanism in fcc metals (Section 2.4) and the overcoming of the lattice resistance (Section 3.2.4) involve energy barriers of heights in the range of 1–2 eV. Thus, dislocation core mechanisms are in general at the origin of thermally activated material properties. Conversely, the overcoming of mesoscopic obstacles by elastic mechanisms involves changes in dislocation line energy that are too prohibitive to be thermally activated (Section 1.3.1).

1.2.2.2 Cross-slip In fcc crystals the Burgers vector of the perfect screw dislocation, $\frac{1}{2}\langle 110\rangle$, is lying at the intersection of two $\{111\}$ planes. Upon gliding on one slip plane, the dislocation can freely cross-slip to the other one, for instance to bypass an obstacle. It can further cross-slip back to its initial slip plane, in which case one has a double cross-slip event. These two mechanisms, which also occur when dislocations are dissociated, are illustrated by Fig. 1.3.

For dissociated dislocations, however, cross-slip requires the overcoming of an energy barrier. As the Shockley partials always have a mixed orientation, they can hardly cross-slip (Section 2.4.1). The minimum energy path consists of a constriction followed by a local recombination of the screw core in its slip plane and its immediate redissociation in the cross-slip plane. At $T = 0$ K, the recombination occurs under stress along the whole screw line. At finite temperatures a critical cross-slipped nucleus is formed and expands sideways along the line. Cross-slip occurs when the motion of a dissociated screw dislocation in its slip plane is obstructed by an obstacle. Thus there are several types of cross-slip mechanisms, each having its own activation energy. During the deformation of fcc crystals, cross-slip induces substantial rearrangements of the dislocation microstructure around a temperature-dependent threshold stress. Then, the transformation of the core involves an energy barrier, which increases with decreasing SFE and approximately ranges from 0.8 to 2 eV in fcc metals (this can be verified from Fig. 2.12). Section 2.4 contains an extended discussion of experiments, modellling and atomistic simulations dedicated to cross-slip in fcc metals.

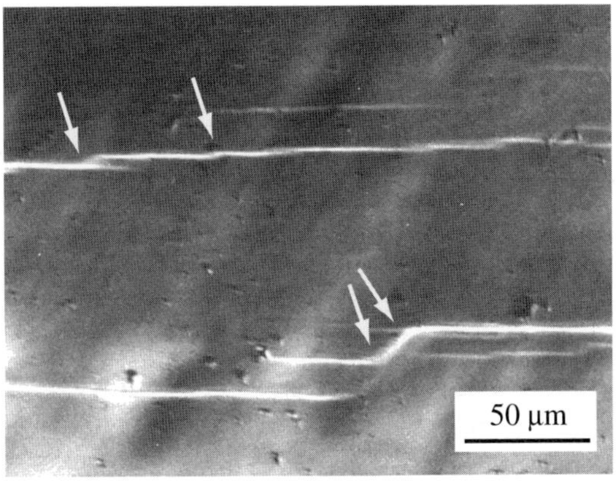

FIG. 1.3. Slip traces at the surface of a copper single crystal deformed at room temperature, showing cross-slip events and a double cross-slip event (arrows). At a higher magnification, these slip traces appear to be composed of finer traces left at the specimen surface by the emergence of dislocations that glide and cross-slip in {111} planes. After Bonneville 1985, courtesy J. Bonneville.

1.2.2.3 Lattice resistance When a dislocation line is parallel to dense atomic rows, its equilibrium position is located midway between these rows in order to minimize the atomic distortions of the core. To move between one stable position to the other, the line has to overcome an unstable high-energy position. The period of this energy profile, the Peierls (or Peierls–Nabarro) energy barrier, is that of the lattice along the direction of motion of the dislocation. At $T = 0\,\mathrm{K}$, the dislocation line moves rigidly under stress until it reaches a core configuration of maximum energy. The corresponding resistive stress, τ_P, is called the *Peierls stress*. The *Peierls energy* is the maximum value of the energy barrier. At finite temperatures and under stress, a small length of line can reach a saddle point with the help of thermal fluctuations. The two atomic-scale defects connecting this configuration to the rest of the line are called *kinks*. Once formed, the *kink-pair* expands sideways, thus transferring the whole line to its next stable position. In that case, the maximum resistive stress is denoted $\tau_P(T)$. In metals, the kinks are extremely mobile and kink-pair formation is the leading process. In covalent materials, it is not so; the migration energy of the kinks is larger than the formation energy of kink-pairs. The term *lattice resistance* is used in what follows to denote in global terms the interaction between dislocations and the crystal lattice.

In fcc metals there is practically no lattice resistance (Section 2.2.1). For this reason, the athermal interactions between dislocations and other defects, including other dislocations, govern plastic flow. In bcc metals, the lattice resistance is strong on screw dislocations (Table 3.1) and quite weak on dislocations of other characters, as well as on kinks. The same holds for some hcp metals (Section 3.3), notably Ti and Zr. In silicon, the strength and strong directionality of the covalent bond induces a high lattice resistance on dislocations of all characters, plus a strong secondary lattice resistance on the kinks (Section 3.4.3). In all these materials the lattice resistance governs the plastic properties below the athermal transition temperature. Chapter 3 is devoted to a discussion of the available and missing connections between dislocation core structures, mesoscale modelling, experimental observations and the mechanical response of bulk crystals.

1.3 Elastic properties of dislocations

As most of the energy of a dislocation is stored outside its core, all the related dislocation properties can be treated quite accurately in an elastic continuum. The elastic theory of dislocations is a very powerful tool, which, in contrast to atomic scale properties, presents the enormous advantage of describing properties that are transposable from one material to the other.

This section recalls the traditional approach to dislocations in linear isotropic elasticity (see, e.g., Hull and Bacon 2001). Displacements, stresses and energies are obtained from closed integrals over dislocation loops of arbitrary shape. The results are fully untractable and methods for deriving analytical forms, for instance for polygonal loops or straight segments, are throughly exposed by Hirth and Lothe (1992). Anisotropic elasticity, which brings further complications, is useful for treating simple specific configurations (Section 4.2.8). The solutions adopted to incorporate the elastic properties of dislocations in DD simulations are discussed in Section 4.2.

1.3.1 Strain energy of a straight dislocation

At rest, each infinitesimal volume of a solid is in mechanical equilibrium and the equilibrium condition is expressed in terms of displacements. We consider a straight, infinite perfect dislocation in a coaxial cylinder of large radius R. Introducing the displacement induced by the Burgers vector, the equilibrium condition yields the displacement field in the cylinder, from which the strain field of the dislocation can be deduced. For instance, the strain field of a screw dislocation, written in polar coordinates with the axis $\mathbf{z}$ collinear to the line, reduces to

$$\varepsilon_{\theta z} = \varepsilon_{z\theta} = \frac{b}{4\pi r}, \tag{1.2}$$

where r is the radial distance to the dislocation. The stress field, or *self-stress*, of the dislocation follows from Hooke's law. In isotropic elasticity, one has

$$\sigma_{\theta z} = \sigma_{z\theta} = 2\mu\varepsilon_{\theta z} = \frac{\mu b}{2\pi r}, \tag{1.3}$$

where μ is the shear modulus. Dislocation fields are long-ranged, so that all dislocation lines in a microstructure mutually interact. During the creation of the dislocation by a Volterra process, the strain energy of an infinitesimal volume element incrementally grows by $\sigma_{ij}\mathrm{d}\epsilon_{ij} = \sigma_{ij}\mathrm{d}\sigma_{ij}/2\mu$, leading to a final value $\sigma_{ij}^2/4\mu$ for each component of the stress tensor. The self-energy is obtained by summing the strain energy components and integrating over the radius r. To avoid the divergency at $r = 0$, an inner cut-off radius r_o is introduced, which excludes the core region. With a large enough outer cut-off radius $R = R_e$, the external surfaces can be considered as being stress-free. In the general case of a straight dislocation of arbitrary character, the *self-energy* per unit length of line, E_{el}, is written

$$E_{el} = E_o(\alpha) \ln \left(\frac{R_e}{r_o} \right), \tag{1.4}$$

where α is the dislocation character, that is, the angle between the Burgers vector and the line, and $E_o = A(\alpha)\mu b^2/4\pi$ is the pre-logarithmic energy factor. The character-dependent coefficient A accounts for the orientation dependence of the line energy. The logarithmic term in eqn. 1.4 is a consequence of the $1/r$ dependency of the stress fields. Another consequence is that the self-energy of a single dislocation line has no upper bound when the cut-off radius indefinitely increases.

One can easily verify that $A = 1$ for a screw dislocation. For an edge dislocation, a calculation similar to the one outlined above yields $A = 1/(1 - \nu)$, where $\nu \approx 1/3$ is the Poisson ratio. Thus, the line energy of an edge dislocation is about 1.5 times larger than that of a screw dislocation. This reflects the predominance in its stress tensor of hydrostatic stress components, as can be guessed from Fig. 1.1. For a mixed dislocation, the orientation factor is obtained by decomposing the dislocation line into two coaxial lines, one of screw character and the other of edge character. By decomposing the Burgers vector into its screw and edge components, $b\cos\alpha$ and $b\sin\alpha$, and summing the two energies, one obtains

$$A(\alpha) = \frac{1 - \nu\cos^2\alpha}{1 - \nu} \tag{1.5}$$

The self-energy per unit length of a straight segment is almost identical to that of an infinite segment (Hirth and Lothe 1992, p. 161). The core radius tends towards r_o with increasing length. For fast estimates this small difference is ignored for most of the time. In order-of-magnitude calculations it is often assumed that $E_{el} \approx \mu b^2$ or $\mu b^2/2$.

A three-dimensional microstructure consisting of connected dislocation segments in total density ρ is characterized by a length scale $\rho^{-1/2}$. Consider for instance a regular distribution of segments of length $\bar{\ell}$, arranged along the edges of a cubic lattice. Within a geometric factor the dislocation density is

$\rho \propto \bar{\ell}/\bar{\ell}^3 = 1/\bar{\ell}^2$. Hence, $\bar{\ell}$ scales like $\rho^{-1/2}$. Irrespective of the type of microstructure considered, this length can be viewed as representing an average segment length or, equivalently, an average distance between segments. One may then assume that the stress field of a dislocation is screened by that of a dislocation of opposite sign located on average at a distance of the order of $\bar{\ell}$. Thus, for estimating a typical value of the elastic energy of a dislocation segment, it seems reasonable to take an outer cut-off radius $R_e \approx \rho^{-1/2}$. This approximation is equivalent to assuming that the logarithmic factor of the line energy weakly depends on the spatial organization of the microstructure. It is sometimes useful for introducing characteristic lengths in continuum models for dislocation densities.

For a dislocation density of $10^{12}\,\text{m}^{-2}$, which is typical of a moderately deformed crystal, $R_e \approx 1\ \mu\text{m}$. With an inner cut-off radius $r_o \approx b$, the strain energy of screw dislocations yielded by eqn. 1.4 is $E_{el} \approx 0.66\mu b^3$ per length b of line. Using the data given in Table 3.1, the elastic energies of screw dislocations in bcc Ta, Mo and α-Fe are estimated to respectively 6, 11.2 and 4.46 eV/b, which can be compared to the computed core energies of, respectively, 0.63, 0.82 and 0.5 eV/b mentioned in Section 1.2.1. Thus, the usual presumption that core energies are about one order of magnitude smaller than elastic energies (Cottrell, 1953) is verified, at least in bcc metals. Since the entropy of dislocations is negligible in metals (Cottrell, 1953), it is justified to calculate the energetics of mesoscopic dislocation processes by means of linear elasticity. Owing to the weak logarithmic dependence of the energy on core radius, the core energy only amounts to about 15% of the elastic energy for $R_e \approx 50$ nm. A more detailed discussion on the approximations made upon neglecting core effects is given in Section 4.2.4. Finally, when reliable atomic-scale calculations of the core radius are not available, which is most often the case, the value of the core radius is in general set to $r_o = b$.

1.3.2 Force on a dislocation

Consider a solid with finite dimensions in mechanical equilibrium, which contains a dislocation and to which an external stress is applied. The total free energy of the solid and the loading device depends on the elastic energy stored by the dislocation in the solid. Hence, it depends on the position of the dislocation line and the associated displacement field—for instance, with respect to a free surface. According to a general principle of thermodynamics, the dislocation tends to move in such a way as to reduce the total free energy of the system. This allows defining what is called a *configurational* force, $\boldsymbol{F}$, which is applied to the dislocation line. This force per unit line is directed along the negative gradient of the total elastic energy. If η is a general coordinate along the line, it is given by

$$\boldsymbol{F}(\eta) = -\nabla_\eta E_{tot} \tag{1.6}$$

The configurational force on a dislocation accounts for forces that are distributed over the whole elastically deformed volume. The stress field that induces dislocation motion may result from an external loading, but it can as well result from any field of internal origin. The book by Phillips (2001) may be consulted for an extended discussion of configurational forces in materials.

A simple example of configurational force is provided by the case of a screw dislocation parallel to and at a distance d from a planar free surface. Its self-energy taken with an outer cut-off radius $R_e = d$ decreases when the dislocation glides towards the surface. Inserting eqn. 1.4 with $r_o = b$ into eqn. 1.6, one has

$$F = \frac{\mu b^2}{4\pi d}, \tag{1.7}$$

where the force is directed towards the surface. This result is traditionally demonstrated by placing on the other side of the surface a mirror image of the screw dislocation, with opposite Burgers vector (Hull and Bacon, 2001). The stress fields of the two screws are of opposite sign and cancel out at the surface, making it stress-free. However the calculation of such *image forces* leads in general to very complex, often untractable, analytical solutions (Hirth and Lothe, 1992). The methods developed for ensuring stress-free conditions in DD simulations of dislocation microstructures are outlined in Sections 4.4.2 and 4.4.3.

We now consider a dislocation line, like the one depicted in Fig. 1.1, which is submitted to a uniform resolved shear stress in its slip plane. A simple geometric reasoning based on the atomic displacements during a virtual shear step allows determination of the Burgers vector and the direction of motion of the line (Hull and Bacon, 2001). For analyses of dislocations in the TEM or in DD simulations it is more convenient to adopt a convention based on crystallography that gives the direction of the Burgers vector in stress-free conditions. In Section C.1, the most usual convention is defined and a general expression for the configurational force induced by a local stress tensor $\boldsymbol{\sigma}$ is derived. The *Peach–Koehler* relation (eqn. C.2), which was first established by Peach and Koehler (1950), is written

$$\boldsymbol{F} = (\boldsymbol{\sigma}.\mathbf{b}) \times \boldsymbol{\ell}_u \tag{1.8}$$

The Peach–Koehler force is a force per unit length of dislocation line. The cross-product expresses that it is always normal to the unit vector along the line direction, $\boldsymbol{\ell}_u$, irrespective of its character. Thus, it has glide and climb components, which are obtained by simple projections (Section C.1). The scalar product expresses the projection of the stress tensor in the direction of the Burgers vector. This relation is fairly general; the crystal may be finite or infinite and elastically isotropic or anisotropic. Equation 1.8 also holds irrespective of the dislocation velocity up to the velocity of transverse shear waves (Mura 1991, see eqns. 39.8 and 39.21). Under a resolved shear stress τ in the slip plane, the glide force is written

$$F = \boldsymbol{\tau}.\mathbf{b} \tag{1.9}$$

As could be expected, the sign of the force depends on the respective directions of the resolved stress and the Burgers vector. For simple configurations of straight dislocations, the interaction forces can be deduced directly from the self-stresses and eqn. 1.9, or from the line energies with the help of eqn. 1.6. As far as dislocation microstructures are concerned, eqn. 1.8 is usually combined with the *superposition principle*. This principle of linear elasticity, which was used above to obtain eqn. 1.5, tells that all displacements, stresses or strains that are superimposed at a given point are additive whatever their origin. It follows that the total Peach–Koehler force on a given dislocation line element can be directly estimated from the sum of all local contributions to the internal stress field on this element.

Considering only dislocation interactions, there are two types of contributions to the Peach–Koehler force on a given line element. The first one arises from the fields of all remote dislocation lines. An integration over these lines provides analytical expressions that are in general too complex to be of practical use. A very useful and quite accurate simplification, which is used in DD simulations, consists in decomposing the curved lines into a succession of straight segments. The interaction stresses can then be calculated analytically (Section 4.2.6). At this step it may be useful to recall that these interaction stresses on dislocations should not be confused with internal stresses, which are defined everywhere in a crystal. Indeed, dislocations do not evenly sample the internal stress field because they tend to reduce the total elastic energy of the crystal by moving towards regions of low internal stress.

The second contribution arises from the self-stress, that is, from elements of the line where the force is calculated. It is obtained on a given line element by integrating on a closed path the stress fields generated by the whole line. The interaction stress of two elements indefinitely increases when their distance along the integration path decreases. It follows that the interaction stress of the given line element with itself diverges. Several procedures have been proposed to regularize this singular behaviour of the self-stress (Section 4.2.5). They all require defining a parameter that plays the role of a core radius.

1.3.3 Line tension

In the absence of applied stress, a dislocation of arbitrary shape straightens out in order to minimize its line length and strain energy. A glide loop spontaneously collapses for the same reason. The notion of *line tension* of a dislocation is based on an analogy with the line tension of a stretched elastic string. The line tension,

Γ, is then defined as the derivative of the self-energy with respect to the line length L

$$\Gamma = \frac{\mathrm{d}E_{el}}{\mathrm{d}L} \tag{1.10}$$

A curved dislocation line is pinned at its ends and bows in its slip plane. We consider a small element AB of this line, of length δs (Fig. 1.4), and ignore its interactions with the rest of the line. Its line tensions are tangential and their components along the normal to the curved element tend to reduce its length. They are equilibrated by a uniform shear stress τ, which applies to the considered element a radial Peach–Koehler force $F = \tau b$ per unit length (eqn. 1.9).

The curvature radius is given by $R = \delta s/\delta\theta$. The Peach–Koehler force is $F = \tau b \delta s$ and the radial components of the line tension apply to the segment a resistive radial force $F_\Gamma = 2\Gamma \sin(\delta\theta/2) \approx \Gamma\,\delta\theta$. At equilibrium, $F = F_\Gamma$, from which we deduce

$$\tau = \frac{\Gamma}{bR} \tag{1.11}$$

For estimating the line tension of the line element, use is made of the self-energy per unit length as given by eqn. 1.4. The simplest approximation consists in ignoring the orientation dependence of the line energy and taking $R_e = R$ (Friedel, 1967; Weertman and Weertman, 1992). The inner cut-off radius is usually taken as that of a straight dislocation, r_o. The dependence of the curvature radius on stress (eqn. 1.11) is then given by

$$\tau \approx \frac{\mu b}{4\pi R} \ln\left(\frac{R}{r_o}\right) \tag{1.12}$$

Like the self-energy, the line tension is often approximated in a rough manner by assuming that the logarithmic term is constant. This leads to $\Gamma \approx \mu b^2$ or $\mu b^2/2$.

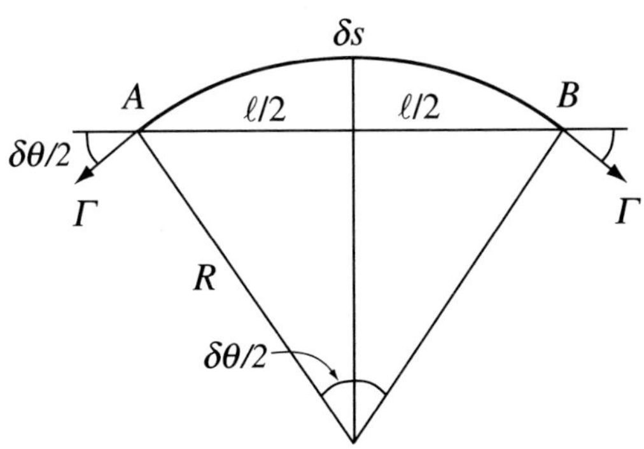

FIG. 1.4. Line tension on a curved element of dislocation AB of radius R.

In eqn. 1.12 the dependence of the line energy on orientation is neglected, and a bowing segment or a dislocation loop assumes a circular shape. This defines the rather crude approximation of the constant, or isotropic, line tension. Anisotropic shapes arise because under an increasing stress the expanding near-screw elements tend to rotate towards the screw direction in order to minimize the total elastic energy. As a consequence, the energy-minimizing shape of a dislocation loop or a curved segment is elongated along the screw direction of minimum energy. It follows that under a given stress the curvature radius of a dislocation loop is the largest for nearly screw orientations, and so is its line tension according to eqn. 1.11.

A more consistent manner of taking into account the orientation dependence of the line tension consists in determining the equilibrium shape that minimizes the contributions of both the line length and the local orientations to the energy (de Wit and Koehler 1959). For small bulges in an initially straight infinite line, a Taylor expansion of the self-energy per unit line is performed to the second order in α (Hirth and Lothe 1992, p. 175). This yields a relation equivalent to the one that expresses the surface tension of a two-dimensional surface.

$$\Gamma = E_{el}(\alpha) + \frac{\partial^2 E_{el}(\alpha)}{\partial \alpha^2}, \tag{1.13}$$

By developing this equation with the help of eqn. 1.5, one obtains

$$\Gamma = \frac{\mu b^2}{4\pi}\left(\frac{1+\nu-3\nu\sin^2\alpha}{1-\nu}\right)\ln\left(\frac{R}{r_o}\right), \tag{1.14}$$

where, actually, the core radius slightly differs from that of an infinite line and becomes identical to it at the limit where the curvature radius R is infinite. According to eqn. 1.14, a weakly bowed segment in equilibrium is $(1+\nu)/(1-2\nu) \approx 4$ times more elongated in the screw direction than in the edge direction. Foreman (1967) compared static 2D simulations of the equilibrium shapes of a curved segment pinned at its ends obtained from eqn. 1.13 and from a regularized self-stress calculation (Brown 1964, see Section 4.2.5.1). The elongation factor was found to decrease in both cases from 4 at low stresses to about 2 at high stresses. In both cases, the critical shear stress at which a segment irreversibly expands is given to a good approximation by the semi-empirical form

$$\tau_c = A\frac{\mu b}{2\pi\ell}\left(\ln\left(\frac{\ell}{r_o}\right) + B\right), \tag{1.15}$$

where ℓ is the length of the segment. Taking $\nu = 1/3$, A is about 1 for edge and 1.5 for screw segments. B is a constant taking into account the fields generated by two semi-infinite side arms. This relation is sometimes used to check critical Frank–Read stresses in DD simulations (see e.g., Weygand *et al.* 2002).

The self-stress of an element of line is non-local, since it depends on the shape of the whole line. In contrast, the line tension is a local quantity that depends only on the curvature radius at the considered position. As it involves no singularity, it provides a means for approximating the interaction stress of an element with itself. Once the interaction stresses with all other elements are included, a non-local approximation for the self-stress of a curved line is obtained.

1.3.4 Line tension strengthening

Figure 1.5 shows a curved segment of a dislocation line pinned by a periodic row of strong obstacles, which is in equilibrium under a uniform shear stress. The obstacles gives rise to a force F on each adjoining segment, which equilibrates its line tension Γ. The total force is $F_{tot} = 2F \sin\theta$.

With increasing stress, a critical stress value τ_c may be reached such that the segments overcome the maximum resistance of the obstacles. The critical angle θ_c characterizes the strength of the obstacles. In the constant line tension approximation with $\Gamma \approx \mu b^2$, the constant curvature radii are given by $R_c = \mu b/\tau_c$ (eqn. 1.11). If d is the distance between obstacles, $R_c = d/2\sin\theta_c$ and the critical stress is given by

$$\tau_c = 2\sin\theta_c \frac{\mu b}{d}, \tag{1.16}$$

where the interactions between curved segments are ignored.

Many models for strengthening by line tension effects are based on more refined versions of eqn. 1.16. Some very strong obstacles cannot be overcome when $\theta = \pi/2$. At this point a segment reaches a semi-circular shape and it can no longer accommodate a stress increase by a uniform reduction of its curvature radius. Hence it becomes unstable and expands under a critical stress $\tau_c \approx 2\mu b/d$. This situation is met, for instance, with non-shearable precipitates; it leads to *Orowan strengthening* (Argon, 2008). In another type of configuration, the very

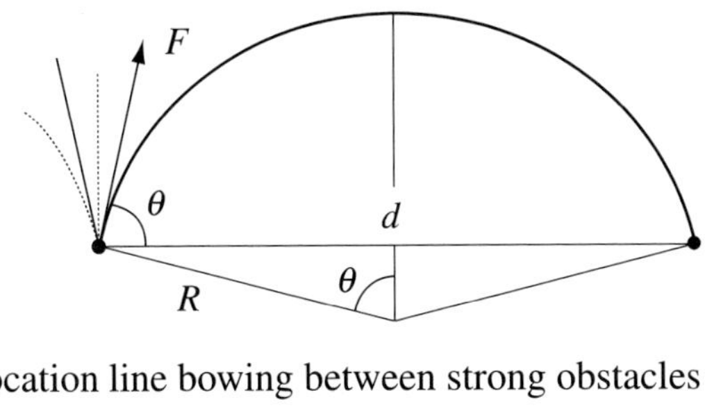

FIG. 1.5. A dislocation line bowing between strong obstacles under stress.

strong pinning points are nodes connecting a segment to a sessile dislocation configuration perpendicular to its slip plane. The segment can then rotate freely around the nodes and emit dislocation loops according to the well-known *Frank–Read source* mechanism. For a bowed dislocation gliding between interfaces separated by a distance $d = R$, the effect of confinement can be very substantial (Section 5.5.3). In copper, with $\theta_c = \pi/2$ and $d = 10\,\text{nm}$, one obtains $\tau_c \approx 2\,\text{GPa}$.

A well-known expression is recovered in the presence of *forest dislocations*, that is, dislocations that pierce the slip plane of mobile dislocations. Although forest dislocations are strong obstacles on average, their critical angle is smaller than $\pi/2$. For a total forest density ρ_f, one has $\bar{d} \approx \rho_f^{-1/2}$ (Section 1.3.1). As a result, one obtains a Taylor-like relation (Taylor, 1934), which is called a *dislocation strengthening* relation (Section 2.5.2)

$$\tau_c \approx 2 \sin\theta_c \mu b \sqrt{\rho_f} \tag{1.17}$$

Forest dislocations are obstacles less strong than source pinning points, but they are involved in numerous contact interactions. Hence, they split dislocation lines into much smaller pinned segments with higher critical stresses. Therefore, in contrast to what is sometimes believed, dislocation sources do not govern plastic flow in bulk crystals (Section 2.5.2). They do govern plastic flow only in a few particular cases, notably in small-scale pillars (Section 5.5.5.2) or in silicon, where the initial density of sources is usually very low (Section 3.4.5).

In summary, the various formulations of the line tension, anisotropic or isotropic, orientation-dependent or not, allow calculating the shape of a dislocation line under stress. Advantage was taken of this possibility in many applications, especially for the modelling of dislocation strengthening mechanisms. In general, a satisfactory compromise is obtained between simplicity and accuracy. Nevertheless, the concept of local line tension is based on an analogy and its physical justification is uncertain.

1.4 Dislocation velocity

This section successively discusses the effective stress on dislocations, which drives their velocity, the main classes of mechanisms governing this velocity and the classical Orowan relation for the macroscopic strain rate.

1.4.1 Effective stress

As was seen in Section 1.3.2, the total resolved force per unit length on a dislocation element, F, is the algebraic sum of Peach–Koehler forces deriving from externally applied or internal stress fields. When dealing with glide velocities it is often more convenient to use the related *effective stress*, which is the net resolved stress in the slip plane projected along the direction of the Burgers vector (eqn. 1.9). This stress is the sum of several contributions, of which some may be resistive. The stress τ_{ext} derives from the externally applied stress tensor and there is necessarily a self-stress τ_{self} for dislocations that bow under stress. In addition, there may be interaction stresses with other dislocation lines, τ_{int}, stresses arising from other obstacles than dislocations, τ_{obs}, and image stresses, τ_{im}. The effective glide stress on a line element, τ_g^*, is written

$$\tau_g^* = \tau_{self} + \tau_{ext} + \tau_{int} + \tau_{obs} + \tau_{im} \tag{1.18}$$

Additional contributions can be added when necessary. The lattice resistance requires a specific treatment (see next section) and should preferably not be included in eqn. 1.18. When the effective stress τ^* on a dislocation is positive, it governs its velocity.

1.4.2 Governing mechanisms

The velocity of a single dislocation is modelled in different manners according to the magnitude of the effective stress and the nature of its interactions with various crystal defects or the lattice itself. Investigating the motion of a straight, infinite dislocation in an elastic continuum leads to a few useful definitions. For a screw dislocation, the conditions of dynamic equilibrium in isotropic elasticity reduce to a wave equation of the form (Friedel, 1967; Hirth and Lothe, 1992)

$$\Delta \mathbf{u} = \frac{1}{c_t{}^2} \frac{\partial^2 \mathbf{u}}{\partial t^2}, \tag{1.19}$$

where $c_t = (\mu/\rho_o)^{1/2}$ is the velocity of the transverse shear waves, which reduces to the sound velocity in isotropic elasticity, and ρ_o is the density of the medium. For large uniform velocities the solutions of eqn. 1.19 exhibit a relativistic-like behaviour. The displacement field undergoes a relativistic contraction, and the total elastic energy diverges like $[(1 - v^2/c_t^2)]^{-1/2}$ when the velocity v approaches the sound velocity. This allows derivation of the *mass of a dislocation* at rest, $m_o = E/c_t^2 \approx \rho_o b^2$ per unit length. This value is equivalent to the mass of a row of atoms per unit length, so that the inertia of dislocations moving with moderate velocities is rather small.

During its motion a dislocation dissipates energy into the lattice by various interactions with the phonons modes. The dissipation mechanisms drastically change around a velocity $v = c_t/2$. Above this velocity a highly dynamic regime prevails (Weertman and Weertman, 1980; Alshits and Indenbom, 1986; Hirth and Lothe, 1992). There are also particular stable elastic solutions of eqn. 1.19, which describe dislocations moving faster than sound (Gumbsch and Gao 1999). Extremely high dislocation velocities can be achieved in shock loading experiments, under high-frequency oscillations or near fast-moving cracks.

In current laboratory tests the sample is imperfect and dislocations interact with various types of obstacles. Therefore, their velocities are much smaller than $c_t/2$. This can be seen from Fig. 1.6, which reproduces a compilation of velocity versus stress responses obtained in different materials.

In Fig. 1.6 the straight intermediate slopes are found at rather low stresses in the alkali metal halides. They are found at higher stresses in metals and semiconductors exhibiting a moderate lattice resistance at the test temperature (Si, Ge, Nb), or in tungsten where the lattice resistance is much higher. The inflections at velocities approaching the $\mathrm{m\,s^{-1}}$ range probably reflect an increased contribution from phonon damping. Other features cannot be interpreted easily. For instance, nickel and copper seem to exhibit velocities that differ by orders of magnitude under stresses of a few MPa, which seems unlikely for fcc metals. This type of measurement, which was very popular in the 1960s, is misleading because dislocation velocities actually depend on effective resolved stresses (eqn. 1.18). Within a physical approach it is preferable to make a distinction between different velocity regimes, with which are associated different types of effective stress versus velocity relations.

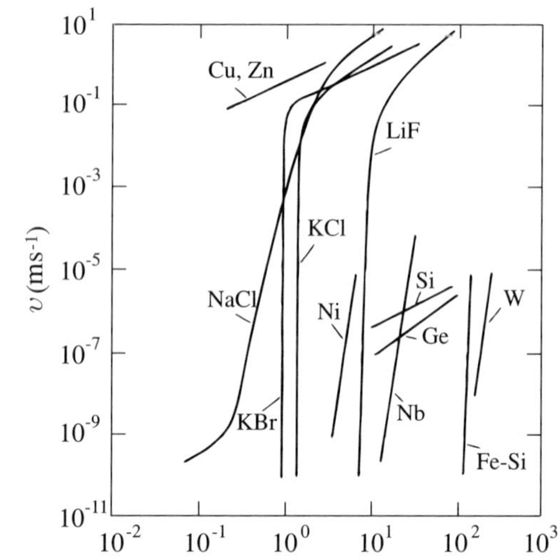

FIG. 1.6. Compilation of experimental data on dislocation velocities at room temperature, except for Ge (450° C) and Si (850° C). The measurements were performed by etching the crystal surfaces to reveal the dislocation positions before and after an applied stress pulse. After Haasen 1996 reproduced by courtesy of Cambridge University Press.

- In the absence of lattice resistance and between two localized obstacles, the free-flight dislocation velocity is governed by phonon drag, except at very low temperatures (Section 2.2.2.3). When the localized obstacles are forest dislocations, the resulting interactions constitute strong obstacles; they govern the average velocity of mobile dislocations. This situation is schematically depicted in Fig. 1.5, where line tension or self-stress effects are predominant. Then, it is essentially the small difference between the stresses τ_{ext} and τ_{self} in eqn. 1.18 that drives the dislocation velocity (Section 4.3.2.1). For long travelled paths, the average velocities can be orders of magnitude smaller than the free-flight velocities. These properties hold in the so-called quasi-static regime, which is defined in the next section. At the microscopic scale, the conventional vision of dislocations moving individually from one obstacle to the other was substantially modified by the recent discovery of dislocation avalanches (Sections 2.7.2 and 5.4.1).

- When the lattice resistance is strong, dislocations move by the kink-pair mechanism and do not bow under stress. As there is no obstacle comparable to the lattice in terms of strength and density, the dislocation velocity is essentially governed by the thermally activated overcoming of the lattice resistance (Section 1.2). Then, the effective stress mainly reduces to the difference between the applied stress and the lattice resistance $\tau_P(T)$. The relation between effective stresses and dislocation velocities for the model materials exhibiting these properties are discussed in Section 3.4.6 for silicon crystals and Section 3.2.4.3 for metallic crystals. Their implementation in DD simulations is discussed in Section 4.3.2.2.

- Real crystals contain, of course, many other types of obstacles which may affect dislocation velocities in different manners. A full coverage of the related strengthening effects is provided by Argon (2008). In earlier reviews, one can find models carried out within an elastic framework (Kelly and Nicholson, 1971), especially for solid solution strengthening (Haasen, 1979; Suzuki, 1979; Neuhäuser and Schwink, 1993) and precipitation strengthening (Gerold, 1979; Nembach, 1996; Martin, 1998). Regarding the interaction between dislocations and grain or interphase boundaries, many different situations have been analysed (Sutton and Balluffi, 1996). DD simulations of several strengthening mechanisms are described in Chapter 5.

1.4.3 Orowan's law

Orowan's law provides a connection between the elementary shears associated with dislocation motion and the macroscopic shear strain rate. In its most commonly used form it implicitly assumes deformation by single slip. By definition, a dislocation that sweeps the entire area A of a slip plane in a sample of height h

and volume V produces a shear $\gamma = b/h = bA/V$. Consider a dislocation segment of length ℓ moving with a velocity v and sweeping an elementary area $\mathrm{d}A = \ell v \mathrm{d}t$ during an infinitesimal time interval $\mathrm{d}t$. The elementary shear is

$$\mathrm{d}\gamma = b\frac{\ell}{V}\, v\mathrm{d}t \tag{1.20}$$

The density of moving dislocations can be considered constant during the time interval $\mathrm{d}t$. If $\mathcal{L}$ is the total length of mobile dislocations, the instantaneous mobile dislocation density is $\rho_m = \mathcal{L}/V$. The integration of eqn. 1.20 at the scale of a slip system in a bulk crystal is then performed by introducing the mobile density and replacing the dislocation velocities by their average value $\bar{v}$. This leads to the familiar form of Orowan's law

$$\dot{\gamma} = \rho_m b\bar{v}, \tag{1.21}$$

where $\mathrm{d}\dot{\gamma} = \mathrm{d}\gamma/\mathrm{d}t$ is the shear strain rate. Orowan's law is valid in conditions of multiple slip provided that adequate geometrical coefficients are introduced to account for the composition of shears in the different slip systems. General relations taking into account the tensorial character of shears and the associated rotations (Section C.2.2) are given by Argon (2008) (see Chapter 2) in component notation and by Zbib and Diaz de la Rubia (2002) in dyadic notation (eqns. 3.16).

In conditions typical of laboratory tests conducted at a constant imposed elongation rate, the total shear rate is about $10^{-4}\,\mathrm{s}^{-1}$. Equation 1.21, taken with $b = 3 \times 10^{-10}\,\mathrm{m}$ and $\rho_m = 10^{12}\,\mathrm{m}^{-2}$, yields an average velocity of $0.3\,\mu\mathrm{m\,s}^{-1}$. Such tests are carried up to strain rates about $10^{-2}\,\mathrm{s}^{-1}$ ($\bar{v} \approx 30\,\mu\mathrm{m\,s}^{-1}$), which approximately marks the upper limit of the quasi-static regime. These typical average values hold for bulk materials irrespective of the mechanism that governs plastic flow. When deformation is predominantly controlled by elastic mechanisms like dislocation intersections, plastic flow occurs, as mentioned above, by a succession of dislocation avalanches. As a consequence, transient velocities can be much higher than the values averaged over a long time interval (Section 5.4.1). For strain rates above the limit of the quasi-static regime, the combined effects of inertia and increased phonon drag drastically modify the microstructure and the mechanical response. DD simulations of the deformation of fcc crystals at high strain rates are discussed in Sections 4.3.2.1 and 5.4.5.

Equation 1.21 is sometimes integrated with respect to time, which yields

$$\gamma = \rho_m b \int \bar{v} dt = \rho_m\, b\bar{L}, \tag{1.22}$$

where $\bar{L}$ is a dislocation path travelled by a constant mobile density. As dislocation lines expand and multiply while they move, the mobile dislocation densities and velocities are not independent variables. Hence, eqn. 1.22 makes sense only in very few specific cases, like the motion of edge segments during the pre-yield stage of pure bcc metals at low temperatures (Section 3.2.1) or the motion of screw dislocations in the channels of persistent slip bands (Section 2.6.6).

1.5 Multiscale modelling

The range of length scales involved in multiscale modelling spans about eight orders of magnitude from interatomic spacings to current sizes of laboratory samples. On the other hand, more than fifteen orders of magnitude separate the characteristic time scale of atomic vibrations ($\nu_D^{-1} \approx 10^{-13}$ s, where ν_D is the Debye frequency) and the typical duration of a conventional laboratory test, typically 200 s for obtaining an elongation of 20% under a typical imposed strain rate of $10^{-3}\,\mathrm{s}^{-1}$. As a result, multiscale modelling carries out information passing through three different domains, each having its specific theoretical approaches and simulation tools, namely solid state physics, physical metallurgy and solid mechanics.

At the smallest scale of investigation one finds first-principles electronic structure calculations (Martin, 2004), which are most of the time based on the density functional theory (DFT). New solutions have been devised to implement boundary conditions accounting for the long-range fields of dislocations in the small simulated volumes (Woodward, 2005). As discussed in Chapter 3, the obtention of realistic screw dislocation cores in bcc metals (Ismail-Beigi and Arias, 2000; Woodward and Rao, 2001) brought a new insight into the Peierls stress as well as new open questions. A new generation of atomistic calculations emerged, which are based on quantum approaches or on potentials fitted to electronic structure calculations. Such simulations are useful for investigating volumes that are not accessible to first-principles calculations. Another method for obtaining first-principles or atomistic information on dislocation cores consists in computing generalized SFE surfaces, or *γ-surfaces*, which are the two-dimensional energy surfaces generated by the rigid shearing of a crystal along all possible directions of crystallographic planes. This method and its applications, in particular to the Peierls–Nabarro model, are discussed in Appendix D.

Atomistic simulations of dislocation processes by molecular dynamics (MD) simulations (see Bulatov and Cai 2006) are well suited for the examination of fast athermal events. Consider a single dislocation in a million-atom MD simulation. The linear dimension of the simulated volume is $d \approx 40$ nm and the mobile density is typically $\rho_m = d/d^3 \approx 10^{15}\,\mathrm{m}^{-2}$. The loading conditions impose a strain rate in the range of $10^8\,\mathrm{s}^{-1}$, and the dislocation velocity predicted by Orowan's law (eqn. 1.21) is about $400\,\mathrm{m\,s}^{-1}$. In current laboratory tests the imposed strain rates are at least ten orders of magnitude smaller, and the average dislocation velocities are typically a fraction of $\mu\mathrm{m\,s}^{-1}$ (Section 1.4.3). Thus, MD simulations operate in highly dynamic conditions and are not well suited for the investigation of quasi-static dislocation mechanisms. This is, for instance, the case for the overcoming of localized obstacles with the help of thermal fluctuations, which becomes less efficient at high dislocation velocities (Section A.2). In consequence, the critical stress for moving dislocations in a MD simulation is the one corresponding to an effective temperature significantly smaller than the nominal one. In general, investigations of thermally activated dislocation processes are preferentially carried out by static methods.

The nudged elastic band (NEB) is a static method which involves no thermal activation but is useful for finding energy minimum paths and saddle points (Jónsson *et al.*, 1998; Lu and Kaxiras, 2005). The initial and final configurations are given and a transition path is first guessed. At discrete positions along this path, replicas of the system are introduced, which are connected by fictitious springs in order to ensure an even sampling of the transition path. Energy minimization is then carried out by allowing the replicas to relax in directions perpendicular to the path. This method has been revisited several times (Henkelman and Jónsson, 2000; Henkelman *et al.*, 2000; Zhu *et al.*, 2007; Gröger and Vitek, 2012). In molecular statics methods, load increments are applied to a configuration and the total energy is allowed to relax between two loading steps.

The kinetic Monte Carlo (KMC) method (see Bulatov and Cai 2006) is used for determining the properties of a dislocation line when they are governed by a set of competing processes with different activation probabilities. Thus, it provides a connection between atomistic and mesoscopic processes. It is typically applied to dislocation motion by the kink-pair mechanism (Sections 3.2.4.2 and 3.4.3) and to the interactions of dislocations with diffusing solute atoms (Section 5.3.1) or irradiation defects (Section 5.3.2).

The other end of the spectrum is represented by solid mechanics. In the present context the strength of this framework resides in its ability to solve rigorously the *boundary value problem*, that is, the conditions for mechanical equilibrium of a body submitted to internal and externally applied fields. The solutions have to be computed numerically. The most widespread computational method is the finite element (FE) method (Dunne and Petrinic, 2005), of which the full development for engineering studies can be traced back to the 1960s. In its classical versions, continuum mechanical modelling does not incorporate length scales. Thus, it cannot account for size effects and the basic scaling properties of dislocation microstructures. At present, the connection with the mesoscale is developed increasingly with the help of FE codes that include a crystal plasticity formulation. In addition to boundary conditions, these codes treat crystallographic slip and can, therefore, accommodate continuous dislocation-based models for plastic flow (Section 2.6.7).

The mesoscale domain is that of the defects microstructure, as observed, for instance, by conventional TEM imaging techniques. At this scale the objective is to unravel the mechanisms that govern three major quantities characterizing the stress versus strain curves. The yield stress is the critical stress at which most of the recorded strain becomes a plastic strain, as a significant density of dislocations starts overcoming obstacles to their motion. The flow stress is the stress needed to obtain sustained plastic flow at a given plastic strain. The strain hardening rate is the derivative of the flow stress with respect to the plastic strain. The main achievement of dislocation theory consists in explaining the yield stress of model metals, alloys and compounds in terms of their microstructure. As far as the relation between microstructure and strain hardening is concerned, the patchwork of models and experimental results accumulated over the years by physical metallurgy contains valuable contributions that are too often ignored.

Three-dimensional DD simulations do not have yet a very long history. Their objective is to fill the gap between atom-scale and fully continuum studies, and to assist theoretical modelling. DD simulations essentially compute the properties of dislocation populations in terms of the elastic theory of dislocations. Core properties are included using the available information that can be obtained from the lower scales, from experiment, and from a few mesoscopic models.

Two other types of simulation methods were applied to mesoscale dislocation dynamics. They provide elegant and efficient solutions for dealing with the elastic properties of dislocations, but are in development as far as other material-dependent properties are concerned. This is the case for the level-set method, which has already produced generic results (Xiang *et al.*, 2003; Quek *et al.*, 2006; Chen *et al.*, 2010). The phase-field method (Khachaturyan, 1983; Steinbach, 2009), which treats coupled elastic and chemical fields, was expanded to investigate micromechanical problems (Wang *et al.*, 2005). A critical discussion of its application to dislocation dynamics is given by Bulatov and Cai (2006). New types of phase-field models may enhance the ability of this method to treat dislocations at all scales (Wang and Li, 2010).

1.6 Introduction to 3D DD simulations

1.6.1 Brief historical sketch

The treatment of the self-stress of a curved segment initiated by Brown (1964) allowed assessment of the accuracy of the most refined line tension models. It was further improved by several authors (Section 4.2.5.1). The analytical treatment of the solutions was, however, too complex to be carried out in the general case. Thus, two-dimensional (2D) simulations of the equilibrium shapes of dislocations segments bowing under stress in their slip plane were initiated by the end of the 1960s (Foreman, 1967; Bacon, 1967). To simplify the calculations the curved segments were decomposed into a succession of straight segments connected by nodes. These early studies substantially contributed to the modelling of the critical stress of Frank–Read sources (Foreman, 1967), the Orowan bypassing of precipitates (Bacon *et al.*, 1973) and the overcoming of various types of obstacles. More references can be found in a further work by Duesbery *et al.* (1992*b*), where several innovations are introduced (Section 4.2.5.1).

Another type of 2D simulation was proposed in the mid-1980s, wherein parallel, infinite dislocations of same Burgers vector are viewed end-on. The simulation by Lépinoux and Kubin (1987) was based on an extension of cellular automata methods (Chopard and Droz, 1998), which further led to three-dimensional (3D) lattice-based simulations. The simulation by Ghoniem and Amodeo (1988) was constructed like a mesoscopic MD simulation. End-on simulations allow investigation of spontaneous forms of organization in a population of dislocations that move under stress and interact via their long-range elastic fields (Fig. 1.7*b*). In spite of the omission or oversimplification of major dislocation mechanisms, 2D end-on simulations may sometimes be useful (Section 5.4.3). 2D in-plane simulations may also be useful to examine dislocation mechanisms and interactions that do not involve cross-slip, like the one shown in Fig. 1.7*a*. They present more interest when the obstacles to dislocation motion are three-dimensionally distributed (Section 5.3.4).

The development of the first 3D DD simulation (Kubin *et al.*, 1992), which was a *lattice-based* simulation (Section 4.2.2), aimed at obtaining more realistic descriptions of individual and collective dislocation behaviour. The simulations that were further developed (Zbib *et al.*, 1998; Schwarz, 1999; Ghoniem and Sun, 1999) were mostly *nodal* simulations (Section 4.2.2). As it takes a lot of time and effort to develop a 3D DD simulation, the number of existing codes is limited. However, several of them are easily accessible (Section 4.5) and are attracting a continuously growing number of users.

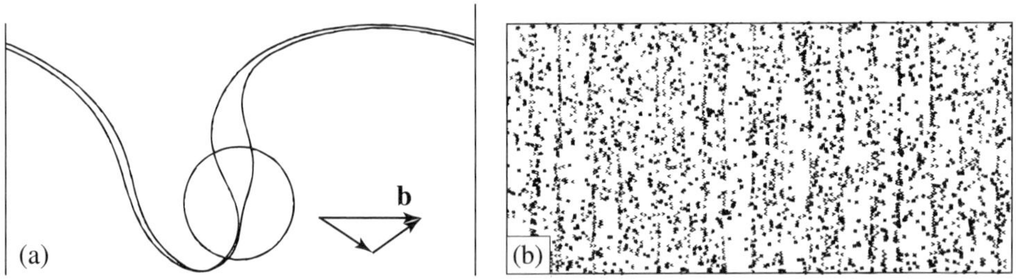

FIG. 1.7. Two-dimensional DD simulations. (a) *In-plane* simulation. Critical shape of a dissociated screw dislocation in an fcc crystal, which shears a periodic linear array of coherent obstacles with diameter 300*b*. From Duesbery *et al.* 1992*b*. (b) *End-on* simulation of cyclic deformation showing the formation of an array of edge dipolar walls. From Hesselbarth and Steck 1992.

1.6.2 Further implementation

Implementing the elastic properties of dislocations with reasonably good accuracy no longer poses serious difficulties. This is not so for material-dependent mechanisms, which are incorporated in the form of constitutive rules adapted to a mesoscopic description. These so-called *local rules* (Section 4.3) deal in particular with thermally activated core properties like cross-slip or the overcoming of a strong lattice resistance. They are based on Arrhenius forms, through which temperature is introduced in the simulations.

Substantial progress was achieved by solving the question of the boundary conditions, for which several solutions are now available (Section 4.4). Periodic boundary conditions allow investigating large volumes. For investigating the properties of finite volumes, DD simulations are coupled to finite element (FE) or equivalent codes, which numerically implement the conditions for mechanical equilibrium.

DD simulations can be used for studies of elementary dislocation configurations and interactions. Such *model simulations* can be refined down to the smallest dimension compatible with the use of linear elasticity, that is, a few lattice spacings. Figure 1.8 shows a comparison of a small-scale configuration of bowed segments in aluminium investigated by molecular statics and DD simulations.

Simulations in volumes representative of the behaviour of bulk materials are called *large-scale simulations* (Fig. 1.9). They are carried out on simulation cells of typical dimensions 10–20 μm containing realistic initial densities (Section 4.1).

A particular feature of DD simulations is the increase of computing load entailed by dislocation multiplication and the increasing number of elastic interactions to be computed. The evolution of DD simulations has benefited from

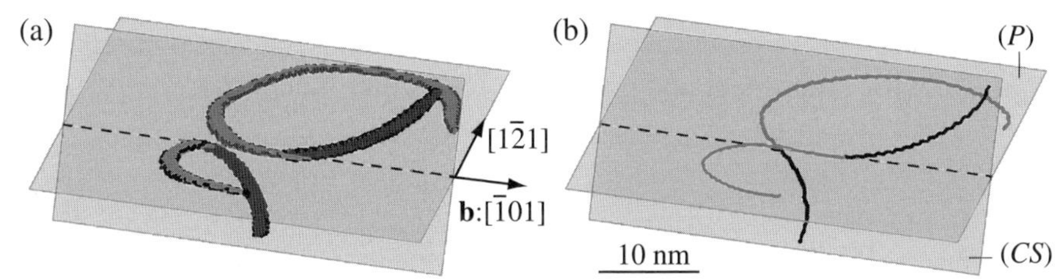

FIG. 1.8. Model configuration of two dislocation lines pinned at their ends in aluminium bowing under a stress of $\approx 1.7\times10^{-2}\mu$. Each line is composed of two segments with same Burgers vector, lying respectively in a primary slip plane *P* (dark lines) and its cross-slip plane *CS* (lighter lines). (a) Static atomistic simulation showing the partials and stacking faults. (b) The same configuration investigated by DD simulations in same conditions, but with perfect dislocations. The two critical configurations look practically identical. After Madec *et al.* 2003.

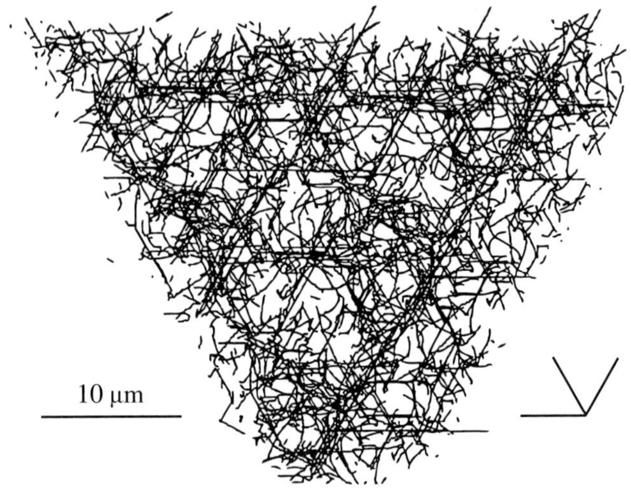

FIG. 1.9. {111} foil of thickness 5 μm extracted from the large-scale simulation with periodic boundary conditions of a copper crystal strained along a ⟨100⟩ axis (the ⟨110⟩ directions are indicated). An early stage of dislocation cell formation can be observed after a tensile plastic strain of $\varepsilon = 4 \times 10^{-3}$ at room temperature under a strain rate of $20\,\mathrm{s}^{-1}$. From Madec *et al.* 2002*b*.

progress in the available computing power and programming solutions such as parallelization and fast multipole algorithms (Section 4.2.7). Validation tests are necessary to define simulations conditions that yield an optimum compromise between computing load and numerical accuracy. This sets an upper limit to the number of time steps and to the maximum plastic strain that can be reached. Such strains vary from fairly small to a few percent, depending upon the problem considered. This is generally sufficient for obtaining relevant numerical data, comparison with experiment and guidelines for further modelling.

2

OBSTACLE-CONTROLLED PLASTIC FLOW

2.1 Outline

This chapter deals with dislocations that experience a negligibly small lattice resistance. In consequence, their free-flight velocity can be high but their average velocity is limited by their interaction with various defects present in the microstructure. In this category of dislocations leading to obstacle-controlled plastic flow, one finds:

- Dislocations in fcc metals, which are dissociated into Shockley partials in close-packed $\{111\}$ slip planes.
- Dislocations in materials with high lattice resistance, like dc covalent materials, at temperatures above their athermal transition temperature (Section A.2). At this temperature, the activation energy for overcoming the lattice resistance is entirely furnished by thermal fluctuations. The dislocation microstructures and the mechanical response are then akin to those found in fcc metals.
- Dislocations in bcc metals, where the strong lattice resistance on screw dislocations vanishes at the athermal transition temperature, whereas non-screw dislocations essentially behave as in fcc metals at all temperatures.
- Dislocations gliding in the basal plane in simple (i.e., divalent) hcp metals such as Be, Mg, Zn or Cd. As in fcc metals, the dislocations are dissociated into Shockley partials in the compact basal plane. The lattice resistance is small in this plane, but stronger in the prismatic planes. The situation is more complex in transition hcp metals (Section 3.3.1). Depending upon the electronic structure, some behave like divalent metals and others exhibit a more or less strong lattice resistance on screw dislocations in all slip planes. The situation is then similar to that of bcc metals.

Some properties of this class of dislocations are examined first at the level requested for further implementation in DD simulations. The emphasis is on pure fcc crystals, which have been the subject of extensive experimental and theoretical investigations. Evidences for a very small lattice resistance are first discussed. Between obstacles, the free-flight velocity of the dislocations is mainly limited by dissipative interactions with phonon modes, from which one can deduce a simple stress–strain velocity law (Section 2.2). Dislocation interactions and reactions are essentially athermal processes, which can be conveniently treated by elasticity theory (Section 2.3). Finally, the cross-slip of screw dislocations is the most important core process in materials with no lattice friction (Section 2.4).

At an upper level of description, elementary mechanisms should ideally be combined to yield an insight into collective dislocation behaviour of bulk crystals and, further, into their mechanical response (size effects in confined fcc materials are discussed in Chapter 5). The relation between flow stress and dislocation density is now well understood in terms of short-range dislocation interactions (Section 2.5). The resolved stress–strain curves of fcc single crystals exhibit orientation-dependent stages of work hardening, which are associated with characteristic dislocation patterns (Section 2.6). The discovery of the intermittent and scale-free nature of elementary slip events, in bulk or confined materials, recently opened a new chapter in the study of collective dislocation processes. In the last ten or fifteen years the traditional modelling of hardening and patterning phenomena has largely benefited from new approaches, theoretical, experimental or by simulations. These problems are, however, complex and still constitute a major challenge for physical approaches to dislocation-based plasticity (Section 2.7).

2.2 Free-flight velocity

2.2.1 The Peierls stress in fcc metals

In fcc metals the Peierls stress and energy are so low that only approximate estimates are available.

The Peierls energy can in principle be obtained from *mechanical spectroscopy* (an advanced type of *internal friction*) techniques, whereby the mechanisms governing the velocity of dislocations are determined from the energy loss of an acoustical wave in a sample submitted to very small oscillating stresses. The stress amplitude is such that no irreversible dislocation motion occurs (see for example Benoit *et al.* 1993). However, the results obtained on pure fcc crystals are paradoxical. The Peierls stresses calculated from the measured Peierls energies are larger by one or two orders of magnitude than the experimental yield stress values at cryogenic temperatures. Possible reasons for this discrepancy were discussed but perhaps not fully clarified (Benoit *et al.* 1993, Nabarro 1997*b*).

The measurement of yield stresses in pure fcc single crystals show that the lowest recorded values are always governed by dislocation interactions and intersections, which can be verified from their dependence on initial dislocation density. In well-annealed crystals, extrapolations to 0 K yield values around $10^{-5}\mu$ (Wang, 1996). This defines an upper limit to the Peierls stress, even for aluminium that may exhibit the highest lattice resistance of all fcc metals because of its extremely narrow dissociation width.

The first theoretical estimates were based on revisited versions of the *Peierls–Nabarro model*, which treats glissile dissociations within a semi-continuum framework (Appendix D). The predicted Peierls stress values were found in the range of $2-4\times10^{-5}\mu$ (Wang, 1996). Methods have been developed for incorporating input from atomic or electronic scale calculations in this model (see Appendix D). As yet, however, the available estimates for Peierls stresses are not fully reliable, which is partly due to difficulties inherent to the matching of continuum and discrete methods in this type of approach (Schoeck and Krystian, 2005). Whatever the method used for determining the Peierls stress in fcc metals, it is also facing a problem of numerical accuracy. Consider a dislocation line moving rigidly by one lattice translation under a stress τ_P. The mechanical work per unit length of line, $\tau_P b^2$, represents the energy change to be estimated, whereas the line energy is of the order of μb^2. The relative energy variation is extremely small and its determination requires an extreme precision.

The only certainty regarding the Peierls stress of moving dislocations in fcc crystals is, therefore, that its upper bound is about $10^{-5}\mu$. The resulting Peierls energies cannot be predicted accurately. However, since mechanical tests show no evidence for a Peierls mechanism at $T = 4.2$ K, the athermal transition temperature should be very low and the Peierls energy should be of the order of 10 meV according to eqn. A.5. Thus, another mechanism governs the free-flight dislocation velocity between obstacles above cryogenic temperatures, namely, phonon damping.

For basal slip in simple hcp metals, yield stress data suggest Peierls stresses of same order as in fcc metals for Zn and Cd, where dislocations are dissociated, and one order of magnitude larger in Be where the dislocation core is compact (Table B.1). For non-screw dislocations in bcc metals, atomistic calculations performed with semi-empirical potentials suggest an upper limit to the Peierls stress of about $10^{-4}\mu$ (Yang *et al.*, 2001; Liu *et al.*, 2004) and a total activation energy for the Peierls mechanism of a few meV (Duesbery and Xu, 1998), this time in agreement with internal friction measurements (Benoit *et al.*, 1993). In this case too, the athermal transition temperature for the lattice resistance should be extremely small.

2.2.2 Phonon drag

2.2.2.1 Mechanisms We consider here dislocations moving in the so-called static domain with slow and moderately fast velocities, $v < c_t/2$, where c_t is the velocity of transverse waves (Section 1.4.2). Beyond this range, the dislocation–phonon interactions become more complex and the time scales are more suited for MD simulations than for DD simulations. The interaction between the elastic field of a moving dislocation and various elementary excitations, principally of conduction electrons and phonons, gives rise to an exchange of momentum and energy dissipation. The resulting effect is a viscous drag on the dislocation, which results in a linear relation between effective force and steady-state velocity. As temperature increases from low values, the thermal population of phonons increases. The resulting contribution to phonon drag increases much faster than the contribution from electrons. As a consequence, the dislocation–phonon interaction predominates over the dislocation–electron interaction at temperatures above $\theta_D/10$, where θ_D is the Debye temperature, that is, the temperature at which all phonon modes are excited in a given material ($\theta_D = 315$ K in copper at 300 K[2]). Phonon drag actually results from multiple dissipation mechanisms. It is estimated mainly by continuum models, which are nevertheless in reasonably good agreement with experiment (Alshits and Indenbom, 1986; Alshits, 1992; Hirth and Lothe, 1992).

In the considered range of dislocation velocities, the most important phonon drag mechanisms arise from second-order (anharmonic) elastic effects close to the dislocation lines. In a simplified picture of pair atomic potentials, anharmonicity is due to the fact that close to equilibrium, the potential energy profile is steeper in the compression side than in the tension side. The scattering of phonons by the anharmonic strain fields of moving dislocations gives rise to a viscous drag through two main mechanisms. There are also other types of interactions between dislocations and phonons. In simple terms, one has the following effects.

- In a coordinate system attached to a moving dislocation line, the flux of incident phonons is not the same ahead of the line and behind it because of anharmonic effects. The phonons scattered by the dislocation line impart to the latter a momentum proportional to the dislocation velocity and opposite to its direction of motion. The asymmetry between these two contributions results in phonon wind, which is a kind of radiation pressure.
- A refinement of models for phonon wind shows that they do not account properly for the dissipation induced by acoustic phonons with short-wave vectors close to a Brillouin zone, which do not have a linear dispersion curve. These phonons move away slowly from the dislocations and the relaxation of slow phonons is responsible for the second anharmonic effect (Alshits and Indenbom, 1986).
- At temperatures $T \gg \theta_D$, thermoelastic effects become important. They are related to the adiabatic heating and cooling of the lattice, which is submitted by passing dislocations to rapid compressions and dilatations. This induces a heat flow through the crystal through additional phonon modes and a thermoelastic dissipation.

[2] θ_D slightly depends on temperature and pressure through c_t.

- Phonon modes set dislocations into oscillation and are scattered by them. The resulting transfer of momentum to dislocations is responsible for a damping effect called the flutter mechanism. This mechanism, contributes to the total damping by one order of magnitude less than anharmonic effects when $T \gg \theta_D/10$.
- In alloys, an additional contribution arises from the excitation of localized phonon modes at solute atoms.

The viscous drag coefficient B, which is temperature and material dependent, is the ratio of the force per unit dislocation line to the resulting steady-state velocity (cf. eqn. 2.8). The contributions of the phonon wind and of the slow phonons suffice to define B to a good approximation in a wide temperature range and for non-relativistic dislocations.

2.2.2.2 *Measurements of the drag coefficient* The drag coefficient is experimentally determined from internal friction measurements and etch-pit measurements of steady-state dislocation velocities. There is usually a large scatter induced by impurities or the lattice friction (for compiled values, see Alshits and Indenbom 1986). Figure 2.1 shows two comparisons between experimental and theoretical determinations of the damping constant in several materials.

A good agreement is obtained, in spite of the complexity of the dislocation–phonon interactions. In copper, which has been the most investigated material, $B(300\,\text{K}) \approx 1.5 \times 10^{-5}$ Pa s. It is interesting to note that because B increases with

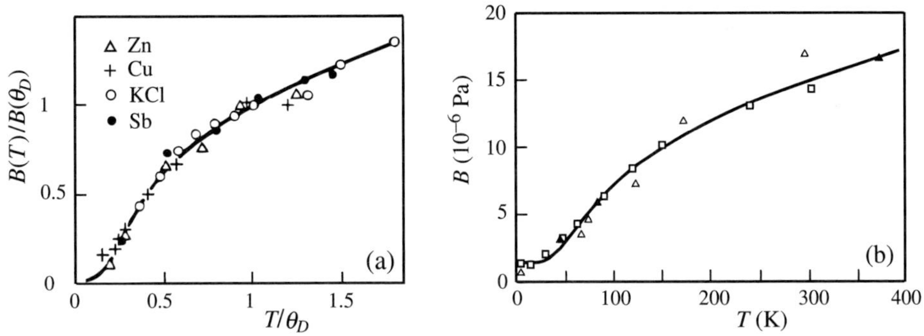

FIG. 2.1. Comparison between theoretical and experimental estimates for the phonon drag coefficient. (a) Temperature dependence of the ratio $B(T)/B(\theta_D)$ for several materials. The full line is the theoretical elastic prediction, which involves only one free parameter, plus an inner cut-off radius for the elastic energy of dislocations. From Alshits and Indenbom 1986, courtesy V. I. Alshits. (b) Temperature dependence of B in pure copper from several experimental sources. The full line is a fit of the theoretical prediction. From Fusenig and Nembach 1993.

temperature, the steady-state dislocation velocity under a given stress decreases with increasing temperature.

The drag coefficient can also be measured by MD simulations from the temperature dependence of steady-state dislocation velocities under stress. To be reliable, classical MD simulations have to be carried out at temperatures around or above the Debye temperature since they do nor account for the Bose–Einstein statistics governing the occupation of phonon states below θ_D. Physically motivated potentials have also to be used, to avoid the interference of artificially high Peierls stresses with phonon damping. For example, with a tight-binding second-moment approximation model, Mordehai *et al.* (2003) obtained values for the drag constant in copper within a factor of only two from those shown in Fig. 2.1*b*. For aluminium, $\theta_D = 390$ K at 300 K. The measured and computed values of the drag constant at 300 K are of same order as in copper (for references, see Bakó *et al.* 2011).

2.2.2.3 Equation of motion and steady state The basic equation of motion of a dislocation line with an average direction parallel to the **x** direction in its slip plane moving along the positive **y** direction (Fig. 2.2), is written

$$m\frac{\partial^2 y}{\partial t^2} + B\frac{\partial y}{\partial t} = \tau_a b - \frac{\Gamma}{R} + \sum_i F_i, \tag{2.1}$$

where all terms are expressed per unit length of line. The first term at the left-hand side accounts for inertial effects and the second one for viscous drag. For velocities much smaller than $c_t/2$, the effective dislocation mass, m, can be assimilated to the rest mass $m_o = \rho_o b^2$, where ρ_o is the density of the material (Section 1.4.2). At the right-hand side, one finds an applied force, $\tau_a b$, a resistive line tension force, Γ/R,

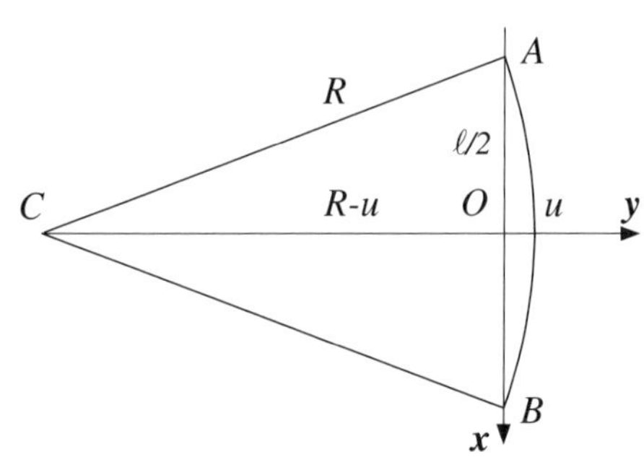

FIG. 2.2. Fundamental vibration mode of a small curved segment of initial length $\ell = AB$, pinned at its ends. R is the curvature radius and $u = y(x = 0, t)$ is a small displacement of the mid-point from the origin O.

which depends on x and y via the curvature radius R, and other forces arising from the interaction of the dislocation with other defects, F_i.

Equation 2.1 is at the base of the *vibrating string model* by Granato and Lücke (1956), which has found many applications in dislocation theory and especially internal friction (see Hirth and Lothe 1992). In general, eqn. 2.1 has to be solved numerically, but simplifications of the line tension term allow obtaining solutions in several specific cases. Static problems can be treated too, as the time-independent solutions yield the equilibrium shape under stress of a dislocation interacting with an obstacle. We now consider several approximations to eqn. 2.1.

The natural vibration frequency of a dislocation line of length ℓ pinned at its ends is obtained by assuming displacements from the straight configuration that are small with respect to ℓ and a small velocity, hence a negligible damping term. For a displacement $y(x = 0, t) = u$ (Fig. 2.2), the relation $R^2 = \ell^2/4 + (R - u)^2$ yields the curvature radius R of the line. When $u \ll \ell$ and $\ell \ll R$, one has $1/R \approx 8u/\ell^2$ and eqn. 2.1 reduces to the traditional equation of a harmonic oscillator

$$m_o \frac{\partial^2 u}{\partial t^2} + \frac{8\Gamma u}{\ell^2} = 0 \tag{2.2}$$

Introducing the velocity of transverse waves, $c_t = (\mu/\rho_o)^{1/2}$ and with $\Gamma \approx \mu b^2$ and $m_o = \rho_o b^2$, the fundamental vibration frequency in eqn. 2.2 comes out as $\nu_\ell = (2\Gamma/\pi^2 m_o \ell^2)^{1/2} = (2^{1/2}/\pi)(c_t/\ell)$. Since the inverse of the Debye frequency $1/\nu_D$ is of same order as the time taken by sound waves to travel one interatomic distance, $\nu_D \approx c_t/b$, one has

$$\nu_\ell \approx \nu_D \frac{b}{\ell} \tag{2.3}$$

A similar derivation is made in mesoscale analyses of thermally activated dislocation processes to define the attempt frequency of a dislocation line in front of an obstacle (Section A.1).

For dislocations moving under stress, the curvature $1/R$ has to be expressed from eqn. 2.1 in terms of $x(y, t)$ and y. To obtain analytical solutions, use is frequently made of an approximate form of the curvature, $1/R \approx \partial^2 y/\partial x^2$, which is valid when $|\partial u/\partial x| \ll 1$. Taking into account that the line tension force is in the direction of $-\mathbf{y}$, eqn. 2.1 is rewritten in the form

$$m \frac{\partial^2 y}{\partial t^2} + B \frac{\partial y}{\partial t} - \Gamma \frac{\partial^2 y}{\partial x^2} = \tau_a b + \sum_i F_i \tag{2.4}$$

When dislocation velocity is governed by viscous drag, dislocation motion can be underdamped (i.e., oscillatory) or overdamped, in which case a steady-state velocity is reached in a non-oscillatory manner. In the range of dislocation velocities considered here, dislocation motion is overdamped as long as the drag constant is not too much reduced by a low temperature (see e.g., Fusenig and Nembach 1993). For a straight dislocation segment moving with a velocity $v(t) = dy/dt$ in an otherwise perfect crystal under a constant stress τ_a, eqns. 2.1 or 2.4 reduce to

$$m_o \frac{dv}{dt} + Bv = \tau_a b \tag{2.5}$$

This equation only has overdamped solutions, which are of the form

$$v = v_s[1 - \exp(-t/t_s)], \tag{2.6}$$

where the steady-state velocity v_s is reached within a characteristic time interval t_s. These two quantities are given by

$$v_s = \tau_a b/B; \qquad t_s = m_o/B = \rho_o b^2/B \tag{2.7}$$

In copper at room temperature, with $B \approx 1.5 \times 10^{-5}$ Pa s, $\rho_o = 8.94 \times 10^3$ kg m^{-3} and $b = 2.56 \times 10^{-10}$ m, a steady-state velocity $v_s = 17 \times 10^{-5}\,\tau_a$ is reached within a time interval $t_s \approx 4 \times 10^{-11}$ s. The distance x_s travelled during this transient is obtained by integrating eqn. 2.6 between $t = 0$ and $t = t_s$, which yields $x_s = v_s t_s/e$. As a rule of thumb, for $\tau_a = 1$ MPa, one has $v_s \approx 20$ m s^{-1} and $x_s \approx 2.5$ nm.

When transient behaviour can be neglected, eqn. 2.5 reduces to

$$\tau_a b = Bv_s \tag{2.8}$$

The use of eqn. 2.8 is restricted to the motion of straight dislocations in materials exhibiting negligible Peierls stresses and to quasi-static deformation conditions. For curved dislocations, use is made of a modified form of eqn. 2.8 where the applied stress is replaced by an effective stress (Section 1.4). The implementation of free-flight velocities in DD simulations is discussed in Section 4.3.2.1.

2.3 Dislocation–dislocation interactions

This section is focused on the discrete mechanisms by which dislocations mutually interact. In pure crystals and when the lattice resistance is negligible, the main obstacle to dislocation motion arises from interactions with other dislocations. Long-range interactions are defined here as non-contact elastic interactions between dislocation segments in a three-dimensional microstructure. An almost equivalent definition takes the average segment length or distance between segments, $\rho^{-1/2}$, where ρ is the total density (Section 1.3.1), as defining the boundary between short- and long-range interactions.[3] Long-range interaction stresses can be computed for each dislocation segment provided that the positions of all other dislocation segments in the microstructure are known. Upon intersections or at short approach distances, stable configurations can be formed that constitute barriers of variable strength to further motion. Three contributions can be distinguished

- Attractive dislocations moving in closely spaced parallel slip planes minimize their total elastic energy by forming stable *dipoles*.
- Upon crossing each other, dislocations moving in non-coplanar (intersecting) slip planes may react to form *junctions*. A dislocation *lock* is a junction that has no slip plane in common with any of its parent dislocations. These configurations minimize the total, elastic plus core, energy of the interacting segments. Owing to their strong impact on the mechanical response, they are discussed is some detail.
- When two dislocations cross each other, an atomic *jog* is formed on each of them.

2.3.1 Short-range interactions in fcc crystals

In fcc crystals, there are four $\{111\}$ slip planes, each containing three $\frac{1}{2}\langle 110\rangle$ Burgers vectors, hence twelve slip systems. Symmetry reduces the number of distinct types of interactions between slip systems to only six. There are four $\langle 111\rangle$ axes with threefold symmetry in cubic crystals and each type of interaction occurs between two slip systems, hence $12^2/(4\times3\times2) = 6$. These elementary interactions are schematically illustrated by Fig. 2.3.

The interaction of dislocations gliding in parallel slip planes is called the *self-interaction* when the Burgers vectors are colinear (Fig. 2.3*a*) and the *coplanar interaction* when they are not (Fig. 2.3*b*). In both cases, the mutual trapping of attractive dislocations may result in dipole formation. Self- and coplanar interactions in a single slip plane may lead respectively to direct annihilation or to the formation of a fully glissile segment having the third coplanar Burgers vector. The probability of occurrence of such configurations is, however, quite low when dislocations glide but cannot climb. Dipoles and multipoles can be destroyed by resolved stresses exceeding a critical value, the *passing stress*. The passing stress τ_p for two edge dislocations of opposite sign gliding in parallel slip planes

[3] In early publications, especially on the Cottrell–Stokes relation (Section 2.5.3), short-range interactions are defined as core interactions leading to a thermal component of the flow stress and long-range interactions are defined as elastic interactions leading to an athermal component.

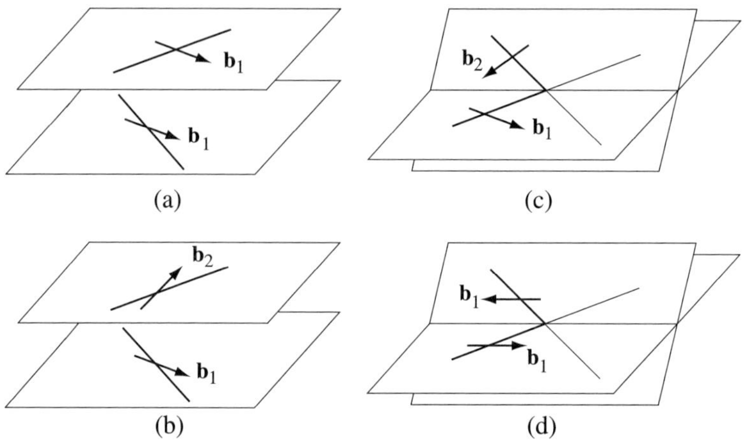

FIG. 2.3. The six elementary short-range interactions of perfect dislocations in the fcc structure. (a) Interaction of a slip system with itself (self-interaction). (b) Interaction between coplanar slip systems (coplanar interaction). (c) Reactions between attractive dislocations. These reactions lead to three possible types of junctions (see text). (d) Interaction between slip systems with parallel or antiparallel Burgers vectors (colinear interaction). The slip plane of each segment is the cross-slip plane of the other. The Burgers vector is parallel to the intersection of the two planes.

separated by a distance h, the dipole height, is a traditional exercise in dislocation theory (Hull and Bacon, 2001). It is given by

$$\tau_p = \frac{\mu b}{8\pi(1-\nu)h} \tag{2.9}$$

The *colinear* interaction (Fig. 2.3*d*) occurs between two dislocations of same Burgers vectors gliding in a slip system and its cross-slip system. When it is attractive, the colinear interaction may lead to partial annihilation of the lines, or, in other terms, to a reaction product with null Burgers vector. This reaction leads to a substantial reduction of the total line energy. As it was extensively studied with the help of DD simulations, its detailed discussion is postponed to Section 5.2.1.2.

In the fcc and related structures, there are three possible types of reactions between non-coplanar perfect dislocations (Figs. 2.3*c* and C.5). Before discussing them, it is useful to visualize interactions and reactions in the fcc structure with the help of the *Thompson tetrahedron* (Fig. 2.4*a*). This tetrahedron is constructed in a cube of edge length one-half the lattice parameter. The edges then represent the six $\frac{1}{2}\langle 110\rangle$ Burgers vectors and the four faces are parallel to the $\{111\}$ slip planes. For

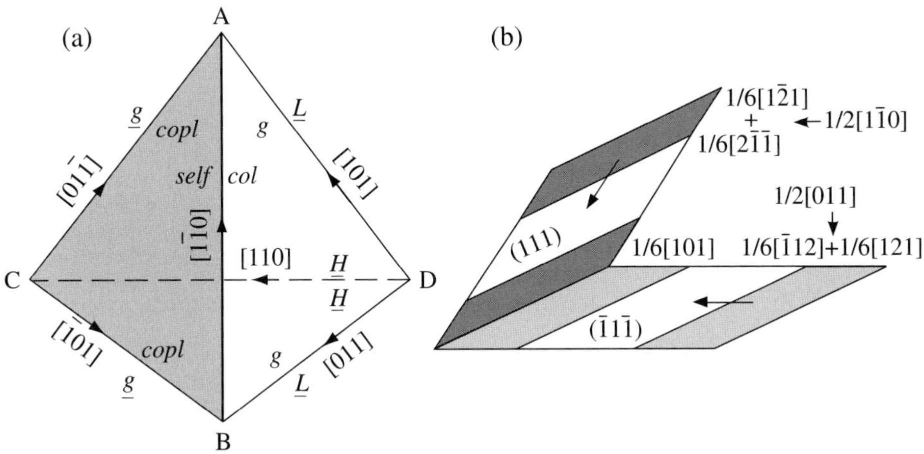

FIG. 2.4. (a) The interactions of a perfect dislocation with Burgers vector along $\pm[1\bar{1}0]$, gliding in the shaded (111) slip plane, with other dislocations in the Thompson tetrahedron. There is one self-interaction (*self*) and two coplanar interactions (*copl*). The notations for the reactions are *L* (Lomer lock), *g* (gissile junction), *H* (Hirth lock) and *col* (colinear annihilation). The underlined interactions occur with dislocations gliding in the slip planes (BCD) or (ACD). (b) A Lomer–Cottrell lock. The sessile stair-rod dislocation of Burgers vector $\frac{1}{6}[101]$ is formed by the reaction of the two leading partials of the attractive dislocations $\frac{1}{2}[1\bar{1}0](111)$ and $\frac{1}{2}[011](\bar{1}1\bar{1})$.

a full discussion of the use of the Thompson tetrahedron, especially for dissociated dislocations, see Hirth and Lothe (1992) and Argon (2008).

One can see from Fig. 2.4*a* that a single slip system interacting with all slip systems forms four glissile junctions, two Lomer locks and two Hirth locks. It also undergoes one self-interaction and one colinear reaction. The Hirth lock has a $\langle 001 \rangle$ Burgers vector. It results from the interaction of non-coplanar dislocations with orthogonal Burgers vectors like **BA** and **DC**. The two other junctions are easily identified from this figure. For instance, consider the reaction between a dislocation with Burgers vectors **BA**, gliding in the plane (ABC) = (111), and an attractive dislocation with Burgers vector **DB**

$$\frac{1}{2}[1\bar{1}0] + \frac{1}{2}[011] \rightarrow \frac{1}{2}[101] \tag{2.10}$$

When the slip plane of **DB** is (ABD) = $(\bar{1}\bar{1}1)$, the line direction of the junction is along the direction of intersection AB of the two slip planes. This junction is glissile because it can glide in (ABD). When the slip plane of **DB** is

(BCD) = $(\bar{1}1\bar{1})$, the two slip planes do not contain the Burgers vector of the junction. The latter is a sessile Lomer lock of edge character. The Lomer–Cottrell (LC) lock shown in Fig. 2.4*b* can be seen as a three-dimensional dissociated form of the previously described Lomer lock. This reaction is frequently observed by TEM during the tensile deformation stage II of fcc single crystals (Section 2.6.3). It serves as a prototypic configuration for explaining the occurrence of strong barriers to dislocation glide in fcc crystals. In aluminium, the reaction occurs between dislocations with very narrow cores, but nevertheless results in the formation of a strong obstacle. Other, less frequent, types of LC locks are discussed by Hirth and Lothe (1992).

2.3.2 Junction formation and destruction

2.3.2.1 Elastic and core energies As seen in Section 1.3.1, the core energy of a dislocation line is a small fraction of its total energy. Further relaxation of a junction by dissociation decreases its core energy by a small fraction of its value. Hence, the gain in total energy upon junction formation can be assimilated to the gain in elastic energy. The destruction of a junction requires a critical stress such that the mechanical work brings the total elastic energy to its initial value. The validity of such elastic calculations was checked by comparing the critical stresses of LC locks by atomistic and mesoscale simulations carried out on identical configurations (Bulatov *et al.*, 1998; Rodney and Phillips, 1999; Shenoy *et al.*, 2000; Shin *et al.*, 2001). An experimental validation also arises from the remarkable constancy of the scaled strain hardening rate of fcc metals in stage II, θ_{II}/μ, as no dependence on SFE is observed (Section 2.6.3).

Assuming that the total elastic energy E of two interacting segments of length $\ell_o = 1\,\mu\text{m}$ is typically reduced by ten percent by their reaction, and with a line energy of 3 eV per length b of line, we have $\Delta E \approx 0.2\mu b^3(\ell_o/b) \approx 2.4 \times 10^3$ eV. Such an energy can in no way be furnished by thermal fluctuations. Hence, the formation and destruction of junctions are essentially elastic and athermal processes, which considerably simplifies their analytical or numerical treatment at the mesoscale.

2.3.2.2 Junction formation The energy balance during junction formation is often discussed in terms of the *Frank's rule*. The self-energy of a dislocation being proportional to b^2 per unit length (eqn. 1.4), the reaction between dislocations with Burgers vectors $\mathbf{b}_1$ and $\mathbf{b}_2$ forming a reaction product of Burgers vector $\mathbf{b}_3 = \mathbf{b}_1 + \mathbf{b}_2$ is energetically favoured when

$$b_1^2 + b_2^2 > b_3^2 \qquad (2.11)$$

This rule can only serve for checking the plausibility of junction formation in the most favourable case where the lines are parallel and react along their whole length. It does not predict whether the total elastic energy is reduced for non-parallel lines (Fig. 2.5), which have an orientation-dependent self-energy, even in isotropic elasticity (Section 1.3.1). In fact, when the Frank rule is fulfilled, only a fraction of the possible attractive configurations react to form stable junctions (Section 5.2.1.1).

Figures 2.5*a* to 2.5*c* illustrate how a junction is formed by the zipping of the interacting lines. Junction destruction, which is discussed in the next section, occurs by the reverse process. The equilibrium configuration of a junction corresponds to the configuration of minimum energy with respect to a small virtual displacement of the triple nodes (*N* and *P* in Fig. 2.6) along the junction direction.

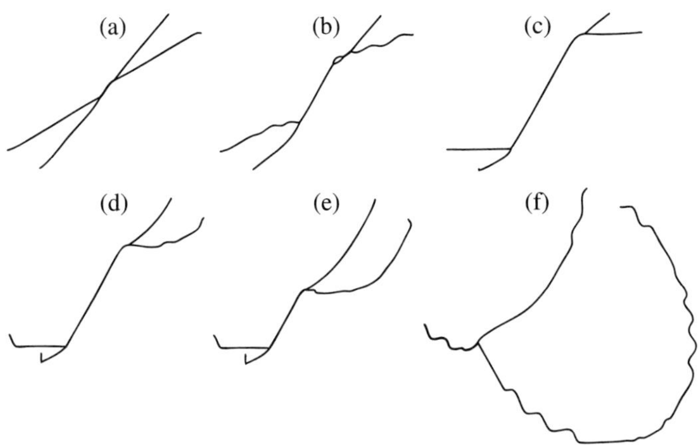

FIG. 2.5. DD simulation of the zipping and unzipping of a junction. (a) Initial configuration under zero stress. The junction zips (b) and reaches its equilibrium length (c). Under stress, one arm of the parent dislocations unzips the junction (d, e), until the initial segments recombine and further expand (f). From Madec *et al.* 2001, courtesy R. Madec.

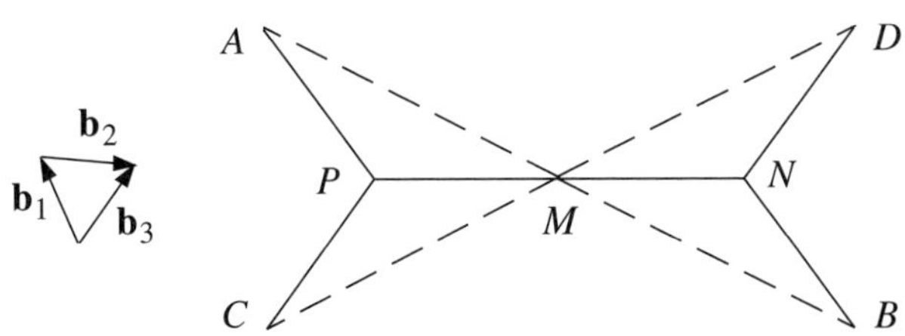

FIG. 2.6. Equilibrium configuration of a junction (schematic). Two segments are pinned in *A*, *B*, *C* and *D*. Their initial positions are shown dotted and the whole configuration is seen in a 2D projection. After reaction, the two triple nodes *N* and *P* have zipped a junction from the intersection point *M* and reached their equilibrium positions.

2.3.2.3 Junction destruction The unzipping of a junction under stress is illustrated by Figs. 2.5*d* to 2.5*f*. Extended sessile barriers like the LC lock constitute extremely strong obstacles only within a 2D vision, where one considers their destruction under the effect of dislocations piling-up in the slip plane containing the stair-rod segment. The unzipping process leads to quite smaller, but still respectable, stresses.

We examine now a simple scaling law for the unzipping stress. For perfect dislocations, the gain in elastic energy upon junction formation in $\Delta E \propto \mu b^2$ per unit length of junction. The mechanical work of the stress, τ, on a parent arm of length ℓ that unzips the junction is $\Delta W \propto \tau b \ell$, also per unit length of junction. All energies are proportional to line lengths and all lengths scale with ℓ_o, the initial length of the interacting segments. Hence, $\Delta W \propto \tau b \ell_o$. The critical stress τ_j for unzipping a junction is then obtained by equating the mechanical work to the gain in elastic energy

$$\tau_j = \beta \frac{\mu b}{\ell_o} \tag{2.12}$$

In this scaling relation, the constant β is a dimensionless measure of the strength of a specific junction configuration. Given the Burgers vector and the normals of the interacting slip planes, the condition for the formation of a junction and the strength of the latter depend on twelve independent geometric parameters, specifically the coordinates of the four end-points of the two parent segments.

2.3.2.4 Junctions in bcc and hcp crystals In the bcc lattice, the Burgers vector is $\frac{1}{2}\langle 111\rangle$. The possible slip planes cannot be always determined without ambiguity but are usually assumed to be the two most compact ones, $\{110\}$ and $\{112\}$, or a combination of both at high temperatures (Section 3.2.1). The reaction products are always dislocations with a $\langle 001\rangle$ Burgers vector. Thus, one has for instance

$$\frac{1}{2}[111] + \frac{1}{2}[1\bar{1}\bar{1}] \rightarrow [001] \tag{2.13}$$

The number of different types of $\langle 001\rangle$ junctions to be investigated depends on the number of slip systems considered. Detailed estimates of the energy gain associated with junction formation were performed by Baird and Gale (1965) in isotropic elasticity and by Püschl (1985) in anisotropic elasticity. Indeed, in bcc metals, the Zener anisotropy ratio, $A = 2c_{44}/(c_{11} - c_{12})$, ranges from 0.51 in niobium to extremely high values for the alkali metals Li (8.4), Na (7.3) and K (7.6). In all the investigated cases, elastic anisotropy was found to increase the energy gain upon reaction without modifying the nature of the reaction product. In hcp crystals, Baird and Gale (1965) examined the possible reactions of basal dislocations with prismatic and pyramidal trees, and Püschl and Schoeck (1984) investigated in isotropic elasticity the reactions between basal dislocations and dislocations with Burgers vectors along the **a**, **c** and **c** + **a** axes. All these calculations were performed for dislocations that bow under stress, as in fcc metals.

2.3.3 Jogs

When two dislocation lines cross each other, each of them is sheared by the other one; it acquires a segment of height equal in magnitude and direction to the Burgers vector of the other line. Although the direction of the lines locally changes, their Burgers vectors remain the same. The shearing of a line by a dislocation with a different Burgers vector creates a *jog*. The glide plane of a jog contains its line direction and the Burgers vector of the jogged line (Fig. 2.7). In contrast to jogs of larger heights or superjogs, produced for example by double cross-slip events, unit jogs created by dislocation intersections have an atomic length and cannot be treated by elasticity theory. Thus, their energy is not the elastic energy of a dislocation segment of length b, but rather its core energy, typically $\mu b^3/10$. When a line is sheared in its own slip plane during an intersection by a repulsive cross-slipped dislocation with same Burgers vector, the small segment is glissile and is called a *kink*.

On a dislocation of edge or mixed character, a perfect jog can move conservatively with the line, as depicted in Fig. 2.7*a*. The component of the jog velocity along the direction of motion of the line decreases as the character of the line goes from edge to screw. On straight screw dislocations, jogs cannot move conservatively with the line; they can only glide sideways until they meet an obstacle (Fig. 2.7*b*). Hence, a small-velocity jog may pin a fast-moving line. This raises the question of jog mobility, especially in materials with average or low SFE where jogs may be either constricted or extended (see Friedel 1967; Hirth and Lothe 1992). Because jogs are atomic-scale defects, their structure and mobility can hardly be predicted by elastic models. For example, in copper, where a complete atomistic study was performed (Vegge and Jacobsen, 2002), jogs on screw dislocations were found to be constricted and extremely mobile along

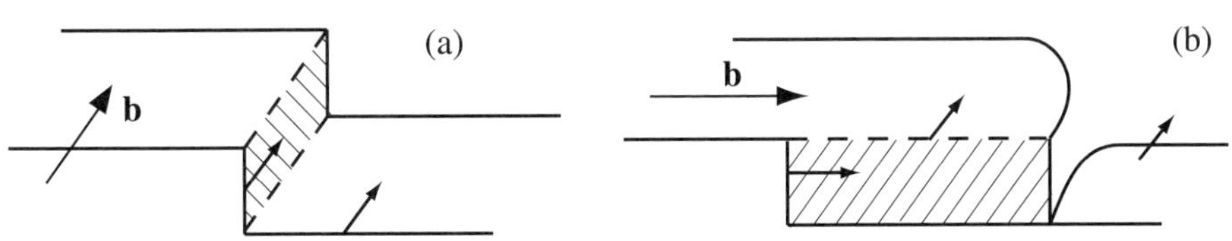

FIG. 2.7. A jog on a perfect dislocation with Burgers vector **b**. The small arrows indicate the directions of motion. The dashed areas represent the glide planes of the jogs. (a) On an edge dislocation, a jog moves conservatively with the line. (b) On a screw dislocation, it can only glide sideways.

the lines, even at very low temperatures. Thus, the dragging of jogs by moving lines, which produces rows of points defects and dislocation debris, seems to be restricted in that case to high dislocation velocities and applied strain rates. Jog strengthening in pure fcc metals deformed in quasi-static conditions is not fully understood (Section 2.5.3). It is usually attributed to the formation energy of jogs upon dislocation intersections.

2.4 Cross-slip in fcc crystals

The cross-slip of a screw dislocation involves a change in core structure and the overcoming of an energy barrier. Although the importance of cross-slip was recognized long ago, modelling efforts were not totally successful and are now essentially carried out at the atomic scale. This section mainly focuses on the Escaig elastic model for cross-slip in fcc crystals. Although treating a dislocation core process by linear elasticity may seem a bit rough, the Escaig model is still useful because some of its predictions were verified by experiment and atomistic simulations. In the last case, some interesting discrepancies were also found. They seem related to the fact that, usually, cross-slip is not occurring spontaneously under a uniform stress field and is assisted by short-range interactions with other defects (Section 2.4.8).

2.4.1 Models for compact cross-slip

There are several types of minimum-energy paths associated with cross-slip mechanisms. They depend upon whether the total stress-free energy of the dislocation is the same or not in the glide and cross-slip planes. In fcc crystals, the dissociation width and the stress-free energy of a dislocation are the same in all close-packed {111} slip planes. This generic type of cross-slip, which is called *compact cross-slip*, is discussed in what follows. *Non-compact cross-slip* involves a variety of cross-slip processes between compact and non-compact slip planes. It is, for instance, illustrated by a study on cross-slip by the kink-pair mechanism in magnesium binary alloys (Yasi *et al.* 2011, 2012). For detailed accounts on cross-slip models and their application to various classes of materials, in particular ordered alloys, the reader is referred to Püschl (2002) and Caillard and Martin (2003).

The modelling of compact cross-slip was initially motivated by the occurrence of a temperature-dependent stage of work hardening in fcc crystals, the dynamic recovery stage, or stage III, during which the strain hardening rate decreases with increasing strain while the three-dimensional arrangement of the dislocation microstructures becomes well-marked. These features are usually interpreted as resulting from the annihilation of attractive screw dislocations by cross-slip (Section 2.6.4).

A Shockley partial dislocation cannot cross-slip without leaving a stair-rod dislocation (Fig. 2.4*b*) at the intersection of the slip and cross-slip planes. Fleischer (1959) proposed a model based on this mechanism and on a further reaction of the trailing Shockley partial with the stair-rod dislocation. This process was, however, shown to require a too substantial activation energy. It is only observed at high stresses, for instance in MD simulations of nanocrystalline materials (Bitzek *et al.*, 2008).

A decisive progress in the search of a low-energy critical configuration was made by Schoeck and Seeger (1955), who proposed that cross-slip is initiated by the formation of a point constriction on the stacking fault ribbon (cf. Fig. 2.8*b*,). In further steps, the screw dislocation was assumed to locally recombine and cross-slip. On this base, Friedel (1957) suggested a less energetic pathway that is now widely accepted.

2.4.2 The Friedel–Escaig mechanism

The Friedel–Escaig mechanism for cross-slip in fcc crystals was modelled by Escaig (1968*a*, 1968*b*, 1968*c*). Figure 2.8 describes a minimum energy path, which involves in a first step the formation of a point constriction on a dissociated screw

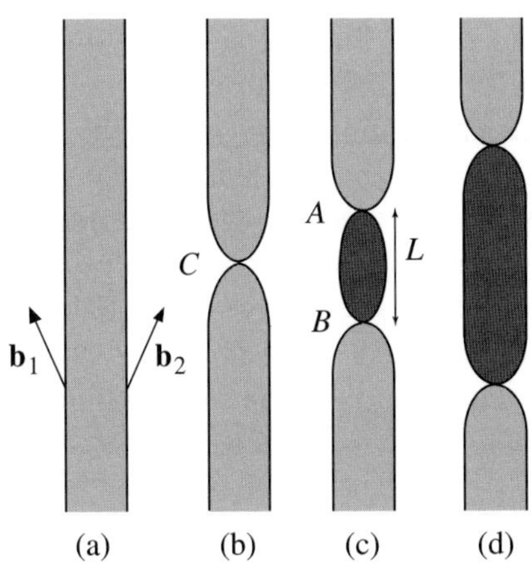

FIG. 2.8. Successive steps of the cross-slip mechanism in fcc crystals according to the Friedel–Escaig model. (a) A screw dislocation is dissociated into two 30° Shockley partials $\mathbf{b_1}$ and $\mathbf{b_2}$. (b) A point constriction C is formed. (c) The constriction splits into two constrictions separating a portion of recombined line that is dissociated in the cross-slip plane. At the energy saddle point, the dissociated loop has a critical length L. (d) Past the critical configuration, the cross-slipped loop expands along the screw direction.

dislocation (Fig. 2.8*b*). Under stress, the point constriction splits into two constrictions (*A* and *B* in Fig. 2.8*c*) separating a portion of recombined screw dislocation line that dissociates in its cross-slip plane as soon as it is formed. This minimizes the activation energy more efficiently than in previous models (Schoeck and Seeger, 1955; Wolf, 1960), where the recombined segment was assumed to dissociate again only after having cross-slipped. For a critical separation between the two constrictions, a saddle point is reached and the cross-slipped loop *AB* expands in an irreversible manner along the screw direction (Fig. 2.8*d*).

As far as the energetics of the cross-slip mechanism are concerned, the shear stresses that matter are those that govern the equilibrium widths of the stacking fault ribbons in the slip and cross-slip planes. The calculations are performed in the approximation of a constant line tension and by integrating the forces to determine the energies. The constriction energy is estimated by analogy with Stroh's calculations of constrictions (Stroh, 1954), with same value b for the inner cut-off radius of the perfect dislocation and the critical distance for recombination of the partials.

The activation energy can be further reduced by assuming that defects on dissociated screws, like jogs produced upon dislocation intersections, may be favourable nucleation sites for the nucleation of constrictions. In such a case, the lines contain a pre-existing constriction and Fig. 2.8*b* represents the initial configuration. Washburn (1965) examined whether cross-slip could be spontaneously initiated in the immediate vicinity of a junction. Half-constrictions, like those shown in Fig. 2.8*b*, are always formed at one of the triple nodes connecting junctions to primary and forest lines. The qualitative mechanism proposed by Washburn discusses the possibility of forming a second half-constriction on a primary segment when it rotates around the triple node under stress and reaches the screw orientation. Provided that the slip plane of the forest segment is the cross-slip plane of the primary one, the whole segment between the two constrictions can relax its energy by cross-slipping. This mechanism may lead to the critical configuration of Fig. 2.8*c* and to athermal cross-slip. However, the condition on the slip system of the forest dislocation restricts the possible reactions to only glissile junctions or colinear annihilations, which are much less commonly formed than LC locks.

2.4.3 The activation energy for cross-slip

In the Escaig model, the activation energy for cross-slip is estimated in a first step as the energy needed to develop a cross-slipped loop from a pre-existing constriction in the primary plane. The contribution of the mean curvature of the split segment under stress is found negligible, which means that the Peach–Koehler force on the perfect screw dislocation plays no role in the cross-slip process. The energy is obtained by solving an equation of the Stroh type for the constriction on a straight dissociated segment. The equilibrium energy under stress is calculated in terms of the distance between the two half-constrictions in the cross-slip plane and the equilibrium splitting width of the two straight portions of partial dislocations. The saddle point configuration is obtained as the maximum of this energy with respect to these two geometric parameters. The calculation of the various terms is detailed by Caillard and Martin (2003) and summarized by Püschl (2002), Bonneville (1985) and Bonneville and Escaig (1979).

The screw and edge components of the Shockley partials are playing distinct roles in Escaig's model. As can be easily seen from Fig. 2.9, these two components have Burgers vectors of different magnitudes, respectively $|b|/2$ and $|b|/2\sqrt{3}$.

Taking into account the mixed character of the Shockley partials and with $\nu = 1/3$, the dissociation width of the perfect screw dislocation in equilibrium is $d_o/b = \mu b/16\pi\gamma$, where γ is the SFE. This width is the same in the slip and cross-slip planes. Under stress, two shear stresses are defined, τ' and τ_s, which are perpendicular to the screw direction in the primary and cross-slip planes, respectively. As the Shockley partials have different Schmid factors, these stresses act on their edge components and modify the widths of the stacking fault ribbons. The latter are denoted d' and d_s; they are given by

$$\frac{d'}{b} = \frac{\mu b}{16\pi(\gamma + \tau' b/2\sqrt{3})} \quad \text{and} \quad \frac{d_s}{b} = \frac{\mu b}{16\pi(\gamma - \tau_s b/2\sqrt{3})} \tag{2.14}$$

The sign convention for these two stresses is that they are taken positive when they favour the cross-slip process by narrowing the dissociation width in the primary plane or widening it in the cross-slip plane.

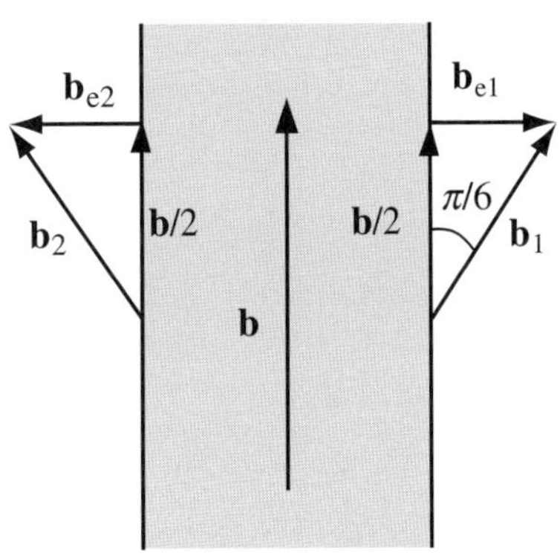

FIG. 2.9. A perfect screw dislocation with Burgers vector $\mathbf{b}$ is dissociated into two Shockley partials, $\mathbf{b}_1$ and $\mathbf{b}_2$. The latter are decomposed into two screw components with same Burgers vector $\mathbf{b}/2$ and two edge components $\mathbf{b}_{e1}$ and $\mathbf{b}_{e2}$ with opposite Burgers vectors.

With a pre-existing constriction the calculation of the saddle point configuration leads to an activation energy for cross-slip in the form

$$E_1 = E_o(d_s)\, w\left(\frac{d_s}{d'}\right) \tag{2.15}$$

The first term at the right-hand side of this equation has the dimension of an energy. It is given by Caillard and Martin (2003) in a form close to the one given by Stroh (1954) for a point constriction energy in the cross-slip plane.

$$E_o(d_s) \propto \mu b^3 \frac{d_s}{b}\left[\ln\left(\frac{d_s\sqrt{3}}{b}\right)\right]^{1/2} \tag{2.16}$$

The function w at the right-hand side of eqn. 2.15 is not analytical. It strongly depends on the two dissociation widths, which are the key quantities driving the activation energy. Figure 2.10 shows the variation of the scaled activation energy as a function of the ratio of the dissociation widths in the cross-slip and primary slip planes, that is, the function $w(d_s/d')$. Using eqns. 2.14, w can also be expressed in terms of the two stresses τ' and τ_s and the SFE.

Under zero stress, $d_s = d' = d_o$ and $w(1) \approx 1.2$. The total activation energy of the cross-slip process with a pre-existing constriction, E_o, is then about $1.2\, E_o(1)$. Thus, eqn. 2.16 taken under zero stress defines a scaling law for the total activation energy of the cross-slip process in fcc materials. Without a pre-existing constriction, a point constriction energy in the glide plane has to be added to E_1.

As the activation energy given by eqn. 2.15 is not analytical, Escaig developed asymptotic forms for high and low stresses. The last one is the most suited for

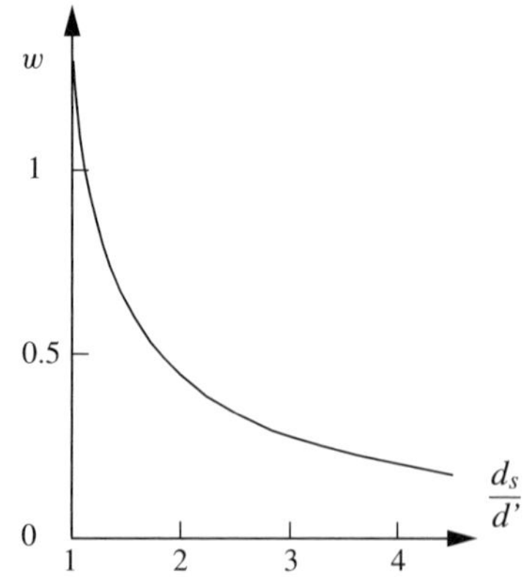

FIG. 2.10. The activation energy for cross-slip with a pre-existing constriction, E_1, scaled by the constriction energy $E_o(d_s)$, as a function of the ratio d_s/d' of the dissociation widths in the cross-slip and slip planes. From Escaig 1968*c*.

comparison with experiment. At low stresses, the equilibrium width of the stacking fault ribbons is expanded to first order in stress. The activation energy is expressed in terms of the resolved applied stress in the primary plane, τ, to facilitate comparison with experiment. The approximations of the low stress regime hold for $\tau \ll \gamma/b$ (≈ 150 MPa in copper). The activation energies with and without a pre-existing constriction are written (Bonneville and Escaig, 1979; Bonneville *et al.*, 1988)

$$E_i(\tau) = 2iAd_o^2\gamma\left[\left(1 - \frac{b_o}{d_o}\right)^2 - \alpha_i\tau b/\gamma\right] \qquad (2.17)$$

The index i takes the values 1 and 2 with and without a pre-existing constriction, respectively. $A \approx 1$ is a slowly varying function akin to the logarithmic term in eqn. 2.16 and b_o is the core width of the perfect screw, which is assimilated to the magnitude of its Burgers vector, b. The coefficients α_i contain two types of contributions. At this step we consider only one, α_i^o, which arises naturally in the model as it incorporates orientation terms relating the stresses τ' and τ_s to the resolved applied stress τ. This coefficient and the second contribution to α_i will be further discussed in Section 2.4.4.

The predicted activation volume, $V = -\partial E_i/\partial\tau$, is a constant in the low stress approximation. It is given by

$$V = 2iA\alpha_i^o bd_o^2 \qquad (2.18)$$

As it is proportional to bd_o^2, the activation volume is very sensitive to the dissociation width. However, as shown in the next section, eqn. 2.18 yields unphysical results.

Saada (1991) revisited the treatment of the point constriction, still in the line tension approximation. He showed that the values assumed for the cut-off radius and recombination distance significantly influence the constriction energy. Duesbery *et al.* (1992*a*) computed more accurately the shape of the critical configuration, but used arbitrary recombination distances and inner cut-off radii. This resulted in rather high activation energies. The most sophisticated calculations were performed by Püschl and co-workers (see Püschl 2002). The inner cut-off and the recombination distances are defined with the help of a Peierls–Nabarro model (Appendix D). Self-energies and self-stresses are estimated from a piecewise decomposition of the lines. The model is constructed as a set of building blocks, and different combinations can be used to investigate more complex cases and various types of non-compact cross-slip. The results are not analytical.

2.4.4 Escaig's effect and Escaig's barrier

The resolved stresses on the edge components of the Shockley partials in the slip and cross-slip planes, τ' and τ_s, are called *Escaig stresses*. The magnitude and direction of these stresses depend upon the orientation and sign of the applied stress through Schmid factors that are incorporated in the coefficients α_i^o mentioned above. The orientation of the loading axis thus affects the dissociation widths and, in consequence, the energetics of the cross-slip process. This is called the *Escaig effect*. The total activation energy is, however, not modified since it is not stress-dependent. The Escaig stresses are non-glide stresses, by which is meant that they have no effect on the stress that governs the motion of a screw segment as a whole. An analogous situation is found in bcc metals, where non-glide stresses significantly influence the core structures of screw dislocations (Section 3.2.3). Figure 2.11*a* shows in a stereographic plot the orientation dependence of the scaled dissociation widths d'/d_o and d_s/d_o in compression.

In the central region B, the stacking fault ribbon contracts in the primary plane and expands in the cross-slip plane, which favours cross-slip. It expands in both planes in region A and contracts in both planes in region C, so that the two variations tend to cancel each other in both cases. All inequalities are reversed in tension and a tension–compression asymmetry is predicted in the central region B. The activation volumes calculated from eqn. 2.18 for copper ($Ad_ob^2 \approx 50$) are

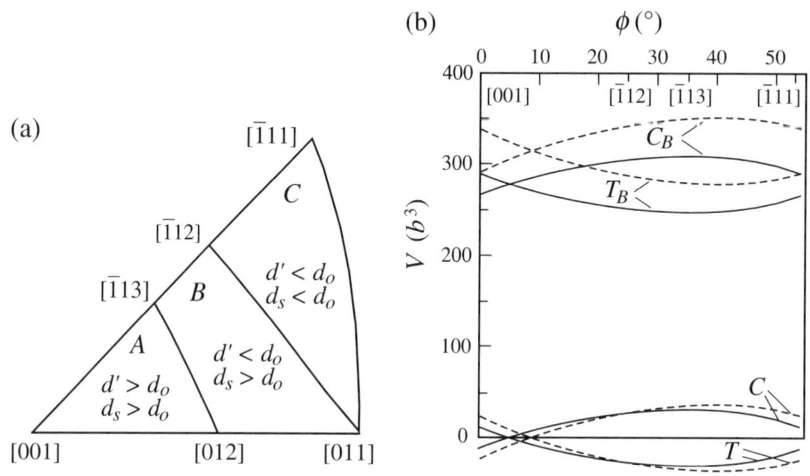

FIG. 2.11. (a) The Escaig effect in compression. Orientation dependence of the dissociation widths in the primary and secondary slip planes, d' and d_s. Three domains are defined wherein the stacking fault ribbons extend or contract with respect to their equilibrium width d_o. From Bonneville and Escaig 1979, © 1979. (b) Predicted orientation dependence of the activation volumes for copper, V, along the [001]-[$\bar{1}$11] zone axis. The orientation is defined by the angle ϕ with respect to [001]. The curves are drawn with and without the Escaig barrier (index B), for tension (T) and compression (C). Full lines: $i = 1$ in eqn. (2.18); dashed lines: $i = 2$.

shown in the lower curves of Fig. 2.11*b* for orientations along the [001]-[$\bar{1}$11] zone axis. There are ranges of orientations, mainly in tension, for which the evolution of dissociation widths is not favoring cross-slip, so that the activation energies increase with stress and the activation volumes are negative. Further, an order of magnitude of the maximum critical stress for cross-slip at $T = 0$ K can be derived by setting d' or $d_s = d_o$ in eqns. 2.14. With $b = b_o$, one obtains extremely high Escaig stresses of about 2.2 GPa for recombining the perfect screw dislocation in copper. Similar values are also expected from the model by Püschl and Schoeck (see Fig. 20 of Püschl 2002).

In the first models for stage III hardening, screw dislocations emitted by Frank–Read sources were supposed to pile-up on LC locks and bypass them by cross-slip. By analogy, and to remedy to the above-mentioned defects, Escaig assumed that cross-slip nucleation is a heterogeneous process occurring under strong stress gradients. As was done previously by Wolf (1960), a rigid, linear barrier to the motion of screw dislocations was introduced in the primary glide plane (Bonneville and Escaig, 1979; Bonneville *et al.*, 1988). The leading partial is blocked by the barrier whereas the trailing partial is pushed against it by the Peach–Koehler force $\tau b/2$ acting on its screw component. A term taking into account the corresponding additional stress is then introduced into the coefficients α_i. It is of the form $\sqrt{3}|\tau|/\tau$, where the absolute value ensures that this term reduces the dissociation width in the primary plane in compression as well as in tension. The factor $\sqrt{3}$ transforms the force applied to the screw component of the trailing Shockley partial into an equivalent force applied to the edge component (cf. Fig. 2.9). The final form of the activation volume is then given by

$$V = 2iA\alpha_i bd_o^2 \tag{2.19}$$

The upper curves of Figure 2.11*b* show that in the presence of this rigid barrier, the activation volumes are in the range of $250-300b^3$. They are modulated by a relatively small orientation dependence. This *ad hoc* barrier, of which the importance is often ignored, plays a decisive role in Escaig's model. Thus, it seems clear that cross-slip cannot occur in an uniform stress field. The nature of the obstacle that is modelled as a barrier is discussed in Section 2.6.4.

2.4.5 Experimental checks

The energetics of cross-slip are rather elusive from an experimental viewpoint. Cross-slip events produce a negligible contribution to the total strain rate and cannot be investigated by conventional testing methods. Attempts at measuring the activation parameters from the temperature dependence of the stress τ_{III} for the onset of dynamic recovery in fcc single crystals yielded mixed results (Escaig, 1968*a*).

To bypass these difficulties, Bonneville and Escaig (1979) devised a two-step deformation procedure analogous to the one used in latent hardening tests (Section 2.5.4). A copper crystal is first prestrained along a $\langle 110 \rangle$ orientation in order to introduce a dislocation forest. Small specimens extracted from this crystal are then further strained in tension or compression along orientations favouring a primary slip plane that has two properties. It was not active during the first step and it is the cross-slip system of one of the previously active slip systems. In such conditions, a burst of cross-slip occurs upon yielding. The yield and flow stresses exhibit an orientation dependence in agreement with the predicted Escaig effect (Fig. 2.11*a*). The low-stress activation volumes are around $280b^3$ in tension, which is the expected value for $i = 2$ (this value was obtained by using a corrected form of the resolved atomic volume, $\omega = b^3\sqrt{3}/2$).

In a subsequent study performed by the same method, the energetics of cross-slip were investigated at different temperatures (Bonneville *et al.*, 1988). The pre-strained copper crystals were deformed in compression along a single slip orientation of region B (Fig. 2.11*a*). The total activation energy for cross-slip was estimated to $E = 1.15 \pm 0.37$ eV. This mean value is intermediate between the two activation energies predicted by Escaig's model with and without a pre-existing constriction (about 0.86 and 1.58 eV, respectively). The activation volume at room temperature was $280 \pm 65b^3$, which corresponds to reasonable dissociation widths around $6.1b$ for $i = 1$ and $6.5b$ for $i = 2$. For aluminium, similar measurements yielded an activation volume one order of magnitude smaller than in copper (Bonneville and Vanderschaeve, 1986), which confirmed the strong dependency of this quantity on the dissociation width. Globally, these experimental results confirm the predictions of the Escaig model but do not allow determining whether cross-slip occurs by a single or double constriction process.

A simpler way to introduce a high density of screw dislocations in a crystal consists in prestraining it below the dynamic recovery stage at a low temperature. Upon restraining at a higher temperature, screw annihilations occur when the dynamic recovery stage is reached (Section 2.5.3). These annihilations are followed by massive athermal rearrangements of the microstructure, which are discussed in Section 2.6.4. Thus, the cross-slip mechanism investigated by Bonneville and Escaig (1979) and the Escaig model seem to have many features in common with dynamic recovery.

2.4.6 Stress-free constriction energies

The total activation energy of the cross-slip process is approximately one or two times the stress-free constriction energy E_o. Thus, this last quantity serves as a benchmark test for values determined by various methods. Figure 2.12 shows a compilation of elastic, experimental and atomistic (see Section 2.4.7) determinations of E_o.

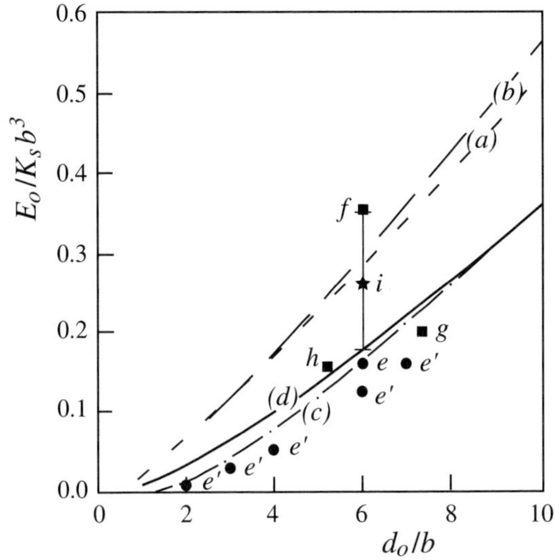

FIG. 2.12. Stress-free constriction energies, E_o, scaled by the energy factor for screw dislocations in anisotropic elasticity, as a function of the scaled dissociation width in equilibrium, d_o/b. Elastic models: (a) Stroh (1954); (b) Duesbery *et al.* (1992*a*); (c) Escaig's model, eqn. 2.16; (d) Püschl and Schoeck (see Püschl 2002); (e, e') Saada (1991) with inner cut-off radii $r_o = b/2$ and b. Atomistic simulations: (f) copper (Rasmussen *et al.*, 1997*a*); (g) copper, activation energy for cross-slip with a pre-existing jog (Vegge *et al.*, 2001); (h) Rao *et al.* (1999) for nickel. (i) Measured activation energy for cross-slip in copper (Bonneville *et al.*, 1988) with its error bar. After Püschl 2002, courtesy W. Püschl.

For atomistic simulations, the dissociations widths indicated are the computed ones. For unconstricted dislocations, the constriction energy is taken as one-half of the total stress-free energy. The models by Stroh (1954) and Duesbery *et al.* (1992*a*) predict too high constriction energies. The elastic models by Püschl and Schoeck and by Escaig are in reasonable agreement except for small dissociation widths. For copper, they predict a constriction energy of $E_o \approx 0.7$ eV. According to Fig. 2.10, the activation energy with a pre-existing constriction should then be $E_1 \approx 0.85$ eV.

2.4.7 Atomistic studies of cross-slip

This section describes results obtained by atomistic simulations on the energetics of the cross-slip process. Lattice statics and the NEB method (Section 1.5) were employed in order to avoid artefacts arising from the high strain rates inherent to MD simulations.

Rasmussen *et al.* (1997*a*, 1997*b*) applied the NEB method to the study of unconstricted screw dislocations in copper, using a many-body potential derived from the effective medium theory. In the prescribed initial and final configurations, the screw dislocation was dissociated respectively in the slip and cross-slip planes. Periodic boundary conditions were employed along the screw direction and other surfaces of the simulated cell were free. The major outcome of these simulations is a confirmation of the minimum energy path assumed in the Friedel–Escaig mechanism (Fig. 2.13).

An important feature, which was predicted by Duesbery *et al.* (1992*a*), was also found. Consider the two constrictions *A* and *B* in Fig. 2.13*d*. The local line direction of the partial dislocations near the constriction tends towards more screw component in *A* (screw-like constriction) and more edge component in *B* (edge-like constriction). As edge lines have a higher self-energy than screw lines, the screw-like constriction should have a lower energy. The shapes of the two constrictions are not substantially different, but the screw-like constriction turns out to have a negative energy (Rasmussen *et al.*, 1997*a*). The total activation energy was found to be 2.7 eV, very close to the sum of the energies of the positive and negative constrictions, respectively 3.8 eV and –1.1 eV.

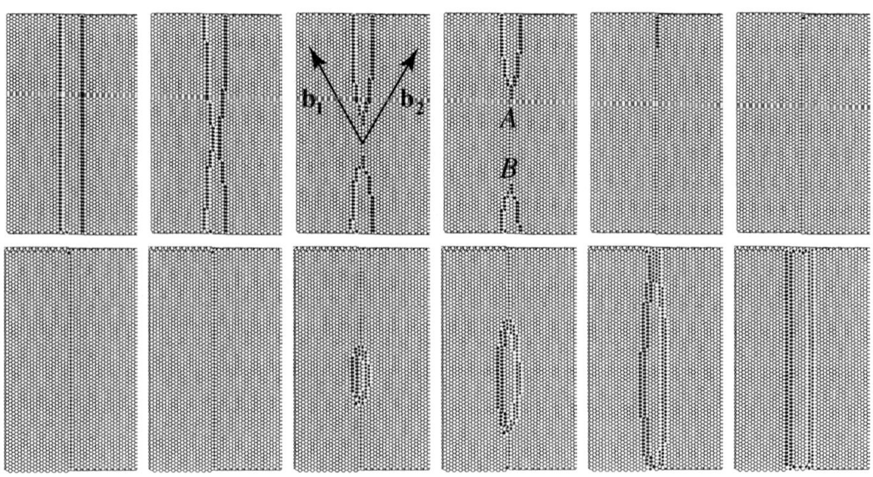

(a) (b) (c) (d) (e) (f)

FIG. 2.13. The minimum energy path for cross-slip in an atomistic simulation of copper (compare with Fig. 2.8). The upper row of frames shows the evolution of the initially non-constricted screw dislocation ($b = \frac{1}{2}[110]$) dissociated into two Shockley partials in a $(1\bar{1}1)$ plane. The lower row shows the parallel evolution in the $(1\bar{1}\bar{1})$ cross-slip plane. (a) and (f) are the initial and final configurations and (d) is the saddle point configuration. From Rasmussen *et al.* 1997*b*, courtesy K. W. Jacobsen.

These studies were further extended to the properties of jogs and kinks on screw lines in copper. Cross-slip of the configuration shown in Fig. 2.13 was simulated with two different pre-existing unit jogs of line directions $a[001]$ and $a/2[\bar{1}01]$. In both cases, the presence of the jog induced a stable, full constriction of the dissociated screw. The formation energy of these jogs seems, therefore, rather large and equal to the constriction energy. In the two cases, the total activation energy for cross-slip was reduced to about 0.9 eV (Vegge *et al.*, 2001; Vegge and Jacobsen, 2002). In these simulations, however, the dissociation width in equilibrium, d_o, is clearly too large (cf. Fig. 2.12), whereas the stress-free constriction energy, 1.6 eV, seems too large in the previous ones.

The critical annihilation distance of two screw dislocations of opposite sign is an important parameter in the modelling of monotonic and cyclic deformation (Section 2.4.8). Static simulations were carried out for short screw dipoles of lengths a few tens of Burgers vectors (Rasmussen *et al.*, 2000) and for similar configurations containing a constricted jog (Vegge and Jacobsen, 2002). In the last case, spontaneous annihilations occurred at $T = 0$ K for stress-free dipoles heights up to 2.4 nm.

Lattice statics simulations with flexible boundary conditions were employed by Rao *et al.* (1999) for investigating the energetics of cross-slip in nickel without a pre-existing constriction. The potentials used for nickel were of the embedded-atom-method type. One of them, Ni(2), had realistic elastic constants, SFE ($\gamma = 119\,\mathrm{mJ\,m^{-2}}$) and screw dissociation width. The Friedel–Escaig mechanism was confirmed by these simulations, as well as the occurrence of a constriction with negative energy. The atomic configurations of the cross-slipping dislocation were examined in some detail. With the Ni(2) potential, the total activation energy was 2.35 eV. According to the elastic scaling law given by the Escaig model (eqn. 2.16), this corresponds to a total activation energy for copper of 1.88 eV, similar to the predicted value without a pre-existing constriction, 1.85 eV.

For estimating the energetics of the cross-slip mechanism, Escaig's stresses were only applied to the partials in the primary plane, in such a way as to constrict the stacking fault ribbon. The activation energies showed a much stronger stress dependence than predicted by the Escaig model and, consequently, a quite small activation volume ($20b^3$ for a stress of $10^{-3}\mu$), compatible with the one predicted without the Escaig's barrier by eqn. 2.18. In agreement with eqn. A.6, this resulted in a too strong temperature dependence of the yield stress. The extrapolated critical stress at $T = 0$ K was above $10^{-2}\mu$, well beyond any measured flow stress value. These discrepancies, which are also found in other simulations (see e.g., Lu *et al.* 2002; Martinez *et al.* 2008), confirmed that the initiation of cross-slip is a heterogeneous process.

The hypothesis proposed by Washburn (1965) of athermal cross-slip nucleation near junctions was revisited using molecular statics simulations and the NEB method by Rao and co-workers (2009, 2010, 2011). Large-scale simulations were

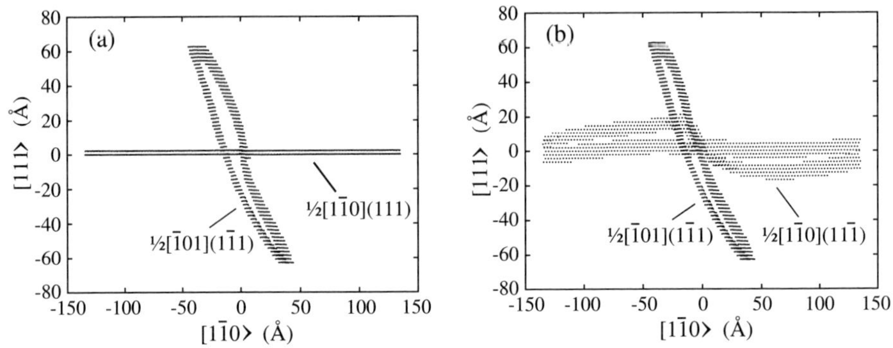

FIG. 2.14. [111] views of a (horizontal) screw segment interacting with an attractive 120° forest tree; the two segments are superimposed. The simulated material is copper. (a) Initial configuration for two segments forming a glissile junction. The screw dislocation is dissociated in its glide plane. (b) The two segments form a LC lock. After reaction, the screw dislocation has cross-slipped. After Rao *et al.* 2011, courtesy S. Rao.

performed with fixed boundary conditions. We focus on the most recent study, which expands the previous ones with improved accuracy. An initial configuration is reproduced in Fig. 2.14*a*, where one can see that the dislocations are constricted at the surfaces of the simulation cell. The slip systems of the attractive forest dislocations do not contain the primary Burgers vector and the combinations of slip systems allow obtaining the three possible types of junctions formed in fcc crystals. The materials investigated are copper and nickel, for which Mishin potentials were used (Mishin *et al.*, 2001; Mishin, 2004). Two other potentials were used for nickel, essentially for the sake of comparison.

Several initial configurations favouring cross-slip were examined, of which some had already been identified (Rao *et al.*, 2009). They lead to a substantial reduction of the energy of the negative constriction in the field of the forest dislocation. The minimum energy path profiles along the reaction coordinate were sampled by varying the initial separation of the screw and forest dislocations. Two successive energy barriers followed by energy minima were found; they correspond to configurations of partially cross-slipped, as well as fully cross-slipped segments (Fig. 2.14*b*). The activation energies for cross-slip deduced from the minimum energy paths are about 0.68 and 0.67 eV for the LC-type configurations in copper and nickel, 0.47 and 0.65 eV for the glissile-type configurations and about only 0.09 and 0.31 eV for the Hirth-type configurations. For both copper and nickel, these values are substantially much smaller than the activation energies predicted by the Escaig model without a pre-existing constriction. The activation volumes for cross-slip are estimated to be $10–20b^3$ in copper and

$20-30b^3$ in nickel. Thus, whereas the total activation energies are reduced with respect to those of the Escaig model, the activation volumes are still very small. The impact of these events on plastic flow in stage II should be quite limited since the strain hardening rate is almost constant and temperature-independent (Section 2.6.3).

In fcc crystals, cross-slip plays a major role in many aspects of the mechanical response. In this section, the modelling part emphasizes the Escaig model. In spite its complexity, which has dissuaded many readers, this rather old model is still the only one that provides a fair quantitative insight into the energetics of thermally activated cross-slip in the presence of strong obstacles. Atomistic simulations of cross-slip seem now able to yield reasonable values for the stress-free point constrictions and total energies. In addition, a whole spectrum of low-energy mechanisms was identified. However, there is still a lack of insight into most of the important processes that are listed in the next section.

2.4.8 The multiple roles of cross-slip

To some extent, cross-slip and climb play a similar role by providing dislocations of respectively screw and edge characters with an additional degree of freedom. The main difference between these mechanisms is that cross-slip events are strongly sensitive to local stresses and can occur at any temperature, including cryogenic ones, whereas climb is less stress-sensitive and becomes effective only at medium homologous temperatures. The most significant effects by which the cross-slip mechanism influences individual or collective dislocation processes, as well as the mechanical response of fcc metals, can be listed as follows.

- Cross-slip relaxes internal stresses by allowing planar debris, small dislocations loops and dislocation tangles to find stable three-dimensional configurations of low energy. The formation of well-defined, three-dimensional, dislocation patterns during plastic flow, like cell structures, approximately coincides with the onset of thermally activated cross-slip (Hirsch, 1975).
- As discussed further in Section 2.6.4, the thermally activated annihilation of screw dipoles by cross-slip is most probably responsible for the occurrence of the dynamic recovery stage (stage III) in monotonic deformation. For low-symmetry orientations, the corresponding critical annihilation distance is usually assumed to be the same as the one measured in cyclic deformation at the same temperature (see next item).
- In the cyclic deformation of fcc and other crystals, extensive investigations were devoted to very regular wall-and-channel patterns, the persistent slip bands (Section 2.6.6). These structures carry the cyclic strain in single slip

under a saturation stress that is strongly temperature-dependent. Like the stress τ_{III} in monotonic deformation, this stress is governed by thermally activated screw dislocation annihilations. For such annihilations, Mughrabi and co-workers (Essmann and Mughrabi, 1979; Mughrabi *et al.*, 1979) mention a critical annihilation distance of about 50 nm under a saturation stress of 28–30 MPa at room temperature.

- Under low applied stresses in single slip, the annihilation of attractive screw dislocations is stress-assisted and almost athermal. TEM observations of small edge (colinear) segments left by such annihilations suggest critical annihilation distances of about 10–15 nm (Section 2.6.2). During this easy glide stage, the continuous production of colinear segments leads to the formation of characteristic dislocation microstructures (Fig. 2.19); it also governs the strain hardening rate until the onset of stage II (Devincre *et al.*, 2007).
- The double cross-slip mechanism allows screw dislocations to bypass repulsive screw dislocations as well as repulsive or impenetrable localized obstacles like precipitates. It also promotes dislocation multiplication when the two superjogs left in the cross-slip plane are immobile or weakly mobile. These mechanisms contribute to slip propagation into non-deformed volumes in a crystal and to the homogenization of plastic flow.
- The initiation of cross-slip at free surfaces, or interfaces, has been frequently observed by X-ray topography and TEM observations. The mechanism was elucidated by Hazzledine *et al.* (1975). A dissociated screw dislocation reaching a free surface is submitted to the combined effects of image stresses and the orientation dependence of the line energy. For a suitable sign of the perfect dislocation, the two partials rotate towards each other and form a screw-like half-constriction of negative energy at the surface. This may result in spontaneous cross-slip and the emission of cross-slipped dislocations into the crystal.

2.5 Flow stress and dislocation densities

This section is devoted to the study of dislocation strengthening in fcc crystals, that is, the relation between flow stress and dislocation density. The flow stress is almost insensitive to temperature and strain rate, except at low temperatures where jog formation during intersections induces a small material-dependent thermal component (Section 2.5.3). In consequence, dislocation strengthening is mostly controlled by the athermal interactions of moving dislocations with other dislocations.

2.5.1 Dislocation strengthening

Whereas the critical stress for overcoming any type of dislocation obstacle can be calculated using linear isotropic elasticity, estimating the macroscopic average strength resulting from all possible local configurations is not simple. Indeed, one has to account for a spectrum of configurations with different strengths, geometries and probabilities of occurrence. Using dimensional arguments, one can define an average scaled critical stress for all types of obstacles, τ_c/μ. This stress is inversely proportional to the scaled mean distance between dislocation segments in a 3D microstructure or their mean length, $\bar{\ell}/b$. Further, one has $\bar{\ell} \propto 1/\sqrt{\rho}$, where ρ is the total density. Thus

$$\tau_c = \alpha \mu b \sqrt{\rho}, \tag{2.20}$$

where α is a dimensionless constant representing an average interaction strength. Equation 2.20 has same form as the classical relation (Taylor, 1934), which gives the long-range interaction stress of a dislocation in a checkerboard pattern of straight parallel edge dislocations with alternating signs.[4] As the meaning of eqn. 2.20 is quite different, we call this last relation the *dislocation strengthening relation* and α is the *strengthening constant or coefficient.*

To calculate the value of α one has to determine a rule defining how the average strength is estimated. Actually, dislocation theory does not provide any general solution to the question of *composition of obstacle strengths*, which appears as a wide open field for DD simulations. It is generally assumed that the critical stresses are additive for obstacles of different nature, whereas the average densities or concentrations are additive for obstacles of same nature and not too different strengths (Kocks *et al.*, 1975; Kocks, 1979). Several composition laws were tested by two-dimensional numerical simulations of dislocations interacting with point obstacles of various strengths (see Brown and Ham 1971, p. 30), namely the additivity of dislocation densities, of stresses and a law of mixtures for stresses. The rule involving additivity of densities, or equivalently of the stresses squared, is the one giving the best results for a wide range of obstacle strengths. It only becomes inaccurate when the strength of a large number of weak obstacles is not negligible with respect to the strength of a small number of strong obstacles. For this reason, the average critical stress is usually written in the form of eqn. 2.20.

The experimental values of α exhibit, however, a large scatter, as shown by a compilation of Lavrentev (1980). The main cause of uncertainty arises from the measurement of dislocation densities. An extended discussion of the most reliable experimental data given by Sauzay and Kubin (2011) leads to the conclusion that the strengthening constant is in the range of $0.25 \leq \alpha \leq 0.5$ for all fcc crystals. In

[4] This relation was actually not explicitly written by Taylor in his article.

spite of this scatter, the dislocation strengthening relation is surprisingly well verified in individual sets of measurements. Equation 2.20 only holds in materials with low lattice resistance where dislocation interactions govern plastic flow. When the lattice resistance is strong, the critical stresses of the two mechanisms are not additive (Section 3.2.5.1). The value of α may also be modified by composition with other strengthening mechanisms, for instance in dilute alloys (Section 5.3.1).

2.5.2 Forest strengthening

During stage II of the deformation of single crystals, the strain hardening rate is remarkably constant and almost temperature-independent (Section 2.6.3). After several unsuccessful attempts at modelling this stage, the first forest models were proposed by Basinski (1959) and by Hirsch and co-workers (see Thornton *et al.* 1962 and references therein). According to these models, the flow stress is mostly governed by the formation of junctions between mobile dislocations and *forest* dislocations piercing their glide planes (Fig. 2.15). Experimental, theoretical and simulation results further confirmed this hypothesis.

The first calculation of the coefficient α in terms of junction strength was performed by Saada (1960) on symmetrical Lomer locks. The value found after averaging over line orientations was $\alpha \approx 0.4$. Baird and Gale (1965) performed

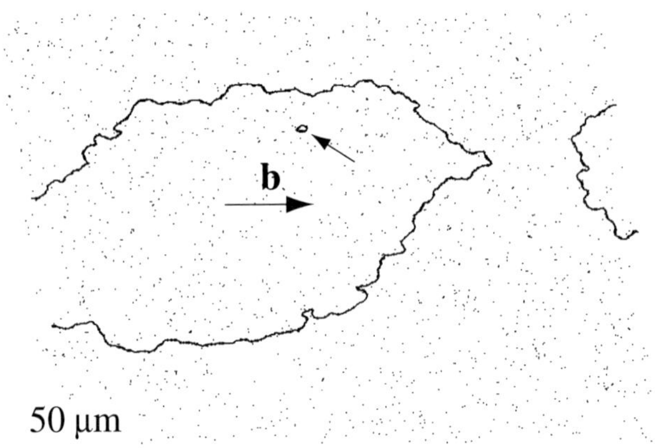

FIG. 2.15. Thin film extracted from a model DD simulation with periodic boundary conditions. A planar dislocation loop expands under stress in a copper crystal. Its intersections with randomly distributed forest dislocations, seen as very short segments, leads to the formation of (non-resolved) glissile junctions. The loop takes a corrugated shape as a result of numerous strong interactions. As a whole, it remains elongated along the screw direction of smaller energy. Notice a small debris left behind by the expanding loop (arrow). Courtesy R. Madec.

quite extensive investigations of the strength of all types of junctions between perfect dislocations in fcc, bcc and hcp crystals. Further studies aimed at reaching more accurate estimates through a more refined treatment of dislocation curvature in hcp crystals (Püschl and Schoeck, 1984) and in fcc crystals. For the latter, Schoeck and Frydman (1972) and Püschl *et al.* (1982) obtained an average junction strength of $\alpha \approx 0.3$.

Figure 2.16 reproduces a compilation of experimental results on flow stress as a function of forest density by Basinski and Basinski (1979), which incorporates data on single crystals and a few polycrystals of copper and silver (rescaled using the shear modulus of copper). The resolved stresses are normalized to room temperature and range from less than 1 MPa to more than 100 MPa. The densities were measured by etch-pit counts at low stresses and by TEM at large stresses. More recent data from DD simulations of copper crystals are included.

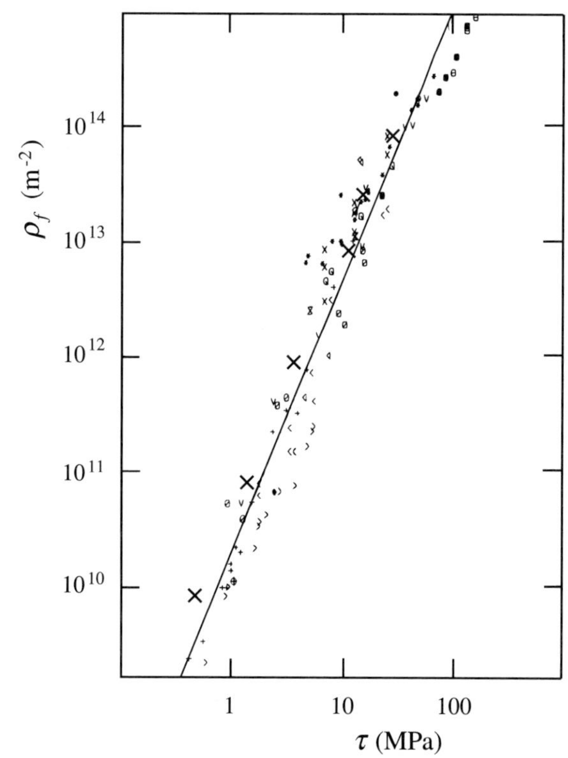

FIG. 2.16. Compilation of experimental results on forest dislocation density versus resolved flow stress in deformed fcc metals. The small symbols refer to sets of results on copper and silver. The full line is a least-squares fit. After Basinski and Basinski 1979. Courtesy S. J. Basinski. The crosses represent data from DD simulations of copper crystals. (Madec *et al.*, 2002*a*).

In spite of the significant scatter in the double logarithmic plot of Fig. 2.16, a linear relation is found to hold over six orders of magnitude in density between the flow stress and the density of forest dislocations. In contrast, a logarithmic plot of primary density versus stress (Basinski and Basinski, 1979) could not be fitted by a straight line. This indicates that dipolar or more distant non-contact interactions do not govern the flow stress.

The linear relation in Fig. 2.16 seems to apply to the low-stress domain of single glide (stage I). This means that, as dipolar hardening does not control the flow stress, a forest mechanism must be involved. A small, unidentified, forest density that amounts to about one-tenth of the primary density was, indeed, detected by etch-pit measurements (Basinski and Basinski, 1964). Its strengthening effect must be very strong in order to be equivalent to the one observed at higher stresses. In fact, most of this forest density is made up of small colinear superjogs lying in the cross-slip plane, which are produced by stress-assisted annihilations of screw dislocations (Section 2.6.2). Thus, dislocation strengthening in stage I is produced by colinear interactions that do not correspond to the activation of a specific slip system. It is convenient to consider this particular mechanism as a self-interaction, as this allows incorporating the primary density into the strengthening relation. For this reason, the latter is actually written in terms of the total density rather than only the forest density.

The exponent affecting the dislocation density in the strengthening relation can be drawn from Fig. 2.16. Its value is not exactly 0.5 but slightly less, 0.43, which is attributed to a small dependence of the line tension on forest density. In eqn. 2.20, μb stands for Γ/b, where the line tension Γ is taken in the simplified form $\Gamma \approx \mu b^2$. When the logarithmic dependence of the line tension (eqn. 1.12) is taken into account, α is no longer a constant. Its value is given by

$$\alpha = \alpha_o \left[\ln\left(1/b\rho_f^{1/2}\right) \Big/ \ln\left(1/b\rho_{fo}^{1/2}\right) \right], \tag{2.21}$$

where ρ_{fo} is a reference dislocation density for which $\alpha = \alpha_o$, and $\rho_f^{-1/2}$ and b represent respectively the outer and inner cut-off radii for the dislocation fields (Section 1.3.1). This logarithmic dependence induces a slow drift of α with increasing forest density, of typically 10% per decade in density. Although this effect may seem tenuous, it matters when wide ranges of dislocation densities are considered. The values of α drawn from the compilation of results by Basinski and Basinski (1979) vary from about 0.5 for the lowest densities to about 0.2 for the highest ones. The fact that the line tension is involved in the strengthening relation constitutes another confirmation that pinning by forest obstacles controls the flow stress.

As mentioned above, Schoeck and Frydman (1972) and Püschl *et al.* (1982) obtained an average junction strength of $\alpha \approx 0.3$, but they did not include the line tension drift. Using as reference the dislocation density implicitly used in their calculations, Gil Sevillano (1993) drew a corrected value of α incorporating the logarithmic drift. In such conditions, the calculated stress values account for almost all (90%) of the experimental flow stresses given by the regression line in Fig. 2.16. The remaining contributions to α can be attributed to repulsive interactions and long-range interactions. As the latter are not affected by a logarithmic drift, their contribution increases with increasing flow stress. DD simulations perform calculations with variable densities that include all contributions to the flow stress except that of jog formation upon intersections. The most reliable estimate yields $\alpha = 0.35 \pm 5\%$ for a reference forest density $\rho_{fo} \approx 10^{12}\ \mathrm{m}^{-2}$ (Madec *et al.*, 2002*a*).

In Fig. 2.16 there seems to be no sensitivity of the flow stress to the various arrangements of the dislocation microstructure formed during the deformation stages. For example, given a high-stress value, the dislocation densities seem to be similar in DD simulations, where the small-strain microstructure is weakly organized, and in real crystals where well-organized dislocation patterns are formed. A possible reason why it is so is as follows. Consider a microstructure containing dislocation-rich walls, characterized by a density ρ_w and a volume fraction f_w, enclosing dislocation-poor cell interiors with a density ρ_c and a volume fraction f_c (see for instance Fig. 2.20). Each of these phases has its own critical stress and the global flow stress can be deduced from a composition rule for these stresses. Kocks and Mecking (2003) showed (p. 260) that when the α-values of the two phases are not too different, which is actually the case (Mughrabi and Ungár, 2002), the difference between the critical stresses of a cell structure and a random distribution of dislocations is within experimental error. Then, the density $\rho = f_w \rho_w + f_c \rho_c$ that enters the strengthening relation (eqn. 2.20) represents both a total and a spatially averaged dislocation density.

In summary, forest strengthening accounts for most of the yield and flow stresses of pure crystals where dislocation interactions govern the mechanical response. Because of the mainly elastic nature of these interactions, the strengthening constant is not much sensitive to temperature, strain rate and dislocation core structure. It is also rather insensitive to patterning effects. As briefly discussed in Section 2.6.6, these properties are shared by the persistent slip bands formed in cyclically deformed fcc crystals.

2.5.3 Jog strengthening

Although the flow stress of fcc crystals is mostly athermal at medium temperatures, it slightly increases at low temperatures. Cottrell and Stokes (1955) attempted to separate the reversible thermal contribution to the flow stress from the irreversible and athermal one. For this purpose, they performed reversible temperature changes at constant microstructure in aluminium polycrystals. The ratio of the two components of the flow stress was found to be approximately a constant that only depends on temperature. This property, which is called the *Cottrell–Stokes law*, was further verified in other fcc crystals. It was interpreted as yielding evidence that the two stresses are governed by the same type of obstacle, specifically forest dislocations. The athermal component mainly arises from short-range reactions between dislocations (Section 2.5.2), whereas the formation of jogs upon dislocation intersections constitute a mechanism that is sufficiently localized to be sensitive to thermal fluctuations. As these two stress components depend on the number of dislocation intersections, they remain proportional to each other during plastic flow.

The Cottrell–Stokes ratio allows estimating the relative sensitivity of the flow stress to temperature changes. It is usually defined from the recorded change in flow stress τ between a temperature T and its linearly back-extrapolated value at $T = 0\,\mathrm{K}$

$$R_{CS} = \frac{1}{T}\frac{\tau(0) - \tau(T)}{\tau(0)} \tag{2.22}$$

Upon deriving this ratio, it is assumed that the flow stress changes are proportional to the temperature changes down to $T = 0\,\mathrm{K}$. When it is so, the Cottrell–Stokes ratio is a material constant.

The Cottrell–Stokes law is valid only during the deformation stage II of single crystals, which is governed by forest interactions (Section 2.6). Beyond this stage, the thermally activated strain hardening rate associated with dynamic recovery comes into play and induces departures from the Cottrell–Stokes law (Mecking and Kocks, 1981). This makes it difficult to determine the temperature at which jog strengthening becomes athermal. Actually, Cottrell and Stokes observed that aluminium specimens strained at low temperatures and restrained at a sufficiently high temperature exhibited a yield drop and lost part of their low temperature strengthening. This irreversible change in flow stress was attributed to a softening mechanism implying some catastrophic release of dislocations. As a consequence, only downwards temperature changes were considered. The recorded instability is now understood as follows (see e.g., Miura and Hamashima 1980). At low temperatures, the specimen is strained in stage II, below the threshold for dynamic recovery and screw dislocation annihilations. The restraining at a higher temperature brings the specimen into the dynamic recovery stage. As a consequence, the annihilations of screw dislocations generated at low temperature and subsequent rearrangements (Section 2.6.4) induce the observed instability.

The formation of jogs during dislocation intersections was discussed by Friedel (1967). During the unzipping of junctions, the maximum stress is reached before the parent dislocations are remobilized. When the parent arms are about to be released (Fig. 2.5*f*), they move with a such a velocity that their kinetic energy is sufficient to provide the formation energy of the jogs. Thus, jog strengthening is thought to arise only from repulsive intersections and the fraction of attractive interactions that do not form junctions. The available models for jog strengthening are based on the elastic theory of dislocations (Seeger *et al.* 1963; Friedel 1967). The main contribution to the activation energy is assumed to arise from a constriction of the lines before the jogs are formed. The model proposed by Friedel (1967) yields a linear temperature dependence of the flow stress in fcc crystals. Jog strengthening is predicted to only matter at low temperatures, in agreement with the available experimental data.

Fig. 2.17 clearly shows that the Cottrell–Stokes ratios increase with the dimensionless SFE (Cousland, 1970). Thus, given the temperature, the ratios $\tau(T)/\tau(0)$ decrease with increasing SFE (eqn. 2.22). This effect may be rationalized in terms of the dependence of the constriction energy on SFE. The measured athermal transition temperatures for jog strengthening are about 100–150 K for Al and 290 K for Ni (Friedel 1967). It follows that the decrease in flow stress between $T = 0$ K and $T = T_a$ is $T_a R_{CS} = 10$–20% in relative value for Al, about 7% for Ni and less for the other fcc metals.

2.5.4 Generalized dislocation strengthening

In the form it was written until now, the dislocation strengthening relation does not contain any information that would allow predicting the number and nature of activated slip systems. Considering all the possible loading orientations, including along zones axes and high symmetry orientations, the six types of dislocations interactions shown in Fig. 2.3 can occur separately or in combination. There is, however, no reason why the interactions between different types of slip systems

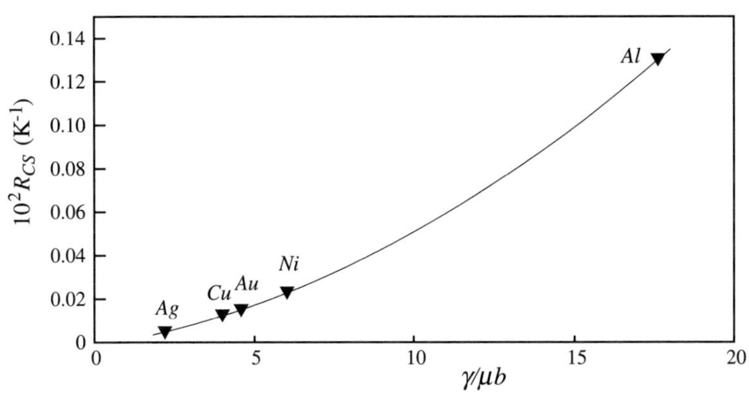

FIG. 2.17. Correlation between the Cottrell–Stokes ratios, R_{CS} for several fcc metals (data from Cousland 1970) and the scaled SFEs (*ab initio* data from Kibey *et al.* 2007).

should exhibit the same strength. To account for these differences, the dislocation strengthening relation has to be replaced by a more detailed expression. More precisely, the athermal critical stress for the activation of a given slip system has to be expressed as a function of the density in each slip system affected by the relevant interaction strength. Such a generalized relation was proposed by Franciosi *et al.* (1980) and is utilized in a model for strain hardening that is discussed in Section 2.6.7. It is written in the form:

$$\tau_c^i = \mu b \sqrt{\sum_j a_{ij} \rho^j}, \tag{2.23}$$

where logarithmic terms are omitted for the sake of simplicity. In this equation, τ_c^i is the athermal critical stress for the activation of slip system (i) and ρ^j is the dislocation density in slip system (j). The summation is carried out over all slip systems including slip system (i). The coefficients $\sqrt{a_{ij}}$ then represent the average strength of the interaction between slip systems (i) and (j) at the scale of a bulk crystal. In fcc crystals, the interaction coefficients a_{ij} are the components of a dense 12×12 matrix. This interaction matrix is diagonal and contains six different coefficients, one for each interaction depicted in Fig. 2.3. The self-interaction coefficient a_{jj} is found along the diagonal and the five other coefficients are distributed symmetrically with respect to it. The interaction matrix of bcc metals is much more complex (Section 5.2.1.3).

An experimental method has been frequently used until the beginning of the 1990s to investigate the strength of the interactions between slip systems. It is based on latent hardening, which is the effect of slip on one slip system on the critical stress of another, inactive, slip system. The experimental procedure is as follows. A specimen is first deformed in tension in single slip on system (i), up to a stress τ_i. A smaller specimen is cut out from the deformed crystal and is further tested in tension along another direction, in such a way as to activate a secondary slip system j that was previously hardened in a latent manner. If the secondary yield stress is τ_j, the ratio τ_j/τ_i is called the latent hardening ratio. This ratio depends on the primary stress or strain, on dislocation densities on the primary and secondary slip systems and on the interaction coefficients a_{ii} and a_{ij}.

The analysis of such tests yielded a semi-quantitative classification of latent hardening ratios by order of increasing value with reference to the self-interaction (Basinski and Basinski, 1979; Franciosi, 1984; Franciosi, 1985; Wu *et al.*, 1991; Bassani and Wu, 1991; Gil Sevillano, 1993). Coplanar slip systems were found to have the smallest ratio and non-coplanar ones the largest ones, especially when glissile junctions and LC locks are formed. No general agreement was obtained regarding colinear slip systems. These experiments cannot always be carried out in ideal conditions and their analysis is not straightforward (Wu *et al.*, 1991). The interaction coefficients can, however, be directly measured when a dislocation-based model for plastic flow is available (Alexander and Crawford, 2000). As shown in Section 5.2.2, the obtained values are then fully consistent with those determined from DD simulations.

2.6 Mechanical response and microstructures

Modelling the mechanical response of single crystals is one of the earliest objectives of dislocation theory, as it represents the first step towards a prediction of the mechanical properties of polycrystals and engineering materials. Although strain hardening is reasonably well understood in terms of elementary dislocation processes, the mechanisms responsible for dislocation patterning (Sections 2.6.7 and 2.7.1) have been and are still the subject of many discussions.

New experimental approaches of intermittent slip events (Section 2.7.2) led to a renewal of interest on the inherently non-uniform nature of dislocation glide, which may in turn provide new insights into the connection between discrete slip events, pattern formation and continuous descriptions of plastic flow. Finally, size effects in small-scale materials are presenting new challenges to the classical dislocation theory; they are reviewed in Section 5.5, in parallel with the contributions of DD simulations to their understanding.

2.6.1 Resolved stress–strain curves

For basic definitions and properties concerning lattice rotations, stable or unstable orientations in tension or compression and the nomenclature of slip systems, the reader is referred to Appendix C.

According to the Schmid law, a slip system is activated when its resolved shear stress becomes larger than its critical stress. These two quantities evolve during plastic flow because of the lattice rotations induced by dislocation glide. As deformation proceeds, the increase of dislocation density in the active slip systems self-hardens them, but also hardens the non-active ones in a latent manner. The interplay between these effects is responsible for the activation of new slip systems and the occurrence of orientation-dependent stress vs. strain exhibiting characteristic stages. In parallel, the dislocation microstructures tend to self-organize in the form of more or less regular spatial arrangements that are similar in single crystals and large grains of polycrystals with same orientation. A common feature of all stages is a reduction in scale of the microstructure with increasing stress, which is discussed in Section 2.6.6.

The first systematic investigation of the deformation stages was performed by Diehl (1956) in a seminal work on pure copper crystals with low-symmetry orientations. The very careful work by Takeuchi (1975), from which Fig. 2.18

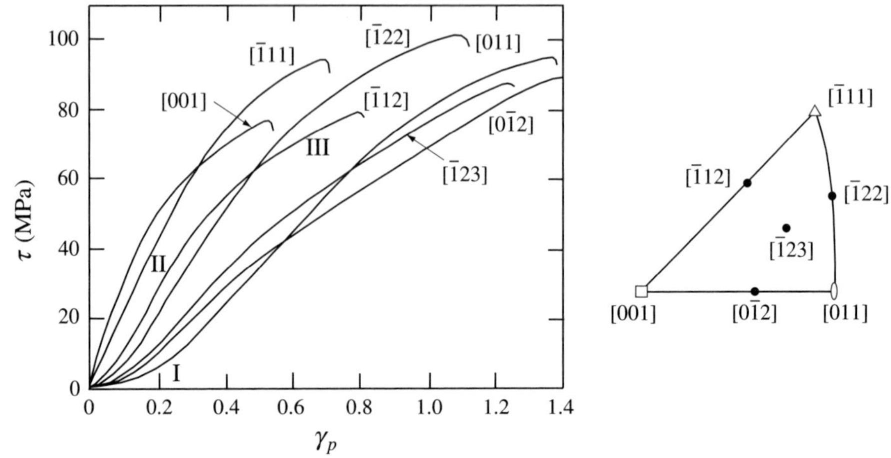

FIG. 2.18. Two- and three-stage shear stress–shear strain curves of copper crystals deformed in tension at room temperature. The initial orientations of the loading axes are shown in a standard stereographic triangle. After Takeuchi 1975.

is drawn, sets a reference for orientations along binary axes and corner orientations. Thcsc investigations were performed in tension, which is the best suited deformation mode for single crystals (Section C.3).

Figure 2.18 shows that some resolved stress–strain curves clearly present three stages, for example those for the low-symmetry orientation [1̄23] and for the unstable high-symmetry orientations [011], [01̄2] and [1̄22]. One can observe successively an initial linear stage with low strain hardening rate, stage I, a second linear stage with a substantially higher strain hardening rate, stage II, and a stage characterized by a pseudo-parabolic strain hardening rate, stage III. For the stable high-symmetry orientations [001] and [1̄11], only the last two stages are observed, whereas the [1̄12] orientation often shows an initial transient stage (Franciosi and Zaoui, 1982). Similar stress–strain curves are found in crystals with other structures than fcc, for example NaCl and Zn (see Gil Sevillano 1993), in pure bcc metals at high homologous temperatures (see Fig. 3.2*a* for Mo and Kumagai *et al.* 1990 for α-Fe), as well as in Si, also at high temperatures (Fig. 3.23).

A considerable amount of experimental work has been devoted to the characterization of the hardening stages. It is summed up in two review articles (Clarebrough and Hargreaves, 1959; Mitchell, 1964) and a recent volume (Argon, 2008). There are also basic texts reporting on extensive TEM investigations of dislocation microstructures (Steeds, 1966; Mughrabi, 1975; Basinski and Basinski, 1979). A thorough discussion of strain hardening mechanisms can be found in a chapter by Gil Sevillano (1993).

2.6.2 Stage I

The *easy glide* stage I is a low-stress stage with a small and approximately constant strain hardening rate, $d\tau/d\gamma = \theta_{\mathrm{I}} \approx 3 \times 10^{-3}\mu$, during which only the primary slip system is active. Near the corner orientations $\langle 001 \rangle$ and $\langle 111 \rangle$, the effect of secondary slips increases the strain hardening rate and reduces the extent of stage I. This extent increases with decreasing temperature and also depends on impurity concentration and specimen surface condition. The shape and cross-sectional dimensions of the samples influence the length of this stage until diameters larger than 1 mm. This effect results from the very large glide distances of dislocations (see Basinski and Basinski 1979 for a review).

A typical dislocation microstructure is shown in Fig. 2.19*a*. In the primary slip plane, dislocations are arranged in the form of dense multipolar bundles perpendicular to the active Burgers vector, which contain a few long edge dislocations strongly interacting with a large density of small prismatic loops (Chiu and Veyssière, 2008; Veyssière *et al.*, 2009). These loops consist of two small edge segments in the primary plane connected by two shorter colinear edge segments in the cross-slip plane.

The quite low density of screw dislocations between the bundles in Fig. 2.19*a* is not a thin foil artefact, as was shown by TEM observations of dislocation microstructures pinned under stress (Mughrabi, 1975). Owing to their large glide paths, mobile screw dislocations have a significant probability for encountering attractive screw segments gliding in very close slip planes. Mutual annihilation may then occur by almost athermal, stress-assisted, cross-slip. Such events leave in the cross-slip plane two edge segments of opposite sign with Burgers vectors colinear to

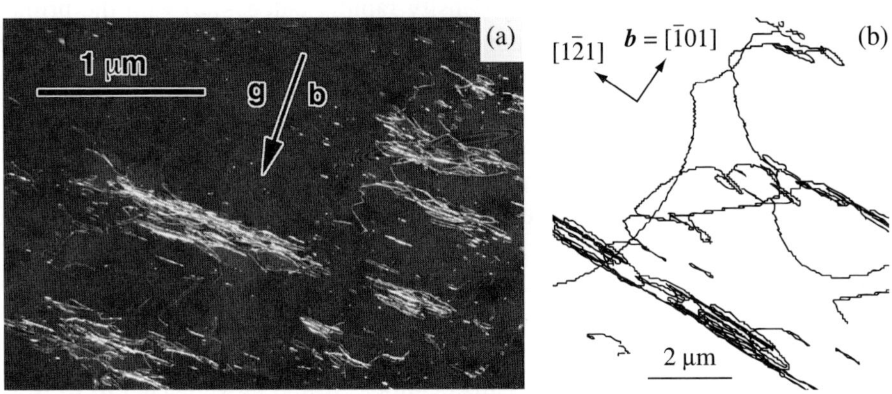

FIG. 2.19. Microstructures of dipolar bundles and small prismatic loops in the primary slip plane during stage I. (a) Aluminium crystal deformed at 77 K (TEM dark field). From Chiu and Veyssière 2008. Courtesy Y.-L. Chiu. (b) DD simulation of a stage I microstructure in copper (see text). The primary slip plane is $\{111\}$. From Devincre *et al.* 2007.

those of the annihilated screws. The heights of isolated prismatic loops and colinear segments outside the bundles correspond to critical annihilation distances in the range of 10–15 nm in Al at 77 K and in copper at 300 K (cf. Fig. 2.19*a* and Steeds 1966).

Further reactions of the primary segments with colinear segments are very frequent and, as a consequence, they govern the evolution of the whole microstructure (Veyssière *et al.* 2009). Figure 2.19*b* shows a DD simulation that reproduces a typical stage I microstructure at a resolved shear strain of 0.8%. As the very large glide distances of screw dislocations cannot be reproduced in DD simulations, the process of stress-assisted screw annihilations was bypassed by introducing a small density (10%) of colinear segments in the initial configuration (Devincre *et al.* 2007). The interactions of these segments with the primary density are responsible for the small strain hardening rate (see Nabarro 1986 and Kubin *et al.* 2008). This is confirmed by Figure 2.16, which shows the occurrence of a forest mechanism at low stresses, and by the absence of long-range internal stresses and dislocation interaction stresses in stage I (Mughrabi, 1975).

2.6.3 Stage II

The transition from stage I to stage II arises from the onset of slip in a secondary system, the conjugate system (Fig. 2.25). Then, the strain hardening rate increases rather abruptly up to $\theta_{\text{II}}/\mu \approx 3 \times 10^{-2}$. This value depends on the number of interacting active slip systems and varies by a factor of about two with orientation (Fig. 2.18); it is also almost insensitive to strain rate and temperature, except for aluminium at low temperatures (Section 2.5.3). For orientations in the middle of the standard triangle, the secondary density rapidly catches up with the primary density. However, the mean free path of moving secondary dislocations is limited by the primary microstructure so that plastic strain is predominantly carried out by primary dislocations (Fig. 2.25).

In the detail, the dislocation arrangements are quite complex. They tend to be crystallographic at low temperatures and more ragged at room temperature. Their evolution during stage II is discussed in some detail by Basinski and Basinski (1979), Prinz and Argon (1980) and Argon (2008). TEM observations reveal a layered, or 'carpet', structure with two types of dislocation-dense slabs approximately parallel to the primary slip plane, the network and bundle regions, which are separated by dislocation-poor regions. The network regions contain extended networks of primary and secondary dislocations, which have reacted to form LC locks. The bundle regions are less dense; they consist of coalesced and evolved edge dislocation bundles inherited from stage I, which are stabilized by interaction with secondary dislocations. Near the end of stage II, the bundle regions evolve to form thick-walled 2D cell structures composed of primary and secondary dislocations (Fig. 2.20*a*).

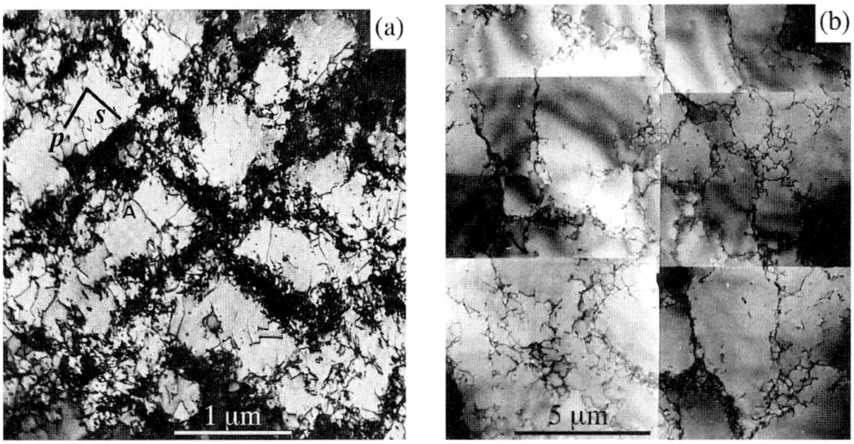

FIG. 2.20. Cell structures. (a) A 2D grid structure with thick walls in copper (end of stage II, $T = 4.2\,\mathrm{K}$, $\tau = 10^{-3}\mu$). The two sets of walls are approximately parallel to the traces of the primary (*p*) and secondary (*s*) slip planes. After Basinski 1974. (b) A well-recovered 3D cell structure with low-density walls in a $\langle 001 \rangle$ aluminium crystal after 5% strain in stage III ($T = 293\,\mathrm{K}$, $\tau \approx 2.5 \times 10^{-4}\mu$). Courtesy X. Huang.

As the critical slip system is only occasionally activated (Grosskreutz and Mughrabi, 1975), the mostly athermal strain hardening rate is attributed to forest interactions with the conjugate system and LC lock formation. The same conclusion holds for the microstructures formed around the two stable corner orientations $\langle 001 \rangle$ and $\langle 111 \rangle$, except that the larger number of interacting slip systems results in the formation of various types of junctions (Fig. C.5).

2.6.4 Stage III

Until the end of stage II, the flow stress is essentially governed by athermal dislocation mechanisms. The transition between stages II and III is not well defined and occurs around a stress τ_{III}. In stage III, the strain hardening rate decreases almost linearly with increasing stress, all the more as temperature increases and the strain rate decreases. Since the relation between stress and dislocation density is mostly athermal, the rate at which the dislocation density accumulates in the crystal is decaying in a thermally activated manner. The process responsible for this effect is called *dynamic recovery*, by opposition to static recovery that occurs in stress-free crystals.

Hirsch (1975) noted that in stage III the cell patterns become well-marked, while the cell walls become thinner (compare Figs. 2.20*a* and 2.20*b*). He attributed these changes to activated annihilations of screw dislocations by cross-slip.

This relation between cross-slip and dynamic recovery can be approximately verified by comparing the experimental athermal transition temperatures for dynamic recovery in fcc crystals to those for cross-slip drawn from eqn. 2.16. In aluminium at room temperature, there is an almost direct transition from stage I to stage III and stage II reduces to an inflection point. This arises from the low athermal transition temperature for cross-slip ($T_a = 320-350\,\text{K}$) induced by the high SFE of this material. A similar behaviour is observed above the athermal threshold in bcc metals (Fig. 3.2*a*), which also exhibit narrow dislocation cores. Finally, the close similarity noted in Section 2.6.6 between the critical stresses associated with thermally activated screw dislocation annihilations in persistent slip bands and the stress τ_{III} leaves little doubt about the nature of the mechanism underlying dynamic recovery.

The critical distance for thermally activated screw dislocation annihilations is usually assumed to be the same as in persistent slip bands, about 50 nm in copper at 300 K for low-symmetry orientations. This is substantially larger than typical values in the 10–15 nm range associated with stress-assisted cross-slip in stage I. As in stage I, dislocation annihilations are accompanied by the formation of colinear superjogs. The latter are extremely reactive and further induce athermal annihilations of screw and non-screw segments (Fig. 5.2). A number of subsequent rearrangements occur in the cell walls (Jackson, 1985), which increase their stability and reduce the internal stress. It follows that dynamic recovery involves a cascade of athermal processes triggered by cross-slip.

The stress τ_{III} looks like a kind of critical stress for cross-slip that is modified by the field of a neighbouring attractive screw segment. Because of this additional complication, it is not certain that all its properties can be captured by the Escaig model. For instance, the dependence of the critical stress on the orientation of the loading axis is well-marked near corner orientations; τ_{III} is rather low near $\langle 001 \rangle$ and rather high near $\langle 111 \rangle$. This feature allows explaining why the stress–strain curves for these two orientations cross each other in Fig. 2.18 (see Section 5.4.2 and Fig. 5.12*b*).

Although the prevailing view about the nature of dynamic recovery was reported in this section, there are still discussions about it. An alternative mechanism for the onset of stage III and subsequent hardening was discussed by Argon (2008, 2009). It does not involve cross-slip but the removal of LC locks in cell walls.

2.6.5 Stage IV

The continuous decrease of strain hardening rate with increasing stress that characterizes stage III is interrupted at large strains by the occurrence of stage IV, which is material and temperature dependent. Stage IV is a stage of sustained hardening that lasts until very large strains. It is characterized by a small, constant strain hardening rate ($10^{-4} < \theta_{IV}/\mu < 10^{-3}$), which depends on temperature and back-extrapolates to stage I hardening.

Strains typical of the end of stage III, or much larger, can easily be obtained in single crystals at medium and high temperatures (Anongba *et al.*, 1993*a*, 1993*b*) or by deforming polycrystals with reasonably large grain sizes (50–300 μm) in torsion or by cold-rolling at room temperature. This allowed performing a quantitative characterization of previously observed microstructures at the end of stage III and in stage IV (see e.g., Bay *et al.* 1992 and Hansen *et al.* 2010). The mechanical response in stage IV, including the beautiful experiments by Alberdi (1984) on polycrystals of copper and aluminium deformed in torsion, is summarized and discussed by Gil Sevillano (1993) and Rollet and Kocks (1994).

The dislocation walls or cells formed in the bundle regions during stage II evolve during stages III and IV. The misorientations across the dislocation walls increase very slowly, through the statistical trapping of glide dislocations (Argon and Haasen, 1993), reaching about one degree for shears $\gamma \approx 1$ (Liu *et al.*, 1998; Hansen *et al.*, 2001). The cell boundaries are therefore called incidental dislocation boundaries (IDBs). In addition, TEM observations show the gradual formation of another microstructure during stage III and its further evolution during stage IV. This microstructure consists of sharp and dense extended walls containing an excess of dislocations of same sign. These dislocation walls accommodate rotations between increasingly misoriented volumes deforming in a different manner; they originate from the network structures formed during stage II and are called geometrically necessary boundaries (GNBs). The average misorientations between GNBs grow from a few degrees to nearly ten degrees for shears $\gamma \approx 1$ (Liu and Hansen, 1995; Liu *et al.*, 1998; Hansen *et al.*, 2001). Thus, the whole microstructure becomes in stage IV a pattern of GNBs enclosing volumes called cell blocks, which contain 'ordinary' dislocation cells or walls (Fig. 2.21*a*).

The small misorientations across IDBs indicate that the cell walls are penetrable to dislocations, so that the same slip systems operate through all the cells of a cell block. The distributions of misorientation angles across GNBs and IDBs was investigated over a very wide range of strains. A universal scaling law was found to hold for the two types of boundaries taken separately when, for each of them, the angles are scaled by their average value (Hughes *et al.* 1997, 1998, Pantleon and Hansen 2001). The average spacings between GNBs and IDBs decrease like the inverse of the average angles, that is, faster for GNBs than for IDBs (Hughes and Hansen, 2000; Hughes, 2001).

A close correlation was established in copper, nickel and aluminium between the dislocation microstructures found in deformed single crystals and grains of polycrystals with same orientations; these results are reviewed by Hansen *et al.* (2010). Three orientation domains corresponding to similar microstructures were defined (Fig. 2.21*b*).

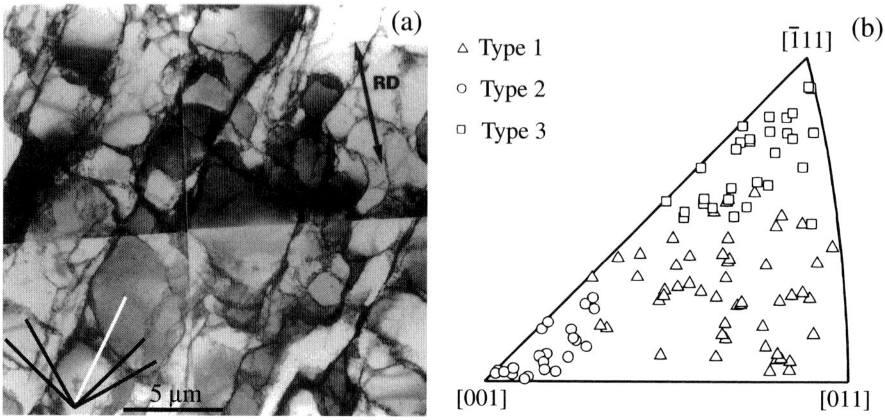

FIG. 2.21. (a) Longitudinal section of a cell block structure in an aluminium polycrystal cold-rolled by 10% at room temperature (the grain size is 300 μm, RD is the rolling direction). The average direction of extended GNBs (white line) is not parallel to the trace of {$\bar{1}$11} slip planes (black lines). Dislocation cells can be seen between the GNBs. After Liu *et al.* 1998, courtesy G. Winther. (b) Grain orientation dependence of the three types of microstructures in polycrystalline copper with a grain size of 90 μm deformed in tension up to true strains of 0.15–0.2. From Huang and Winther 2007.

- The type 1 structure is found in the middle of the standard triangle. It consists of extended crystallographic GNBs nearly parallel to a slip plane, which define cell blocks containing dislocation cells.
- The type 2 structure is found for orientations around [001]. It consists of ordinary dislocation cells with non-crystallographic walls (Figs. 2.20*b*, 2.22 and 2.24) and almost no net Burgers vector in the cell walls. As discussed in the next section, these microstructures exhibit a fractal structure. GNBs were however observed in a commercial aluminium alloy strained at 250 °C (Cizek, 2010), owing to very active dynamic recovery at this temperature.
- The type 3 structure is found in an extended neighbourhood around [$\bar{1}$11]. It is similar to the type 1 structure, except that the planes of extended GNBs are non crystallographic and significantly deviate from {111} orientations (Fig. 2.21*a*). Although IDBs can be observed, it is often difficult to distinguish ordinary cells inside the cell blocks, especially close to the [$\bar{1}$11] orientation (Kawasaki and Takeuchi, 1980; Huang, 1998).

The scaling laws for microstructural parameters mentioned above apply to both stages III and IV (Hughes, 2001; Hansen *et al.*, 2001; Hansen *et al.*, 2010). Thus, the spacings between IDBs and GNBs continuously refine with increasing stress or strain, while the average misorientation angles increase. In stage IV, the microstructure is dominated by the cell block structure, in which an increasing fraction exhibits misorientations of a few tens of degrees and becomes impenetrable to dislocations. It follows that strain hardening is governed by the evolution of dislocation arrangements in GNBs.

Stage IV is less well understood than the previous stages. The successive models proposed to explain strain hardening were reviewed by Gil Sevillano (1993), Rollet and Kocks (1994), Kocks and Mecking (2003) and Pantleon (2004). In broad terms, one can distinguish two main categories of models. In the first category, the evolution of a heterogeneous two-phase structure of misoriented GNBs and dislocation cells is assumed to generate long-range dislocation interactions that account for the persistence of strain hardening during stage IV (Argon and Haasen, 1993; Argon, 2008). In the second type of models, dislocation storage and recovery are still active during stage IV, but are modified by an additional density that cannot be totally annihilated by dynamic recovery. This density is assumed to consist of dislocation debris (Rollet and Kocks, 1994), of small prismatic loops, by analogy with stage I microstructures (Gil Sevillano, 1993), or by excess dislocations stored in GNBs (Pantleon, 2004). These last models can be formulated as modifications of the storage–recovery framework that is discussed in Section 2.6.7.

Stage IV is followed by a final domain of decreasing strain hardening rate that eventually leads to stress saturation or fracture. Depending upon the authors (see Kocks and Mecking 2003), this domain is considered or not as a distinct stage, stage V.

2.6.6 Similitude and self-similarity

Scaling laws involving averaged microstructural parameters provide simple relations that are frequently used for the dislocation-based modelling of plasticity. As they smoothen out probability distributions and small-scale features, they are somehow equivalent to what could be obtained from a mean-field approach of dislocation patterning. The strengthening relation between flow stress and the square root of total, spatially averaged, dislocation density (eqn. 2.20) is a typical example of such scaling laws. It is also a common observation that the average characteristic dimension of dislocation patterns refines with increasing stress. Figure 2.22 allows visualizing this property in a GaAs wafer, which was submitted to small non-uniform stresses during cooling after solidification.

Experimental measurements, yield an inverse relation between the resolved flow stress and the average dimension $\bar{d}$ of dislocation patterns.

$$\frac{\tau}{\mu} = K\frac{b}{\bar{d}} \tag{2.24}$$

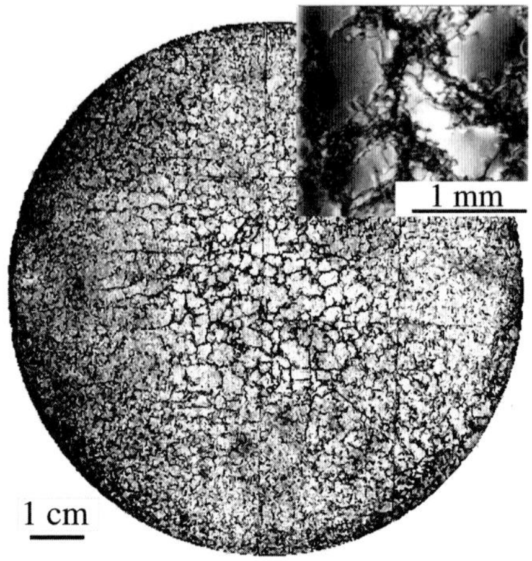

FIG. 2.22. Giant dislocation cells and similitude in an as-grown GaAs wafer with ⟨001⟩ axis. The cell dimensions, as revealed by etching, are in the range of $10^{-3} - 10^{-2}$ m. They decrease from the centre to the periphery, whereas thermomechanical stresses that occurred during cooling increase from typically 0.1 MPa to 2 MPa. The inset shows an X-ray synchrotron topograph of cells at a larger magnification in another GaAs crystal. After Rudolph 2005, topograph by T. Tuomi. Courtesy P. Rudolph.

Like the strengthening coefficient α, K is a dimensionless coefficient common to many materials. Equation 2.24 does not apply to cell blocks (Section 2.6.4) and to materials with a strong lattice resistance. The scaling relation 2.24 is often called the similitude principle (Kuhlmann-Wilsdorf, 1962). It is, however, purely empirical and has never been convincingly modelled. Hence, we simply call it the *similitude relation*.

A compilation of experimental results on similitude in monotonic and cyclic deformation, can be found in a review article by Sauzay and Kubin (2011). The dislocation densities were mostly estimated by TEM on cell structures; they are not very accurate and yield similitude coefficients in the range of $K = 5$–10. The lower values derive from measurements of the diameters of cell interiors, usually in single crystals (see e.g. Ambrosi *et al.* 1974). Values derived from the true size, which also includes a cell wall thickness, are more reliable, because they are numerous and more accurate. A huge compilation of results by Raj and Pharr (1986) on subgrains and cells formed in many materials, mainly polycrystalline, at various temperatures, in tensile and creep deformation, yields $K \approx 23$ in terms of axial stresses. These stresses can be converted into shear stresses using an average Schmid factor $1/M$, where $M \approx 3.06$ is the Taylor average for fcc crystals (Hosford 1993). As a result, $K \approx 7.5$ (Kocks and Mecking, 2003). An earlier work on copper polycrystals deformed at a fixed strain and different temperatures by Staker and Holt (1972) yielded an often quoted value of $K = 10.5$. The analysis

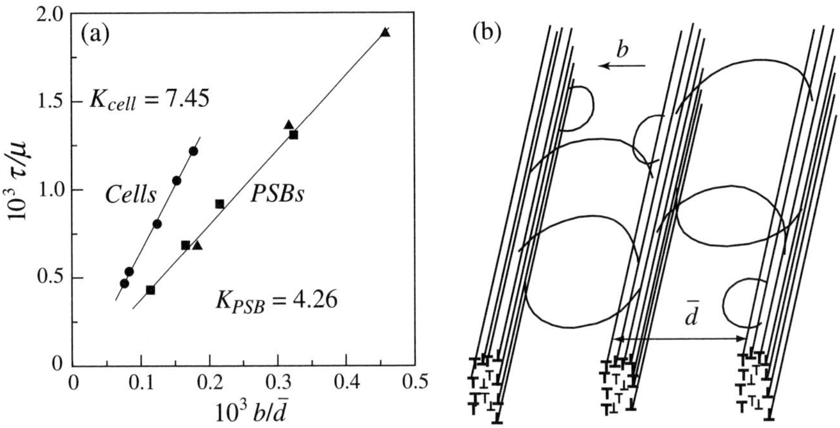

FIG. 2.23. (a) The similitude coefficient K in copper. (a) Scaled resolved flow stress in tension or cyclic saturation stress for PSBs versus the scaled wavelength of the microstructure, $\bar{d}/b$. The bullets represent revisited data from Staker and Holt (1972) for dislocation cells in tension between 300 and 823 K. Black triangles and squares represent PSB data from Basinski and Basinski (1992) and Holzwarth and Essmann (1993, 1994). (a) Schematic dislocation microstructure of PSBs in fcc metals (see text). From Mughrabi 2009.

of the results is, however, inaccurate, in particular because of the use of an average Schmid factor of 0.5. Using again the Taylor factor, one obtains $K = 7.45$ (Fig. 2.23*a*). Taking into account a few single crystal data, corrected when necessary to include a cell wall thickness, the most reliable values of the similitude constant seem to be in the range of $K = 7.5 \pm 0.5$. The error margin may be larger because, as will be discussed in what follows, the thickness of cell walls does not fully follow similitude. Furthermore, a dependence of cell size on single crystal orientation is not excluded (Section 5.4.3). The few available experimental data are, however, too uncertain to allow checking this possibility.

Another experimental evidence for similitude arises from the cyclic deformation of fcc single crystals oriented for single slip (Laird, 1996; Mughrabi, 2009). Under a low plastic strain amplitude per cycle, the maximum resolved stress of the cycles saturates at a constant value. This saturation stress corresponds to the localization of plastic flow in primary bands containing a very regular structure of walls and channels (Fig. 2.23*b*), the persistent slip bands (PSBs). The peculiarity of PSBs is to be active in dynamic steady-sate conditions both in the walls and the channels. These steady-state conditions can persist for extremely large numbers of cycles. A large part of the cyclic strain is carried out in the channels by the back and forth motion of screw dislocations. The latter are produced by segments bowing out from the walls; they mutually annihilate by thermally activated cross-slip. As a consequence, the saturation stress exhibits a strong temperature dependence; actually, both its values and its temperature dependence are very close to those of the stress $\tau_{\rm III}$ for the onset of the dynamic recovery stage (see Fig. 29 of Sauzay and Kubin 2011). As the PSB widths are also temperature-dependent, the similitude coefficient can be measured by varying the temperature. Fig. 2.23*a* shows that a similitude relation is obtained, but with a constant K_{PSB} that is significantly smaller than in monotonic deformation. The value of the strengthening constant α is also reduced. These effects may be attributed to the partial reversibility of cyclic deformation (Sauzay and Kubin, 2011).

As yet, the origin of the similitude relation is not understood, probably because it is related to another unsolved problem, that of dislocation pattern formation (Section 2.7.1). Combining eqns. 2.20 and 2.24, one obtains

$$\bar{d}\sqrt{\rho} = \frac{K}{\alpha} = const. \tag{2.25}$$

With $\alpha = 0.3 - 0.4$ (Section 2.5.2), $K/\alpha \approx 20$. Since the strengthening coefficient α is governed by dislocation intersection processes, the same property should apply to dislocation patterns (Section 5.4.3). In a coarse-grained dislocation pattern with a total density ρ, we consider the average length of interacting segments, $\bar{\ell} \propto \rho^{-1/2} \propto 1/\tau$. Then, eqn. 2.25 shows that $\bar{d}/\bar{\ell}$ is also a constant. This last relation can be considered as the basic postulate of similitude; it was experimentally verified in copper crystals deformed in stage II (Prinz and Argon, 1980; Argon, 2008). It also follows from eqn. 2.25 that the similitude coefficient may exhibit, like the coefficient α, a small line tension drift. The latter cannot, however, be evidenced by lack of sufficiently accurate data.

Cell patterns actually have two average characteristic lengths; one is associated to the wall thickness and the other to the diameter of cell interiors. Full similitude would imply that both lengths are inversely proportional to stress with the same similitude coefficient. However, the onset of dynamic recovery in stage III should partially break similitude in the cell walls, all the more as temperature is high. It was, indeed, noted by Staker and Holt (1972) and Kocks and Mecking (2003) that the thickness of cell walls decreases with increasing temperature and decreasing stress for a given strain. There are, however, experimental evidences that in copper at room temperature the volume fraction of walls is proportional to the volume fraction of cell interiors until large strains (Knoesen and Kritzinger 1982; Neuhaus and Schwink 1992). Thus, similitude may hold in an approximate and weakly temperature-dependent manner in stage III.

Similitude implies that under a varying stress a microstructure remains self-similar, that is, similar to itself. This property applies to periodic or pseudo-periodic patterns where the average cell size is well-defined. The cell structures of type 2 discussed in Section 2.6.5 are more complex, as they exhibit a

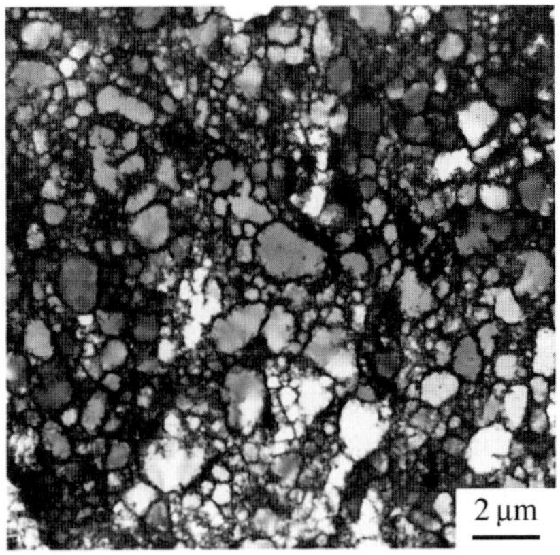

FIG. 2.24. Type 2 dislocation cell structure in a copper polycrystal deformed in tension by $\varepsilon \approx 0.2$ at room temperature. The thin foil is perpendicular to the tensile axis of a grain. From Hansen *et al.* 2010.

fractal structure (Fig. 2.24). A fractal microstructure appears exactly or statistically identical to itself at all magnifications. It is then self-similar in space and exhibits a full range of length scales. This range is, however, always bounded, which allows determining an average length scale. The experimental average length scale is found to follow similitude (for a discussion, see Zaiser *et al.* 1999; Zaiser and Hähner 1999).

TEM examinations on copper polycrystals cold-rolled at room temperature suggested that dislocation cells exhibit a fractal structure (Gil Sevillano *et al.*, 1991), which disappears at large axial strains. Deformation-induced cell structures formed in $\langle 001 \rangle$ Cu and Cu–Mn crystals were thoroughly analysed by Hähner *et al.* (1998), Zaiser *et al.* (1999) and Zaiser and Hähner (1999), who showed that cell interiors have a fractal structure with a stress-dependent fractal dimension. These evidences were obtained in sections perpendicular to the stress axis similar to the one shown in Fig. 2.24. In sections parallel to the stress axis, the cell structure is anisotropic and elongated along the straining direction (Huang, 1998; Kawasaki and Takeuchi, 1980). The domain of occurrence of such fractal structures seems to be restricted to type 2 dislocation cells. In that case, it was often noted upon measuring the similitude coefficient that the distribution of sizes was rather wide. No convincing evidence for a fractal dimension was obtained for other orientations, especially around $\langle 111 \rangle$.

2.6.7 The storage–recovery model

Ideally, models for strain hardening should be based on a small number of relevant mechanisms and microstructural features extracted from the existing wealth of experimental observations. As a simplified vision of the dislocation microstructures has to serve as a starting point, controversies have persisted for a long time, for instance about the respective weights of short- and long-range interactions or the nature of the dislocation mechanisms responsible for the deformation stages. A critical discussion of the traditional microstructural approaches to strain hardening is given by Nabarro (1986). Various models are also collected in a set of chapters assembled by Nabarro and Duesbery (2002) and in the book by Argon (2008).

The storage–recovery framework proceeds differently. It is based on a rather simple set of scalar, continuum equations with length scales (Kocks, 1976). Further developments are reviewed by Kocks and Mecking (2003). This approach treats the evolution with time or strain of the total dislocation density in a uniform microstructure, under the influence of three generic mechanisms, dislocation strengthening, storage and dynamic recovery. Such a formulation is particularly suited for multiscale modelling because of its kinematic character and its potential ability to reproduce stress–strain curves and their deformation stages in a seamless manner. A generalized version that treats uniform dislocation densities per slip systems was proposed by Teodosiu *et al.* (1993). Its output is much richer than that of the scalar model because it predicts the nature and behaviour of the active slip systems. The price to pay is, however, a number of well-defined but poorly known parameters. Figure 2.25 shows the transition between stages I and

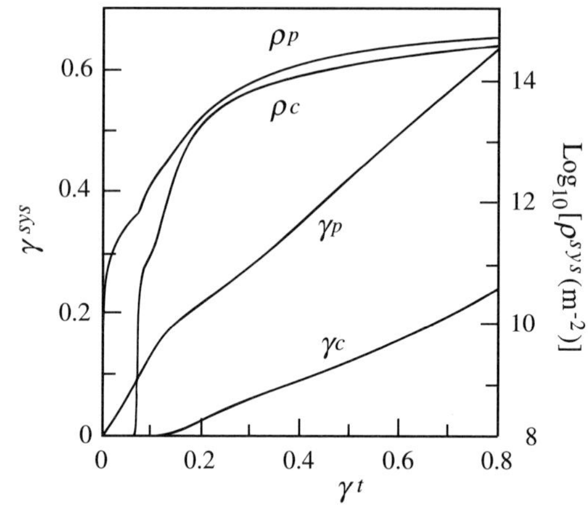

FIG. 2.25. The transition between stages I and II, as yielded by a generalized storage–recovery model. Evolution of plastic shear strains (γ^{sys}) and densities (ρ^{sys}) per slip system as a function of the total shear strain γ^t. The indices p and c refer respectively to the primary and conjugate slip systems. There is no free parameter. From Devincre *et al.* 2008.

II in copper at room temperature, as obtained from a generalized storage–recovery model revisited with the help of DD simulations (Section 5.4.2).

In what follows we describe a basic generalized storage–recovery model that is devised for *isotropic hardening*, here strain hardening in tension, which is the component of strain hardening that is insensitive to changes in strain path. It is first assumed that the dislocation microstructure mostly consists of dislocations immobilized (stored) by interaction with other dislocations. In each slip system (i), the stored density ρ^i is much larger than the instantaneous mobile density and is assimilated to the total density. It is also assumed that there are always enough mobile dislocations to carry the imposed strain rate and compensate for losses by storage and annihilation. Thus, the continuous storage of mobile dislocations is accounted for, whereas the mobile density does not explicitly appear in the model. These simplifications appear to be justified by the intermittent nature of plastic flow at a fine scale (Sections 2.7.2.2 and 5.4.1). The first equation of the model is the generalized dislocation strengthening relation, eqn. 2.23, which relates the critical resolved stress τ_c^i to the stored uniform densities ρ^j per slip system.

The most original part of the model requires defining the mean free path of dislocations, which is done first in scalar terms. Consider a segment of average length $\bar{\ell}$ moving by an incremental step δx. In a volume V, it produces a shear strain $\delta\gamma = b\bar{\ell}\,\delta x/V$. If it is immobilized, the stored density is incremented by $\delta\rho = \bar{\ell}/V$. The mean free path Λ is the average length travelled by mobile segments during their flight time. In the present case, the storage probability is $\delta x/\Lambda$ and the storage rate is $(\delta x/\Lambda)(\delta\rho/\delta\gamma) = b/\Lambda$. As the length of a dislocation segment changes when it moves, the mean free path is a virtual quantity that has to be defined for infinitesimal steps of motion.

In the general formulation for fcc crystals, a mean free path Λ^{int} is associated to each of the six types of interactions between slip systems. For the self-interaction, use is made of the definition given in Section 2.6.2. The storage rates arising from these interactions are additive, so that the inverse of the mean free paths are also additive. The net storage rate in slip system (i) during tensile deformation is then expressed in terms of the resolved shear strain rate $\dot{\gamma}^i$ by the differential form

$$\frac{d\rho^i}{d\gamma^i} = \frac{\dot{\rho}^i}{\dot{\gamma}^i} = \frac{1}{b}\left(\sum \frac{1}{\Lambda^{int}} - y_{hkl}\rho^i\right) \tag{2.26}$$

The first term at the right-hand side is the sum of all contributions to the storage rate. The next term, where y_{hkl} is an orientation-dependent length, accounts for dynamic recovery through a negative contribution of system (i). As shear strain rates are proportional to mobile densities (eqn. 1.21), the product $d\gamma^i\rho^i \propto \rho_m^i\rho^i$ represents an event that leads to annihilations of mobile and stored densities in system (i). This event is, therefore, assimilated to the annihilation of screw dislocations by thermally activated cross-slip and subsequent rearrangements.

In this formulation, strain hardening is represented by a matrix with coefficients

$$\theta_{ij} = \frac{d\tau_c^i}{d\gamma^j} = \frac{d\tau_c^i}{d\rho^j}\frac{d\rho^j}{d\gamma^j}, \tag{2.27}$$

where the terms at the right-hand side are obtained from the derivative of eqn. 2.23 with respect to ρ^j and from eqn. 2.26. The model is set in closed form by an equation relating the ratio of applied to athermal shear stresses to the shear strain rate in each slip system. For active slip systems, this ratio is close to one and accounts for jog formation at low temperatures (Section 2.5.3).

Simple solutions are obtained for the scalar version of the model (Kocks and Mecking, 2003). This scalar version is obtained by replacing eqn. 2.23 by eqn. 2.20, removing all indices from eqn. 2.26 and lumping all mean free paths into a single length Λ. Then, the mean free path follows similitude (Kubin *et al.*, 2008) and $\Lambda = K_\Lambda \mu b/\tau_c$, where K_Λ is a mean free path coefficient. In stage II, dynamic recovery is not active and it can be verified easily that the model yields a constant strain-hardening rate $\theta_{\mathrm{II}}/\mu = \alpha^2/2K_\Lambda$. Upon reintroducing dynamic recovery, the same set of equations is easily solved as well, yielding the strain hardening rate in stage III

$$\theta_{\mathrm{III}} = \theta_{\mathrm{II}}\left(1 - \frac{\tau_c}{\tau_{\mathrm{III}}^s}\right); \qquad \tau_{\mathrm{III}}^s = \alpha^2 \mu b/yK_\Lambda, \tag{2.28}$$

In this equation, the stress τ_{III}^s is a virtual saturation stress obtained by extrapolation to $\theta_{\mathrm{III}} = 0$. This relation, where the strain hardening rate linearly decreases with increasing stress, is called a Voce law. Other obstacles than dislocations can be implemented in the general and scalar models. For instance, in a polycrystal with grain size d_g, the mean free path of dislocations is proportional to $1/d_g$. It is easily checked that when storage at grain boundaries predominates, the scalar form of the model yields a strain-dependent strengthening of the Hall–Petch type.

As it is formulated in an adequately scaled form, the storage–recovery model successfully recovers strain hardening in stages II and III without incorporating a description of the dislocation microstructures. It is mainly devised for the modelling of plastic flow and textures in polycrystals, but has some limitations. It does not apply to large strains beyond stage III and it does not account for *kinematic hardening*. Kinematic hardening is the contribution to strain hardening that arises from a change in strain path, for instance a strain rate reversal from tension to compression or a change in the orientation of the loading axis. These deficiencies are currently overcome by incorporating into the model various phenomenological complements (see e.g., Peeters *et al.* 2001*b*).

The constitutive equations of the generalized model are usually solved by crystal plasticity FE codes, which are particular FE codes dealing, in addition to boundary conditions, with the crystallographic aspects of plastic flow (Dunne and Petrinic, 2005; Roters *et al.*, 2010). This establishes a connection with the continuum mechanical framework at the level of slip systems, which allows going much further than classical models. For instance, it is no longer necessary to make use of the traditional simplifications of continuum mechanics for determining the active slip systems and plastic strains inside polycrystal grains (Teodosiu *et al.*, 1993).

Because of its flexibility, the storage–recovery model is widely used in mechanical engineering. For an overview of the methods and their applications one may consult Roters *et al.* (2010), as well as a few representative publications (Zikry and Kao 1996; Peeters *et al.* 2001*a*, 2001*b*; Arsenlis *et al.* 2004; Ma *et al.* 2006). The situation is, nevertheless, not totally satisfactory because a number of free parameter values are drawn from experiment, especially in the generalized formulation. Truly predictive modelling implies removing phenomenology as much as possible and reducing the number of free parameters. Progress in this direction was achieved by combining dislocation modelling and DD simulations (Sections 5.2.2 and 5.4.2).

2.7 Collective dislocation behaviour

This section is dealing with two topics related to collective dislocation phenomena. Dislocation patterning is a long-standing problem that is still raising new questions and new debates. In the past years, experimental studies on dislocation avalanches yielded a new insight into the inherently heterogeneous nature of plastic flow at a fine scale. Collective dislocation behaviour is now treated by models that introduce into the field of crystal plasticity methods of non-linear dynamics and statistical mechanics for dissipative systems driven far from equilibrium. A general review of these methods is given by Ananthakrishna (2007).

2.7.1 The modelling of dislocation patterns

Traditional models for dislocation patterning lack a framework that would allow treating the collective behaviour of dislocations in physical terms. Various explanations were proposed for the formation of ordinary cell structures (see Kubin 1993 for a review). The similitude relation is the benchmark test for all models; it was, however, assumed without proof most of the time and derived only once, using an incorrect demonstration (Holt 1970). Arguments based on the hypothesis that patterns are in thermodynamic equilibrium are also frequently found in the literature (see for instance Kuhlmann-Wilsdorf 2002). Such an approach, however, does not apply to the metastable dislocation microstructures formed during plastic flow. A kinetic approach was used in the percolation and statistical storage models for mobile dislocations interacting with forest dislocations (see Kocks and Mecking 2003). The latter can somehow be viewed as an early anticipation of the not-yet-discovered avalanche mechanisms.

In the mid-1980s it was fully realized that the strongly dissipative and irreversible nature of plastic flow precludes any analysis in terms of equilibrium or close-to-equilibrium thermodynamics (Section 1.1). A generation of models then appeared, which were based on concepts borrowed from non-linear dynamics. The coupled evolution of dislocation populations in time and space was modelled under stress and the instability of their uniform solutions against spatial fluctuations allowed obtaining preferred wavelengths for non-uniform solutions. The first attempts in this direction were purely deterministic and focused on the patterns formed during cyclic deformation. The models by Walgraef and Aifantis (1985) and Aifantis (1986) and by Kratochvil and co-workers (Kratochvíl and Libovický, 1986; Franěk *et al.*, 1991) realistically reproduced, respectively, the successive patterns that lead to the formation of PSBs in fcc crystals (Section 2.6.6) and the patterns observed during the cyclic deformation of Fe–Si alloys. These models were, however, not sufficiently specific in terms of dislocation mechanisms to capture the experimental length scales and their material dependence (see Kubin *et al.* (2002) for discussion and additional references). The review article on dislocation self-organization processes by Malygin (1999) shows on many examples that all the power of this new approach has not yet been exploited.

The latest set of models is based on a statistical mechanics framework and postulates that fluctuations of long-range internal stresses (or interaction stresses) are responsible for the formation of dislocation patterns. The original model was proposed by Hähner (1996*b*) for monotonic deformation (see Hähner 1996*a* for cyclic deformation). It was further expanded by Hähner and Zaiser (1997) and Zaiser and Seeger (2002). Rather large strain rate fluctuations are predicted in fcc metals, where the strain rate is highly stress-sensitive. The origin of these fluctuations is attributed to collective dislocation glide events propagating through the microstructure. These dislocation groups remain correlated provided that the interaction stresses that stabilize them are larger than the stress fluctuations caused by their short-range interactions with forest dislocations. This condition defines a critical correlation length for fluctuations that was further assimilated to a cell diameter. The similitude relation was then recovered within the assumption that long-range interaction stresses between dislocations govern the flow stress. There is, however, ample evidence that this condition is not fulfilled (Sections 2.5.2 and 5.4.3).

The dislocation field model (Fressengeas *et al.*, 2009), which is briefly discussed in Section 2.7.2.4, gives a flavour of the present evolution in the modelling of dislocation populations. The next generation of models has quite ambitious objectives and may emerge in the next years. It aims at solving numerically in time and space a physical model for plasticity based on statistical mechanics foundations and incorporating the classical continuum theory of dislocations (see Kröner

2001 and references therein). The latter is expanded to include evolutionary laws for dislocation populations in time and space as well as discrete dislocation mechanisms. A truly predictive description of strain hardening may then be reached with a minimum of assumptions by performing successive steps of integration or grain-coarsening up to the level of continuum mechanics. Dislocation avalanches and dislocation patterns should also appear naturally in these numerical models. The difficulties to solve are typical of multiscale approaches; they are discussed by Dimiduk *et al.* (2006*a*) and other authors in the same special issue. The present state of the most advanced model is discussed by El-Azab (2006) and Deng and El-Azab (2009, 2010).

2.7.2 Dislocation avalanches

2.7.2.1 Slip traces in deformed crystals In a review chapter that summarizes a considerable amount of work, Neuhäuser (1983) recalls that the study of slip traces at the surface of deformed crystals and the identification of their crystallographic nature goes back to the end of the nineteenth century. These studies and those that followed made it clear that plastic flow is not uniform at a fine scale. The term *dislocation avalanche* may have appeared before the beginning of dislocation theory, although the repeated yielding that was sometimes recorded in stress–strain curves (for instance by Becker and Orowan in 1932) should probably be attributed to a plastic instability specific to dilute alloys and impure materials, the Portevin–Le Chatelier effect (Bharathi *et al.* 2002). At the beginning of the 1950s, it is understood that the rather abrupt formation of slip traces results from a collective mode of dislocation motion (see Cottrell 1953, p. 178). Slip is then described as a succession of avalanches produced by Frank–Read sources that emit dislocation loops during very short time intervals separated by periods of inactivity (Fisher *et al.*, 1952).

An important distinction has to be made between slip traces in pure fcc materials and planar slip in alloys containing strong and localized obstacles to dislocation motion like small shearable precipitates, short-range order or irradiation defects (Section 5.3.2). The shearing of these obstacles by mobile dislocations locally reduces the strength of the alloy. This process induces the formation of dynamic dislocation pile-ups in the strain-softened glide bands and leads to what is called a Lüders-like behaviour. Single slip is initiated with the help of stress concentrations, usually near a specimen head, and propagates all through the gauge length, leaving coarse slip bands at the surface. There is no strain hardening, as the active slip bands propagate into the initial microstructure. Thus, most of the time and in contrast to what is sometimes believed, planar slip in alloys is not caused by a low SFE that impedes cross-slip (Olfe and Neuhäuser, 1988; Gerold and Karnthaler, 1989). Lüders-like behaviour is also observed in single crystals containing a dense dislocation microstructure (Neuhäuser, 1983; Luft, 1991).

2.7.2.2 Avalanches in bulk crystals A new insight into the properties of dislocation avalanches was obtained by applying methods for the statistical analysis of non-equilibrium systems to recordings of acoustic emission (AE) events during plastic flow (see Zaiser 2006 for a review). The abrupt release of stored elastic energy in a stressed solid gives rise to the emission of acoustic waves, whereas smoother events only produce a continuous signal. The analysis of AE bursts in crystalline materials is typically based on models for sudden steps of motion of one or a few dislocation lines (Rouby *et al.*, 1983). In 2001, Miguel *et al.* analysed the probability distribution of slip bursts energies, $P(E)$, during basal slip in ice crystals crept in compression at -10 °C (Fig. 2.26). Irrespective of the applied stress, this probability was found to exhibit a power-law distribution, $P(E) \propto E^{-\eta_E}$ with $\eta_E \approx 1.6$. Intermittent and scale-free (i.e., containing an extended range of scales) behaviour was characterized over more than six decades in energy.

An analysis of AE peak amplitudes, which are proportional to the maximum strain rate, also yielded power-law distributions in crept ice crystals, with an exponent $\eta_A \approx 2$ (Richeton *et al.*, 2005*a*; Weiss *et al.*, 2002). Further analyses were performed at room temperature and under strain control on copper crystals and hcp crystals of cadmium and Zn-0.08%Al, deformed in single, double or multiple slip. In all cases, power-law distributions were obtained with exponents similar to those found in ice crystals, $\eta_E \approx 1.5-1.6$ (Richeton *et al.*, 2006*a*, 2007) and $\eta_A \approx 2$ (Weiss *et al.*, 2007). These power-law exponents are therefore the same for all the tested materials and crystal orientations, in load and strain

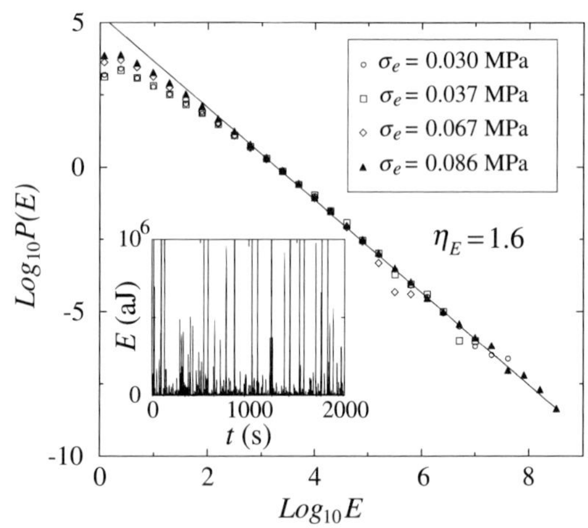

FIG. 2.26. Probability distribution $P(E)$ of the energy E of acoustic bursts in an ice crystal deformed in compression under several applied stresses. The insert illustrates the intermittent nature of strain burst amplitudes. At low energies, the power-law distribution merges into the background noise. From Miguel *et al.* 2001, courtesy M. C. Miguel.

control. The amplitude distributions exhibit an upper cut-off in double and multiple slip conditions, which raises questions about the origin of this truncation and, more generally, about the mechanism that limits the propagation of dislocation avalanches (see Section 5.4.1).

Avalanches emerging at a specimen surface produce very fine slip traces with small but measurable heights (Mader, 1957; Basinski and Basinski, 1979). If such shears are representative of dislocation avalanche sizes in the volume, they should also exhibit scale-free behaviour. The probability distribution functions of step heights were measured in compressed KCl and LiF crystals by atomic force microscopy and scanning white light interferometry by Schwerdtfeger *et al.* (2007, 2010). Power-law distributions of sizes with an exponent $\eta_s \approx 1.5$ were found in KCl and as-grown LiF. In irradiated LiF crystals, the statistical properties of slip heights were found to significantly differ from those of as-grown crystals (Schwerdtfeger *et al.*, 2010), probably because of the above-mentioned occurrence of strain softening effects in irradiated materials (Section 2.7.2.1).

In crept ice crystals, Weiss and Marsan (2003) constructed a three-dimensional mapping of the location of AE sources as a function of time and position on the specimen free surfaces by recording and 'triangulating' their arrival times at five transducers. A correlation analysis revealed that dislocation avalanches are clustered in space in the form of a fractal pattern. A spatio-temporal coupling of avalanches was also evidenced; the more two avalanches are close in space, the more they are close in time. Further examination of this clustering of events by Weiss and Miguel (2004) revealed that an avalanche may induce secondary avalanches in its vicinity. The number of such aftershocks increases with the size of the primary avalanche. As in real avalanches, this phenomenon is attributed to local stress redistributions. In these compression experiments carried out in single slip, the stresses induced by the lattice rotation of the sample with respect to the compression platens is accommodated by kink bands (Section C.3). The latter develop in most of the sample (Fig. 1 of Weiss and Marsan 2003) and may affect the recorded long-range spatial correlations.

2.7.2.3 Avalanches in small-scale pillars Micro- and nanopillars are small-scale deformation specimens (Fig. 2.27*a*) with diameters ranging, respectively, between about one to a few tens of microns and a few tens of nanometres to one micron. Owing to their small dimensions, these specimens exhibit a size effect, which is discussed in Section 5.5.5. Micropillars contain a significant initial dislocation density (Fig. 5.20); they should not be confused with *whiskers* (Dehm, 2009; Bei *et al.*, 2007), which are essentially dislocation-free microcrystals, natural or artificial, with almost perfect surfaces. The maximum yield stress of the smallest whiskers is close to the theoretical strength (Brenner 1956); the introduction of surface or volume defects reduces their strength (Brenner, 1956; Shim *et al.*, 2009).

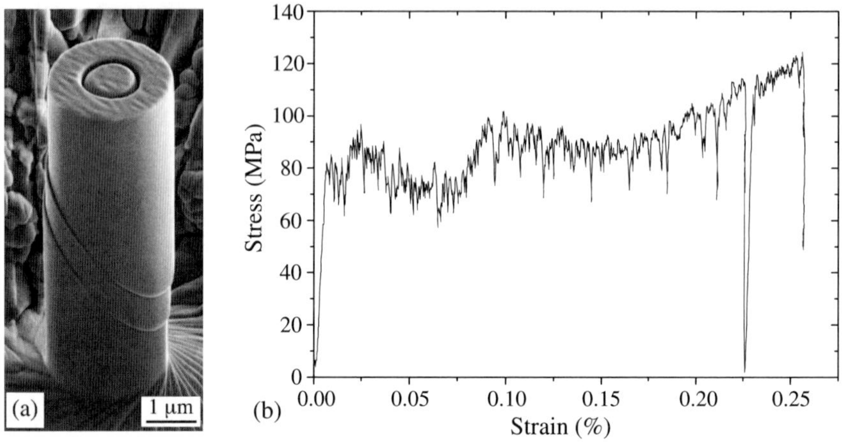

FIG. 2.27. Deformation of micropillars. (a) Nickel crystal of diameter 2 μm deformed in compression by 3% along a ⟨269⟩ axis. Courtesy M. Uchic. (b) Tensile stress–strain curve of a ⟨234⟩ copper crystal deformed in situ in strain control mode in a scanning electron microscope. The dimensions of the specimen are $3 \times 3 \times 8\,\mu m^3$. The large stress drop is an unloading-loading sequence. From Kiener *et al.* 2008, courtesy D. Kiener.

The size s of an avalanche is defined as the area swept by dislocations, δS, or the corresponding shear and axial strains, $\delta\gamma = b\,\delta S/V \propto b/L$ and $\delta\varepsilon = \mathcal{F}\delta\gamma$, where V is the volume of the deforming specimen, L is its length and $\mathcal{F}$ is the Schmid factor. The stress–strain curves of the bulk materials discussed so far are smooth, as the strain bursts produced by avalanches are too small to be recorded. In small-scale pillars, the high surface-to-volume ratio induces larger strains for a given swept area. The stress–strain curves are no longer smooth below a certain diameter, but stepped or serrated, thus revealing individual strain bursts (Figure 2.27*b*). During an avalanche in strain control mode, the applied stress may relax and this unloading contributes to slow down avalanches, reduce their size and increase their number. For this reason, the statistics of strain bursts are much more accurate in strain control than in load control.

The first analysis of repeated microyielding events was performed by Dimiduk *et al.* (2006*b*) on single crystalline nickel micropillars with diameters in the range of 20–30 μm. Specimens with a single-slip ⟨269⟩ orientation were loaded in compression by a nanoindenter with flattened tip. Deformation was carried out in a mixed strain-load control. The distribution of axial elongation bursts was shown to be intermittent and scale-free with an exponent $\eta_s = 1.5-1.6$. A similar scale-free behaviour was recorded in aluminium micropillars (Ng and Ngan, 2008*b*), with a power-law exponent $1.47 \leq \eta_s \leq 1.67$ (in load control), and in gold nanopillars (Brinckmann *et al.*, 2008) with $\eta_s = 1.5$.

Small-scale pillars of ⟨100⟩ bcc crystals (Mo, Nb, Ta, W) were deformed in compression (Schneider *et al.*, 2009) and in tension (Kim *et al.*, 2010). For diameters smaller than about 1 μm, strain bursts are recorded in all the tested crystals, although there is clear evidence that the specimens retain some lattice resistance. Statistical analyses of strain bursts were performed in molybdenum by Brinckmann *et al.* (2008) and Zaiser *et al.* (2008). A power-law distribution was recorded in both cases, with exponent values of, respectively, $\eta_s \approx 1.5$ and $1.34 \leq \eta_s \leq 1.76$. Several mechanisms were suggested to explain this unexpected behaviour of bcc crystals (Greer *et al.*, 2008). In particular, Cai and Weinberger (2008) proposed on the base of MD and DD simulations a mechanism by which screw dislocations nucleated at free surfaces can self-multiply in the volume.

Further experiments on nanopillars show that the statistics of slip events exhibit same exponent values in the regimes of plasticity controlled by dislocation sources and dislocation nucleation (see Section 5.5.5.4). Friedman *et al.* (2012) report on the deformation of cylindrical nanopillars with diameters ranging from 75 nm to 1 μm. The materials tested are fcc Al, Cu and Au and bcc Nb, Ta, Mo and W. In all cases the distributions of axial burst sizes at constant stress exhibit an exponent $\eta_s \approx 1.5$. The size distributions and their cut-off are, however, observed to be stress-dependent with a characteristic exponent of about 2.

Prestraining defect-free whiskers allows exploring their microstructure and mechanical response over a wide range of initial densities. Such an investigation was carried out by Bei *et al.* (2008) and Sudharshan Phani *et al.* (2011) on ⟨100⟩ nanofibres of bcc MoAl obtained by selectively etching a eutectic NiAl-Mo alloy (Bei *et al.*, 2007). Specimens with diameters of about 500 nm were tested in compression. After a small prestrain (4 %), the axial yield stress is about 2 GPa, well below that of the as-grown whiskers (10 GPa $\approx \mu/10$). The stress–strain curves exhibit strain bursts and look very similar to those of molybdenum crystals with same diameter. After higher prestrains (11 and 16%) the stress level is in the GPa range and the stress–strain curves are smooth. The effect of an increased dislocation density on intermittent behaviour is, thus, similar to that of an increased diameter since it softens the material. This type of mechanical response, which is typical of submicron-sized pillars (Fig. 5.20*a*), is discussed in Section 5.5.5.

The values reported above for the various critical exponents are mutually consistent. This can be checked by making simple assumptions about the temporal profiles of strain bursts (Fressengeas *et al.*, 2009) or from the predictions of statistical models for systems exhibiting a depinning transition (see Zaiser 2006, pp. 10–15). In particular, one should have $\eta_s = \eta_E \approx 1.5{-}1.6$ and $\eta_A \approx 1.8{-}2$. Only LiF micropillars were found to behave differently, as rather large exponents ($\eta_s = 1.8{-}2.9$) were measured in both irradiated and as-grown specimens (Dimiduk *et al.*, 2010). In as-grown specimens this large value is at variance with the one obtained by Schwerdtfeger *et al.* (2010). In polycrystalline micropillars, the distribution of avalanche amplitudes exhibits two power laws with different exponents, the single crystal exponent $\eta_A = 2$ and another exponent, $\eta_A = 1.35$, which invades the distribution as the grain size decreases (Richeton *et al.*, 2005*b*). The upper cut-off of the distribution also decreases with decreasing grain size. These results further led to a re-examination of grain-size strengthening and the Hall–Petch relation in polycrystals (Richeton *et al.*, 2006*b*).

2.7.2.4 Avalanches as depinning transitions Power-law distributions appear in a number of natural and man-made phenomena; they have drawn the attention of physicists for many years. In a review article, Newman (2005) provides many examples, of which some are quite surprising, of systems exhibiting scale-free distributions. Methods for analysing such distributions and some important mechanisms by which they can arise are also discussed. Among the latter, one finds a number of driven, non-linear dynamical systems that 'crackle'—that is, respond to a slow variation of the driving force by abrupt and intermittent events over a wide range of sizes. This behaviour is characteristic of critical phenomena and depinning transitions, for which mean-field theories predict the experimentally recorded values of the critical exponents (Fisher, 1998; Zapperi *et al.*, 1998). In this category of systems, one finds (real) avalanches, earthquakes, the Portevin–Le Chatelier effect in dilute alloys, the depinning transition of sliding charge density waves in conductors, irreversible martensitic phase transitions, fracture in amorphous materials, the Barkhausen effect in ferromagnetic materials and many others.

The modelling of these phenomena is discussed in broad terms by Sethna *et al.* (2001). It is possible to construct continuous models, valid at large time and length scales, which predict the critical exponents, the average or maximum event sizes as well as generic scaling laws. All phenomena to which such predictions apply are then said to belong to a class of universality. Because the scale-free distributions extend over several decades in size, microstructural details are of little importance in such coarse-grained models. As a consequence, one can also devise a number of simplified models exhibiting some generic properties without caring too much of the specific underlying mechanisms. Constructing specific models is a more difficult task. As far as dislocation-based models are concerned, it is useful to recall that there are no long-range internal or interaction stresses during easy glide (Mughrabi, 1975; Grosskreutz and Mughrabi, 1975). Further, in multiple glide, the flow stress is governed by short-ranged forest interactions until large strains (Section 2.5.2).

Both mean-field and simplified models were developed for dislocation avalanches (Zaiser, 2006). The simplified models were most of the time based on an analogy with non-equilibrium phase transitions involving a yielding transition. They were applied to the easy glide of two-dimensional systems of parallel, straight edge dislocations (Miguel *et al.*, 2001; Miguel and Zapperi, 2006; Laurson and Alava, 2006; Miguel *et al.*, 2008). Koslowski *et al*, (2002, 2004) made use of a two-dimensional phase-field simulation to examine different configurations, in which flexible dislocations shear fixed obstacles representing forest dislocations. Although the method poses a conceptual problem (see Zaiser 2006), this model has the merit of incorporating line tension effects and short-range interactions of dislocations. Intermittency was obtained in all models with reasonable critical exponent values.

A generic model based on a fully continuum mechanical framework was adapted to reproduce the main features of strain bursts by Zaiser and Moretti (2005). Space-dependent local stresses and strains are related by a constitutive equation under a uniform externally applied stress. Spatial fluctuations of the local glide resistance are introduced and long-range stresses ensuing from local strain non-uniformities are taken into account. The model correctly predicts the statistical properties of strain bursts, slip patterns and surface morphologies, as well as the scaling properties of the upper cut-off (Zaiser and Nikitas, 2007). Figure 5.10*a* shows the collapse of size probability distributions obtained by DD simulations into the predicted generic curve. Another mean-field model, leading to a different generic curve is also proposed by Friedman *et al.* (2012). It is based on a micromechanical model for slip avalanches (Dahmen *et al.*, 2009) and accounts for the observed stress dependence of the size probability distributions in nanopillars.

As in the case of dislocation patterning, revisited formulations of the continuum theory of dislocations potentially provide more neutral approaches at more refined scales. The field dislocation model (or FDM, see Fressengeas *et al.* 2009 and references therein), formally treats excess or *polar* dislocations, also known as GNDs, via the Nye density tensor. It makes use of a transport equation derived by Mura (1963) for the evolution and transport of the polar density. The latter is coupled to phenomenological evolutionary laws for the statistical mobile and forest dislocation densities. As the density of excess dislocations depends on the area of the Burgers circuit on which it is measured, variable spatial resolutions can be selected in order to treat plasticity problems that involve variable contributions of statistical and polar dislocation densities. Fluctuations naturally occur in the model and intermittent, scale-free behaviour is recorded. The FDM was used to analyse high-resolution extensometry experiments at the surface of $\langle 001 \rangle$ copper crystals deformed in tension (Fressengeas *et al.* 2009) and long- and short-range correlations in ice crystals deformed in torsion (Chevy *et al.* 2010). In the last case, the results seem to challenge the emphasis that is often put on long-range correlations (Weiss and Montagnat 2007).

3D DD simulations of dislocation avalanches also yield $\eta_s = 1.5-1.6$ (Section 5.4.1). Such simulations are ideally suited for establishing connections between dislocation mechanisms, collective dislocation properties and the mechanical response (Section 5.4.2).

3

LATTICE-CONTROLLED PLASTIC FLOW

3.1 Outline

In many crystalline materials a lattice resistance arises from the structure of screw dislocation cores or from the non-metallic nature of bonding. The mechanical response directly reflects the low dislocation mobility resulting from the interaction of dislocations with the lattice. As dislocation motion is controlled by the core structure, it occurs by a specific thermally activated process, the *kink-pair mechanism* (Appendix E). In such conditions the mechanical properties differ from those of fcc crystals in a domain of low homologous temperatures limited by the athermal transition temperature (Section A.1) of the kink-pair mechanism. Three types of elemental materials are successively discussed in this chapter (Sections 3.2, 3.3 and 3.4):

- Body-centred cubic metals, in which the lattice resistance arises from the core structure of $\frac{1}{2}\langle 111\rangle$ screw dislocations. The single kinks are highly mobile and the kink-pair mechanism is governed by the generation of kink pairs.
- Hexagonal close-packed metals that preferentially deform by prismatic slip, in particular titanium and zirconium. The lattice resistance arises from the core structure of screw dislocations with the basal Burgers vector $\frac{1}{3}\langle 11\bar{2}0\rangle$. In this case too, the kinks are highly mobile.
- Silicon, a model covalent material in which the directional character of atomic bonds is responsible for a strong lattice resistance. Single kinks on gliding $\frac{1}{2}\langle 110\rangle$ dislocations have to break covalent bonds during their sideways motion. As a consequence, the kink-pair mechanism must account for both kink-pair nucleation and kink propagation.

The current state of the art is examined in terms of the structure of the dislocation cores, their relation to dislocation velocity and the mechanical response of bulk crystals. Although significant progress was achieved in the determination of dislocation core structures, the connection to mesoscale properties is far from being realized, and there remain major open questions for each class of material.

3.2 The lattice resistance in bcc metals

3.2.1 Deformation properties of bcc metals

Elemental bcc materials include principally the transition metals of columns V (V, Nb, Ta) and VI (Cr, Mo, W) of the periodic table of elements,[5] α-Fe and alkali metals such as Li, Na, K and Ba. Small amounts of impurities, especially interstitial solutes, induce a significant strengthening (Section 3.2.5.2). The archetypal example of such extrinsic strengthening is that of carbon and nitrogen in α-iron. Over the years, deformation tests and TEM observations performed on increasingly pure materials showed, however, that there is a quite substantial intrinsic strengthening at low temperatures.

The occurrence of a strong lattice resistance in bcc metals at low temperatures was suspected for some time before being rationalized by Hirsch (1960). As the $\langle 111 \rangle$ screw direction is lying along a threefold symmetry axis, Hirsch predicted that screw dislocations should be extended in a threefold manner in dense $\{110\}$ or $\{112\}$ planes. This hypothesis was the subject of intensive experimental investigations on bcc single crystals during the period 1960–1980. These studies and further ones by Seeger and co-workers, which are summarized by Hollang (2002), constitute most of the presently available data and observations. They have been reviewed by Kubin (1976), Sěsták (1979), Christian (1982), Kubin (1982), Suzuki *et al.* (1985), Duesbery (1989) and Taylor (1992). The specific properties of alkali metals were reviewed by Christian (1982), Kubin (1982), Duesbery (1989) and Taylor (1992). Potassium was further investigated by Duesbery and Basinski (1993). The first *in situ* straining experiments in the TEM were carried out on α-iron by Saka *et al.* (1970) in a high-voltage electron microscope. Further developements of this technique (Saka *et al.* 1973, see Kubin and Martin 1979 for a review), revealed a wealth of mechanisms specific to the strong interaction of screw dislocations with the lattice. Such experimentation is now pursued on standard electron microscopes (Caillard and Martin, 2003).

Atomistic simulations confirmed Hirsch's prediction and yielded screw dislocation cores that are not dissociated into partials but extended in an anisotropic manner. *Ab initio* calculations corrected this view by showing that the stress-free cores are much more isotropic than was thought before (Ismail-Beigi and Arias, 2000; Woodward and Rao, 2001).

The athermal transition temperature T_a marks the upper bound of the domain where the thermally activated velocity of screw dislocations affects the yield stress. Above this temperature, the stress–strain curves of single crystals exhibit dislocation stages (see e.g., Richter 1970*a* for Mo and Spitzig and Keh 1970 and

[5] Before the unification of nomenclatures, these columns were labelled VA and VIA or VB and VIB.

TABLE 3.1. Material constants and data on the Peierls mechanism for screw dislocations in common bcc metals.

	Nb	α-Fe	Ta	Mo	W
T_a (K)	340	340	440	465	800
ΔH_o (eV)	0.69	0.80	1.04	1.10	1.89
μ_{111} (GPa)	47.13	70.67	62.87	134	157
b (nm)	0.286	0.248	0.286	0.273	0.274
$\mu_{111}b^3$ (eV)	6.87	6.76	9.19	16.95	20.19
τ_P (MPa)	350	390	360	730	≈ 820

Kumagai *et al.* 1990 for α-Fe). In parallel, the dislocation microstructures present many similarities with those of fcc metals. Some data on athermal transition temperatures (after Suzuki *et al.* 1999, except for Nb and W) and related quantities are given in Table 3.1. The values of T_a are not so well defined, as they correspond to a rather smooth transition that depends on strain rate and impurity content. They are given for ultra-pure materials and conventional strain rates (10^{-4}–$10^{-3}\,s^{-1}$). The transition temperatures and the total activation enthalpies ΔH_o must approximately verify eqn. A.5, that is, $\Delta H_o \approx 25\,k_B T_a$ (Section 3.2.4.3). The Peierls stresses are typical of orientations with a high Schmid factor on the primary $\{110\}$ slip plane. The isotropic shear modulus along the $\langle 111 \rangle$ screw direction, $\mu_{111} = (c_{11} - c_{12} + c_{44})/3$ (Section B.2), is frequently used as a scaling factor for the Peierls stress τ_P. The quantity $\mu_{111}b^3$ is also a scaling factor for energies.

A certain number of characteristic properties, which will be discussed further here, appear at temperatures below $T = T_a$ and many of them become more pronounced as temperature decreases. To some extent, these properties are modulated by the position of the considered bcc metal in the periodic classification, hence by the electronic structure.

The yield and flow stresses markedly increase at low temperatures. In the purest crystals, the extrapolated Peierls stresses are found in the range of $\tau_P/\mu = 5 - 8 \times 10^{-3}$ (see Takeuchi *et al.* 1982 for Nb, Brunner 2010 for W and the compilation by Suzuki *et al.* 1999). For potassium, which is a 'weak' bcc metal, $\tau_P/\mu \approx 2 - 3 \times 10^{-3}$ (Suzuki *et al.*, 1999). Figure 3.1 shows the temperature dependence of the yield stress obtained in two different studies on ultra-pure tantalum. From Fig. 3.1*a*, one has $\tau_P/\mu \approx 5 - 6 \times 10^{-3}$, in agreement with measurements by Takeuchi and Maeda (1977) between 0.7 and 40 K.

There are no stages in the stress–strain curves; they are pseudo-parabolic (Fig. 3.2*a*) and the flow stress significantly depends on strain rate (Fig. 3.1*b*). At low temperatures, the activation volumes decrease to values of the order a few b^3, which are characteristic of very localized obstacles to dislocation motion.

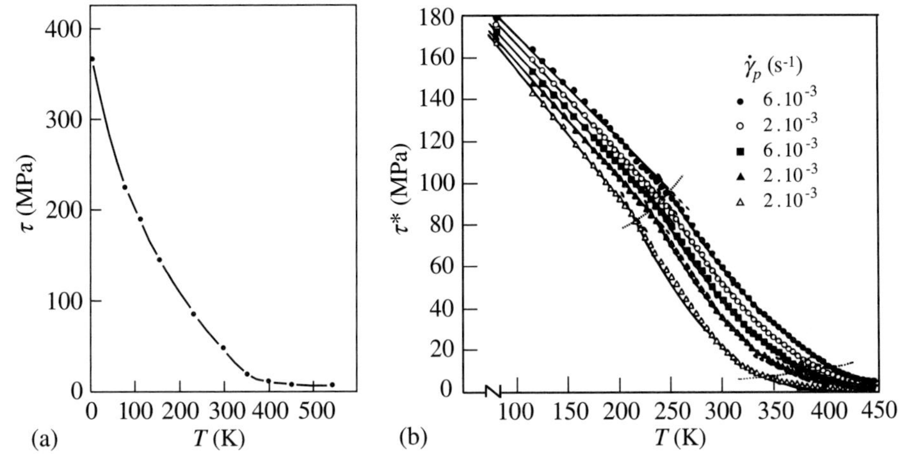

FIG. 3.1. Temperature dependence of the yield stress in ultra-high purity Ta crystals oriented for single glide. Stresses are resolved on the primary ⟨111⟩{110} slip system. (a) Critical resolved shear stresses in non-prestrained crystals. From Smialek *et al.* 1970. (b) Effective resolved flow stresses for several imposed shear strain rates on crystals pre-cycled to saturation at 400 K. A well-marked knee occurs around 250 K and a less well-marked one occurs near the athermal transition temperature. From Werner 1987.

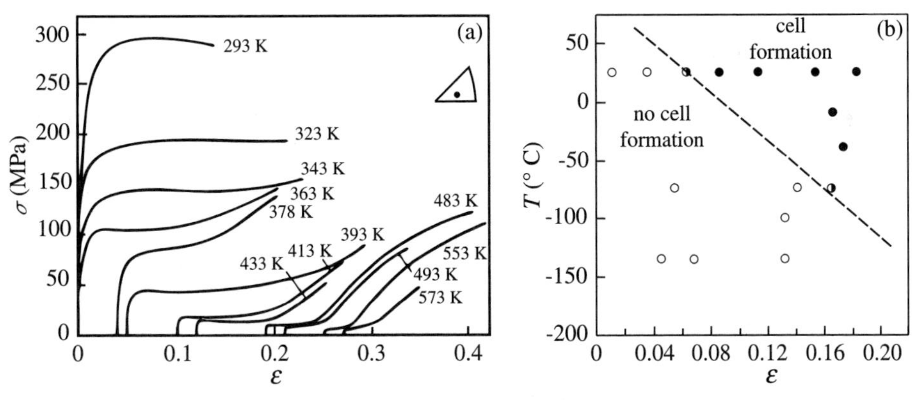

FIG. 3.2. Transition between high and low temperature regimes in bcc metals. (a) Axial stress–strain curves of Mo single crystals oriented for easy glide. Under an imposed strain rate of $7\times10^{-6}\,s^{-1}$, the hardening stages start disappearing around 450 K. From Richter 1970*a*. (b) Transition between cell structure formation and a uniform microstructure of screw dislocations in α-Fe polycrystals as a function of temperature and plastic strain. From Keh and Weissmann 1963.

No dislocation cell formation is observed below a critical strain value that increases with decreasing temperature. This is illustrated by Fig. 3.2*b* in the case of α-iron polycrystals. Cell formation occurs at very small strains above the athermal transition temperature ($T_a \approx 340$ K). At lower temperatures the interaction between screw dislocations and the lattice delays cell formation until strains such that dislocation strengthening prevails over the lattice resistance (Section 3.2.5.1). The dashed line in Fig. 3.2*b* represents the crossover between these two mechanisms. The strain dependence of this transition also manifests itself in Fig. 3.2*a* by an earlier disappearance of the first deformation stages with decreasing temperature.

The critical resolved shear stresses (CRSSs) cannot be rationalized in terms of the Schmid law (Christian, 1982; Duesbery, 1989; Duesbery and Vitek, 1998). This effect manifests itself in two different manners and is all the more pronounced as temperature decreases. In bcc metals, twinning occurs on {112} planes along a well-defined sense of the ⟨111⟩ direction. When {112} planes bear the maximum resolved shear stress (MRSS), the CRSS in the twinning sense is substantially smaller than in the reverse, or anti-twinning, sense (Fig. 3.6). In addition, there is an asymmetry between tension and compression for most orientations of the loading axis. The asymmetry of the CRSSs at 77 K in groups V and VI bcc metals was compiled by Duesbery (1989); it served as a basis for further compilations by Duesbery and Vitek (1998) and Argon (2008).

The slip geometry of bcc metals, as inferred from slip trace analyses, also departs from the prediction of the Schmid law. It is not always well documented and was reviewed by Duesbery (1989) and Duesbery and Vitek (1998). The global trends are similar for transition metals within some differences between metals of groups V, VI and α-Fe. Near the athermal transition themperature, the slip traces are usually wavy and tend to follow the MRSS plane. This so-called *pencil glide* was discovered by Taylor and Elam (1926) in α-Fe crystals (see also Honeycombe 1968; Duesbery and Richardson 1991). With decreasing temperature, MRSS slip persists but the slip traces become progressively less wavy and deviate less from crystallographic orientations. Macroscopic slip occurs on {110} planes and also on {112} planes, preferably in the twinning sense, when they bear the MRSS. This {112} slip as well as MRSS slip may be in many cases a composition of elementary steps in {110} planes (Section 3.2.2). In the detail, the slip geometry is quite complex, as exemplified by studies on Mo crystals by Richter (1970*b*), Irwin *et al.* (1974) and Guiu (1969). At the lowest temperatures, {110} slip exclusively occurs in Fe (Christian, 1982) and Mo (Guiu, 1969) and predominates in Ta (Takeuchi *et al.*, 1972; Shields *et al.*, 1975). The alkali metals behave in a quite different manner (Duesbery 1989; Christian 1982; Kubin 1982). Li and Na undergo martensitic transformations at low temperatures and do not exhibit a bcc-like behaviour. In K and Li-Mg alloys, the bcc phase remains stable and slip predominantly occurs on {112} planes in the twinning and anti-twinning directions.

A convenient method for comparing the slip geometry predicted by the Schmid law to the experimentally observed one was devised by Taylor (see Argon 2008). Owing to the crystal symmetry, it is sufficient to consider a single stereographic triangle, for example [001]-[011]-[$\bar{1}$11], where the primary slip system is [111]($\bar{1}$01). When a pure shear τ is applied in the [111] slip direction, the (111) plane contains the normals to all possible slip systems (cf. Fig. 3.6). The angle of these normals with [$\bar{1}$01], which is denoted χ, goes from $\chi = -30°$ for [$\bar{1}\bar{1}$2], where the shear is in the twinning sense, to $\chi = +30°$ for [$\bar{2}$11], where the shear is in the anti-twinning sense. Thus, a reversal of the applied shear from [111] to [$\bar{1}\bar{1}\bar{1}$] exchanges the twinning and anti-twinning senses, which is equivalent to changing the sign of χ. As the shear direction is colinear to the slip direction, the Schmid factor is $\cos(\chi)$, and $\tau = \sigma\cos(\chi)$ where σ is the applied axial stress. The experimentally found slip plane normals are defined by their angle ψ with [$\bar{1}$01]. On $\psi(\chi)$ curves, the diagonal $\psi = \chi$ corresponds to MRSS slip, the relation $\psi(-\chi) = -\psi(\chi)$ no longer holds in the presence of slip asymmetry and $\psi(\chi) = 0$ corresponds to slip on the ($\bar{1}$01) plane (examples of such curves are given by Christian 1982, Duesbery 1989 and Argon 2008).

The deformation of group-V bcc metals at low temperatures is characterized by the observation on a wide range of orientations of coarse crystallographic slip traces produced by slip on an unexpected {110} plane (Christian, 1982). The occurrence of such *anomalous slip* was investigated in detail in Nb and Ta by Wasserbäch (1995), who showed that it is favoured by a high purity and suppressed by a prestrain. The anomalous slip plane has the peculiarity of containing the primary and conjugate Burgers vectors (Fig. 3.3) and, hence, their reaction product. The reaction between primary and secondary screw dislocations with Burgers vectors $\frac{1}{2}\langle 111\rangle$ and $\frac{1}{2}\langle 1\bar{1}\bar{1}\rangle$, gliding respectively in ($\bar{1}$01) and (101) planes is given by eqn. 2.13. The junction has a Burgers vector [100] and the whole configuration may be glissile in the (0$\bar{1}$1) slip plane. Coplanar double slip in the anomalous plane is though to be assisted by elastic interactions between the junction and its parent arms (Matsui and Kimura, 1975). This implies that [100] dislocations should exhibit a smaller lattice resistance than screw dislocations. An atomistic simulation of coplanar double slip in the anomalous plane using a Finnis–Sinclair potential indicated that it may be so in Mo (Bulatov and Cai, 2002). However, anomalous slip was observed only once in Mo and at strains smaller than 0.5% (Matsui and Kimura, 1976). Anomalous slip was never observed in α-Fe (Aono *et al.*, 1982).

In situ TEM deformation shows that the ratio of non-screw to screw dislocation velocities increases with decreasing temperature. Highly mobile non-screw segments start moving along the edge direction at low stresses (Saka *et al.*, 1973; Kubin, 1976), even on slip systems with small Schmid factors. These segments trail elongated dipoles of immobile screw segments over extremely large distances in ultra-pure materials strained at low temperatures. As the instantaneous

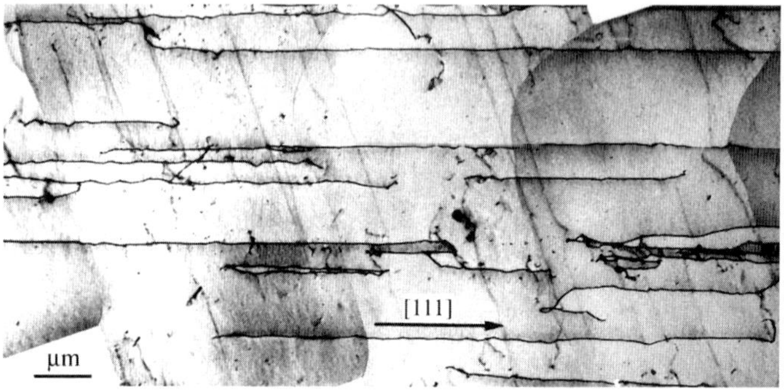

FIG. 3.3. TEM observation of elongated primary screw dislocations in a niobium crystal strained at 50 K. The nearly vertical faint lines are secondary screws that are almost out of contrast. The foil plane is parallel to the anomalous $(0\bar{1}1)$ slip plane. After Louchet and Kubin 1975.

mobile density is too small to accommodate the imposed strain rate, a considerable hardening is recorded, which may amount to a fraction of the shear modulus. Near the end of this pre-yield stage, or stage 0, some screw dislocations start moving and the yield stress is reached when they carry a large fraction of the imposed strain rate. The significant densities of screw dislocations accumulated in many slip systems during stage 0 have often been mistakenly interpreted as reflecting extensive slip activity on these systems. This stage rapidly disappears with increasing impurity content, owing to to the strong interaction between non-screw dislocations and solute atoms.

Beyond stage 0, TEM examinations show the predominance of straight, elongated screw segments (Fig. 3.3). *In situ* deformation in the TEM also shows that the screw dislocations move as straight lines and in a smooth manner, all the more as temperature is low (Section 3.2.5.1). This constitutes a signature of their interaction with the lattice. More recent observations carried out on α-iron with an almost nanometric resolution (Caillard, 2010*b*) reveal, however, a more complex behaviour (Section 3.2.4.4).

3.2.2 Core structure of screw dislocations

In cubic lattices, the ternary symmetry around $\langle 111 \rangle$ implies that these directions are at the intersection of three $\{110\}$ and three $\{112\}$ low-index planes (Fig. 3.4). This is why Hirsch (1960) postulated that the high strength of pure bcc metals at low temperature arises from the simultaneous spreading of screw dislocations cores in these planes. Such a spread core cannot move in a conservative manner. To become glissile, it has to be recombined under stress into a planar configuration of higher energy, which moves by successive atomic jumps between stable, non-planar configurations.

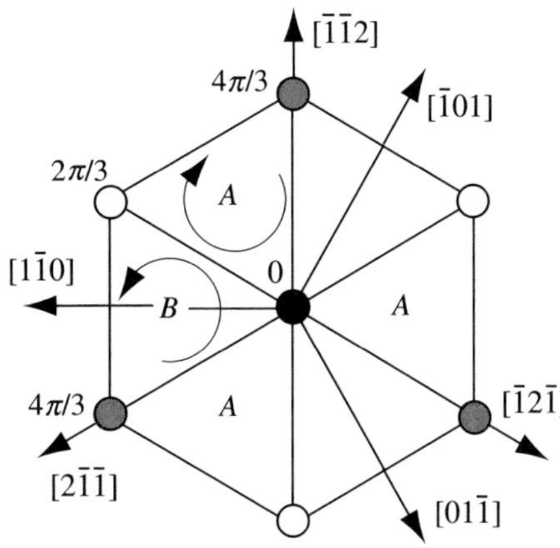

FIG. 3.4. [111] view of three consecutive atomic planes in the bcc lattice. The atomic layers are materialized by black, grey and white atoms. *A* and *B* are two possible sites for the geometric line of a screw dislocation. Around these sites, the heights of nearest-neighbouring atoms are represented by phases that correspond to steps of $b/3$.

Figure 3.4 shows a [111] view of the bcc lattice containing three superimposed (111) atomic planes. These slip planes are separated by distances that are half their value in the fcc latice, $a/2\sqrt{3} = b/3$. As the stable positions of a screw dislocation line minimize the atomic displacements, they are located at sites such as *A* or *B*, at equal distances from the nearest atomic columns. Around these sites, the atomic positions wind up by steps of $b/3$ in a clockwise or anti-clockwise manner. In Fig. 3.4, phases in rotation are attributed to these helical steps. The introduction of a screw dislocation in site *A* induces additional shifts of $\pm(0, 2\pi/3, 4\pi/3)$, in the same or opposite sense depending upon the sign of the Burgers vector. In the latter case, the final phases are $(0, 0, 0)$ and the atoms are brought to the same plane. This corresponds to a 'hard', high-energy configuration. In the former case, the displacements of the atomic columns add up and the phases become $(0, 4\pi/3, 2\pi/3)$; the 'easy', helical configuration of neighbouring atoms is preserved, but its sign is changed.

The review articles by Duesbery (1989), Duesbery and Richardson (1991), Duesbery and Vitek (1998), Moriarty *et al.* (2002*b*) and Cai *et al.* (2004) summarize results of many atomistic simulations on screw dislocation cores carried out with empirical and semi-empirical potentials. *Ab initio* calculations effectively account for the anisotropy of electronic bonding in bcc transition metals with a partially filled *d* band. They are, however, computationally intensive and require boundary conditions adapted to the treatment of very small simulated volumes (see Woodward 2005 for a discussion). *Ab initio* simulations were performed first on tantalum and molybdenum, the model materials for bcc metals of groups V and VI (Ismail-Beigi and Arias 2000; Woodward and Rao 2001, 2002; Segall *et al.* 2003) and further on α-iron (Frederiksen and Jacobsen 2003; Domain and Monnet 2005; Ventelon and Willaime 2007, 2010).

Electronic structure calculations reveal stress-free core structures that differ from the ones obtained in most earlier atomistic studies. This is illustrated by the differential displacement maps of Fig. 3.5 for the type-B 'easy' core configurations in Ta and Mo. For tantalum, the threefold symmetrical core, which does not exhibit a sixfold symmetry as sometimes mentioned, is compact and almost isotropic. This non-degenerate configuration is obtained by *ab initio* calculations (Fig. 3.5*a*) and was previously yielded by some atomistic simulations (Fig. 3.5*c*). For molybdenum and group-VI bcc metals, atomistic simulations predicted a different core structure characterized by an anisotropic spreading on three $\{110\}$ planes (Fig. 3.5*d*). This core is doubly degenerate, as the asymmetric spreading of the cores in the $\langle 110 \rangle$ planes breaks a symmetry operation of the bcc lattice.

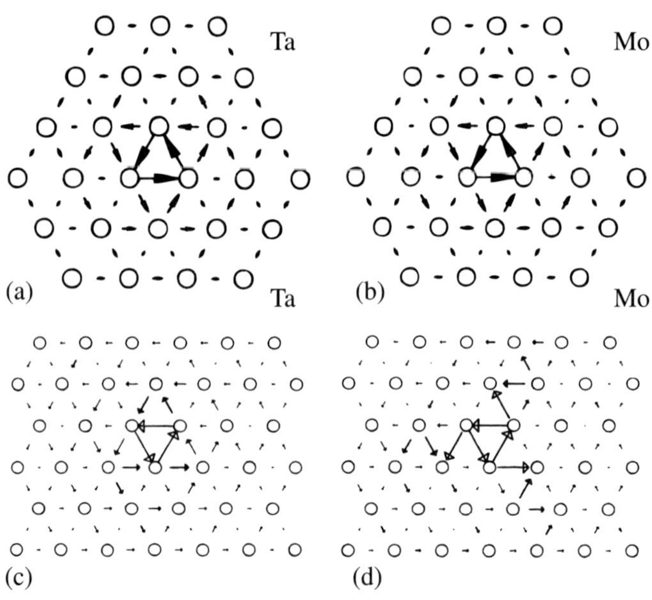

FIG. 3.5. Differential displacement maps for the stress-free cores of $\frac{1}{2}\langle 111 \rangle$ screw dislocations in Ta and Mo. The projection is the same as in Fig. 3.4. The magnitude and sense of the relative vertical (screw) displacements induced by the dislocation between two neighbouring atomic rows are represented by arrowed lines normalized by the interatomic distance |b/3|. (a) and (b) *Ab initio* calculations. After Woodward and Rao 2001. (c) and (d) Atomistic calculations using a Finnis–Sinclair potential. After Ito and Vitek 2001.

However, *ab initio* calculations also predict a non-degenerate isotropic core for molybdenum (Fig. 3.5*b*). Thus, the non-degenerate easy core seems to be the most stable one in all cases. The hard configuration is unstable in Mo (Ismail-Beigi and Arias, 2000) and relaxes to the easy configuration. It may be weakly metastable in Ta (Segall *et al.*, 2003) and is found either weakly metastable or unstable in α-Fe (Domain and Monnet, 2005; Ventelon and Willaime, 2010; Itakura *et al.*, 2012).

The structure of the easy cores is very similar to the one predicted by linear anisotropic elasticity for a perfect screw. Therefore, the transformations of the core under stress control the lattice resistance to a much larger extent than the structure of the stress-free core. The screw dislocations can be displaced by applying a pure shear in $\{110\}$ or $\{112\}$ planes along the direction of the Burgers vector. In *ab initio* calculations, the screw cores always move from an easy type-A site to a next-neighbouring site of same type. This motion occurs in $\{110\}$ planes and along $\langle 112 \rangle$ directions (Fig. 3.4). In Mo, the successive steps of motion occur in the same directions and predominant $\{110\}$ slip is obtained. In Ta, $\{112\}$ slip in the twinning sense is produced by alternating motion on $\{110\}$ planes. This is globally consistent with previous atomistic calculations (Duesbery and Vitek, 1998; Ito and Vitek, 2001). As expected from the available experimental results reported in Section 3.2.1, $\{110\}$ slip is also obtained in Ta (C. Woodward, personal communication). At finite temperatures, elementary steps of motion on $\{112\}$ planes are most probably excluded because kinks have an elastic energy three times larger than in $\{110\}$ planes owing to their larger height. Thus, macroscopic $\{112\}$ slip or MRSS slip is likely to result from a composition of atomic steps on $\{110\}$ planes.

New interaction models were devised in order to treat screw dislocation properties that are beyond the reach of *ab initio* calculations. Two of them are quantum-based potentials accounting for the directional bonding of bcc transition metals with a partially filled *d* band, which reproduce the non-degenerate screw dislocation cores. The bond-order potentials (BOPs) are based on the tight-binding theory. They were developed for molybdenum, tungsten and magnetic α-Fe by Mrovec and co-workers (2004, 2007, 2011). The optimized MGPT potential is based on a model generalized pseudo-potential theory (Yang *et al.*, 2001; Moriarty *et al.*, 2002*a*; Moriarty *et al.*, 2006).

The degree of threefold spreading of a screw dislocation core can be characterized by a polarization index p that varies continuously from $p = 0$ for a non-degenerate core to $p = 1$ for a fully degenerate core (Wang *et al.*, 2003*a*; Ventelon and Willaime, 2010). Such a continuous evolution was obtained by Yang *et al.* (2010), who used optimized MGPT potentials to simulate the core structures of screw dislocations in V, Ta and Nb at high pressure.

Empirical potentials based on the embedded-atom method (EAM) only treat radial interatomic interactions. The modified EAM potential (MEAM), which explicitly contains angular terms, was fitted to *ab initio* data on tantalum by Segall *et al.* (2003). Improved EAM potentials fitted to *ab initio* data were developed for α-iron by Mendelev *et al.* (2003), Ackland, Mendelev *et al.* (2004) and Dudarev and Derlet (2005). Only the first two ones reproduce non-degenerate cores and correctly predict {110} slip for most orientations, as experimentally observed (Section 3.2.1). The γ surfaces (Appendix D) and energy barriers for displacing a screw dislocation between two stable configurations were investigated for modified EAM-type potentials and compared to *ab initio* results by Wang *et al.* (2003*a*), Chaussidon *et al.* (2006) and Ventelon and Willaime (2007, 2010). The energy barriers of the EAM potentials exhibit a camel-hump shape, with a well-marked minimum midway of the activation path. The metastable configurations correspond to split cores that were investigated by Gordon *et al.* (2010). Their occurrence tends to reduce the Peierls stresses with respect to *ab initio* values.

3.2.3 Non-Schmid effects and Peierls stresses

The generality of the Schmid law was already questioned before its second publication in 1928 by Schmid and Boas (see Duesbery 1989; Duesbery and Vitek 1998 for references). Deviations from its predictions were experimentally found in α-Fe and β-brass alloys (with B2 bcc-like structure), leading Taylor to propose in 1928 an explanation for these discrepancies in terms of the symmetries of the bcc lattice. The Schmid law implicitly assumes that the CRSS does not depend on the sense of the resolved shear in the slip plane. For instance, in a bcc lattice, a change of slip direction from [111] to $[\bar{1}\bar{1}\bar{1}]$ in a $(1\bar{1}0)$ plane does not modify the CRSS by reason of symmetry. However, only {110} planes have this property. All other planes of the $\langle 111 \rangle$ zone are non-centrosymmetric, so that positive and negative shears are not equivalent. The intrinsic asymmetry of {112} planes is reflected in a preferential sense for twinning and, as mentioned in Section 3.2.1, a smaller lattice resistance in the twinning sense than in the anti-twinning sense.

The breakdown of the Schmid law has a second cause, the origin of which is of same nature as the Escaig effect in fcc crystals (Section 2.4.4). There are small edge displacements in the screw dislocation cores, which globally cancel out. These edge contributions can be rationalized in terms of fractional dislocations (Appendix D), that is, dislocations with irrational Burgers vectors separated by ribbons of constant, unstable SFE. Non-glide components of the applied stress tensor that do not affect the perfect screw dislocation, induce Peach–Koehler forces on the fractional dislocations. These forces distort the core and modify the CRSS in a very complicated manner (Duesbery, 1989). As a result, the CRSS of a single crystal depends on the orientation of the stress axis not only through the resolved shear stress, but also through non-glide stress components. The experimentally observed tension–compression asymmetry of the CRSS results from this effect and from the twinning–anti-twinning asymmetry.

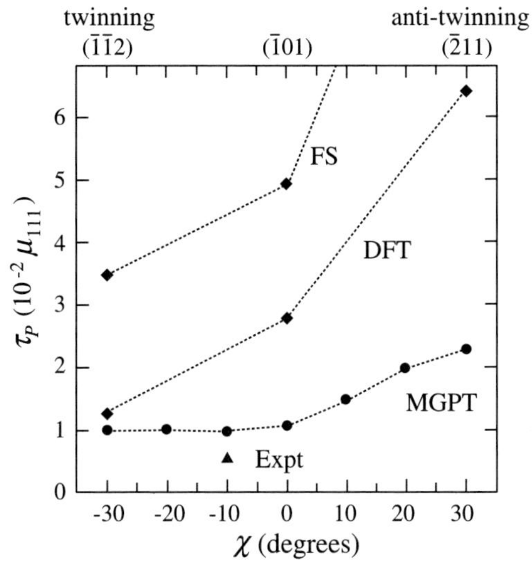

FIG. 3.6. Orientation dependence of the Peierls stress in Ta, in units of the shear modulus μ_{111}. χ is the angle between the MRSS plane and the reference ($\bar{1}01$) plane (Section 3.2.1). The calculations are performed with a central-force Finnis–Sinclair (FS) potential (Ito and Vitek, 2001), an *ab initio* method (Woodward and Rao, 2002) based on density functional theory (DFT) and an optimized MGPT potential (Yang *et al.*, 2001). A typical experimental value ($5\times10^{-3}\mu_{111}$) is given for $\chi = -10°$. From Moriarty *et al.* 2002*b*. Courtesy J. Moriarty.

The Peierls stress of a rigid screw dislocation submitted at $T = 0\,\mathrm{K}$ to a pure shear stress along $\langle 111 \rangle$ depends on the orientation of the MRSS plane. This orientation dependence was determined in Ta and Mo by several atomistic simulations and by *ab initio* calculations (Woodward and Rao, 2002). Fig. 3.6 is extracted from a review article summarizing these results (Moriarty *et al.* 2002*b*, see Gröger *et al.* 2008*a* for further simulations on Mo). It shows the evolution of the Peierls stresss in Ta for orientations going from $\chi = -30°$ (twinning orientation) to $\chi = +30°$ (anti-twinning orientation).

The Peierls stress increases when going from $\chi = -30°$ to $\chi = +30°$ and reaches rather high values for slip in the anti-twinning orientation. The twinning–anti-twinning asymmetry can be characterized by the ratio $r = \tau_P(+30°)/\tau_P(-30°)$. The DFT results yield $r \approx 6$, whereas the Schmid law predicts $r = 2$ for $\{112\}$ slip. A similar trend is observed in Mo with less steep variations. In that case, slip is known to exclusively occur in $\{110\}$ planes and the Schmid law predicts $r = 1$. The DFT-computed ratio is $r = 1.8$, which is consistent with the only existing measurement of this quantity by direct application of a pure shear, $r \approx 1.5$, between 77 and 200 K (Guiu, 1969). Thus, a significant non-Schmid behaviour is evidenced in spite of the poor quantitative agreement between DFT and atomistic calculations.

Fig. 3.6 shows that the MGPT and DFT Peierls stresses for $\chi = -10^\circ$ are larger than the experimental one by a factor of two and four, respectively. The problem is quite general and was noted very early (Duesbery, 1983*b*). All simulations performed to date predict too high Peierls stresses for all orientations, with only one exception (the MGPT stress for $\chi = 0^\circ$ in Mo is close to the experimental value). For instance, the Mendelev potential yields a Peierls stress of about $1.7 \times 10^{-2}\mu$ for α-iron at $\chi = 0^\circ$ (Chaussidon *et al.*, 2006), whereas the experimental value is about $5 \times 10^{-3}\mu$. Although the NEB method may overestimate the height of Peierls barriers by about 30% (Gröger and Vitek, 2012), this does not solve the problem. Several explanations were proposed for this puzzling phenomenon.

Bulatov and Cai (2002) suggested that anomalous slip is responsible for a significant softening with respect to normal slip at low temperatures. However, as indicated in Section 3.2.1, observations of anomalous slip are essentially restricted to group-V bcc metals.

Gröger and Vitek (2007) proposed a temperature-dependent softening mechanism based on the postulate that Frank–Read sources emit fast-moving dislocation loops that transiently expand as in fcc metals without experiencing any lattice resistance. This postulate contradicts a wealth of *in situ* observations (Section 3.2.1) performed over a large time span (Saka *et al.*, 1973; Caillard, 2010*a*).

Using an atomistic potential, Rodney (2007) investigated the kink-pair mechanism on the Lomer dislocation in aluminium. A comparison of dynamic and static simulations suggested that the critical stresses for kink-pair nucleation at very low temperatures could be typically two times smaller than the Peierls stress. However, further simulations showed that the two stresses are very close to each other.

In a set of experiments carried out at very low temperatures (see Section 3.2.4.3) Takeuchi *et al.* (1982) observed a progressive breakdown of the classical Arrhenius rate equation below 50 K and, by extrapolation, the occurrence of a residual strain rate sensitivity at $T = 0\,\mathrm{K}$. This last phenomenon was attributed to a quantum effect induced by zero-point vibrations and modelled in a heuristic manner. A revisited model based on quantum transition state theory was proposed by Proville *et al.* (2012). It successfully accounts for most of the discrepancy between computed Peierls stresses, which do not account for such quantum effects, and experimental values.

3.2.4 Kink-pair mechanisms and models

The history of the kink-pair mechanism can be found in a review article by Seeger (1984). In 1956, Seeger proposed that thermally activated dislocation motion over the Peierls relief occurs at finite temperatures via the formation of a saddle point configuration of finite size under a stress $\tau(T) < \tau_P$. Beyond the critical position, this configuration expands in an irreversible manner through the sideways motion of two kinks of opposite sign. The early approaches outlined in Section 3.2.4.1 make use of linear elasticity to investigate the formation and properties of kink-pairs on screw dislocation cores.[6] Atomistic simulations further provided a more consistent approach of the energetics of kink-pairs, but a full connection with the mesoscale properties of screw dislocations is still missing.

3.2.4.1 Phenomenological models In the earliest models, no connection is made between screw dislocation cores and the lattice resistance. The critical configurations of a perfect dislocation in equilibrium under stress is calculated in a non-deformable 'Peierls potential' of assumed shape (Section E.1), usually in the approximation of a constant line tension. The wavelength of this potential is the distance h_o between $\langle 111 \rangle$ atomic rows in the slip plane. At finite temperatures, there are two types of critical configurations (cf. Fig. E.1 and Fig. 3.7). At high and medium stresses, the critical configurations consist of bulges with a height smaller than h_o and a curvature radius approximately proportional to the inverse of the resolved shear stress. At low stresses, the saddle point configuration consists of more or less well-formed geometric kink-pairs of height h_o. These models are reviewed by Kocks *et al.* (1975) and Caillard and Martin (2003). The methods used for calculating the saddle point configurations and the related activation energies are detailed in Section E.2. Such analytical solutions can be useful for predicting simple generic forms or scaling laws for the energetics of kink-pair processes (Suzuki *et al.*, 1999; Caillard and Martin, 2003).

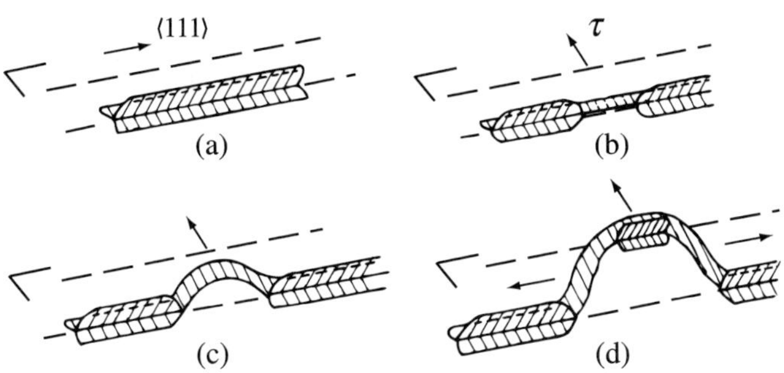

FIG. 3.7. Kink-pair formation on a dissociated screw dislocation. (a) Stress-free dissociated configuration on three $\{110\}$ planes. (b) Recombination of two partials under stress leading to a glissile configuration. (c) Critical configuration of bulge-type. (d) Critical geometric kink-pair re-dissociated over a portion of its length. From Kubin 1976.

[6] Following Seeger's convention, the critical configuration is called here a kink-pair; a double kink is then a kink of height twice the elementary height.

The most sophisticated model was developed by Seeger and co-workers (Seeger 1984, see Argon 2008 for a simplified version of this model and a comparison with experiment). It divides the temperature range between the athermal transition temperature $T = T_a$ and $T = 0\,\mathrm{K}$ into three regimes separated by knees or humps that are more or less visible on the experimental stress–strain curves (Fig. 3.1*b*). The adjustment of the model to experimental data on ultra-pure bcc single crystals loaded along low-symmetry orientations yields excellent fits. In particular the rather simple analytical form derived for the low-stress regime (Seeger and Schiller, 1962) is particularly well suited for comparisons with atomistic simulations (Section 3.2.4.2).

Hirsch's hypothesis led to the development of other types of models, in which doubly-degenerate cores are depicted in terms of screw dislocations dissociated into partials. A typical dissociation in three {110} planes occurs for instance by the reaction (Escaig, 1968*d*)

$$\frac{1}{2}[111] = \frac{1}{8}[110](1\bar{1}0) + \frac{1}{8}[101](\bar{1}01) + \frac{1}{8}[011](01\bar{1}) + \frac{1}{4}[111], \qquad (3.1)$$

where the $\frac{1}{4}[111]$ screw partial accommodates the shears at the intersection of the three stacking faults. Other dissociation modes are reviewed by Duesbery (1989). In such models, the extended core transforms under stress into a glissile recombined configuration that forms critical kink-pairs (Fig. 3.7).

The energetics of this process is calculated within an elastic framework. There are two free parameters, the SFE and the recombination distance.

3.2.4.2 Kink-pairs at the atomic scale We consider kink-pairs in {110} planes and in ⟨112⟩ directions. In the bcc lattice, the {111} planes do not possess mirror symmetry. As a consequence, there are two non-equivalent families of kinks with different energies on the ⟨111⟩ screw lines, plus symmetry-related kink-pairs for lines moving in opposite directions (see e.g., Ventelon *et al.* 2009). The situation is far more complex when the screw dislocation cores are doubly-degenerate (Cai *et al.*, 2004). There are six different types of kinks depending upon whether the kinks connect lines with same or different core structure. Further, the two core configurations can coexist along a screw line; they are then separated by antiphase defects, or 'flips', which may locally modify the formation energy of kink-pairs.

The first detailed discussion of the various types of kinks, kink-pairs and their energies was given by Duesbery (1983*a*, 1983*b*) with reference to α-iron and potassium. Similar studies were further performed on α-iron by Wen and Ngan (2000), using an EAM potential and the NEB method (Section 1.5), and by Yang *et al.* (2001) and Moriarty *et al.* (2002*a*) in tantalum using a MGPT potential and static methods. Such atomistic simulations provided the first informations about the kink-pairs with lowest energy. The total activation energies are reasonable and, although the Peierls stresses are too high (Section 3.2.3), the scaled values

τ/τ_P exhibit a reasonably realistic temperature dependence. These simulations also show that the barrier opposed to kink migration, or secondary Peierls stress, is typically one or two orders of magnitude smaller than the primary Peierls stress. This confirms that the nucleation of kink-pairs is the rate-controlling step for screw dislocation motion.

The atomistic simulations of kink-pair properties performed by Moriarty *et al.* (2002*a*, 2006) on an almost isotropic screw dislocation core in tantalum provide a first method for approaching mesoscopic screw dislocation velocities in bcc crystals. The formation energy of the kink-pair with lowest energy is $E^f_{kp} = 0.96\,\text{eV}$, close to the experimental value for the total activation enthalpy of the kink-pair mechanism ($\Delta H_o \approx 1.04\,\text{eV}$). At low stresses, the configuration investigated is a well-formed geometric kink-pair. For very small stresses and kink-pair widths $w \approx 20b$, the computed activation energy asymptotically approaches its maximum value, E^f_{kp}. The low stress regime holds for kink-pair separations $w \geq 10b$ and stresses $\tau \leq 0.2\,\tau_P$. In this range, the atomistic results are found in rather good agreement with the prediction of the continuum model by Seeger and Schiller (1962), which is detailed in Section E.3. For simulations in the high stress regime, it is no longer possible to guess a definite initial shape for the saddle point configuration and its determination becomes more difficult. Figure 3.8 shows the

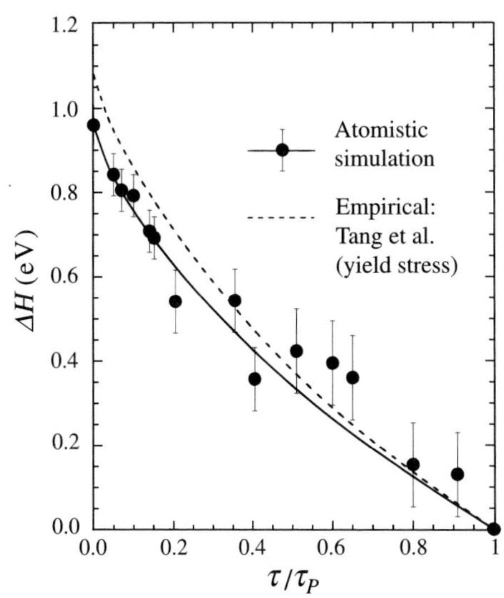

FIG. 3.8. Atomistic calculation of the stress dependence of the activation enthalpy, ΔH for kink-pair nucleation in tantalum. The stress τ is scaled by the Peierls stress τ_P drawn from the simulations. The regression line drawn through the data is compared to a fit (dashed line) drawn from experimental data on ultra-pure tantalum (Fig. 3.13). After Moriarty *et al.* 2002*a*, courtesy J. Moriarty.

full dependence of the activation enthalpy on the scaled stress $\tau(T)/\tau_P$ for two different sets of data. The simulated results reproduce well a fit drawn from experimental data on high-purity tantalum (Tang *et al.*, 1998). Hence, the functional form $\Delta H(\tau/\tau_P)$ is not affected by the too high value of the Peierls stress in atomistic simulations. Hence, the only quantity that is affected is the temperature dependence of the activation enthalpy.

The dynamics of screw dislocation motion was investigated by MD simulations in α-iron by Marian *et al.* (2004) for degenerate screw dislocation cores and by Chaussidon *et al.* (2006) for non-degenerate cores. With increasing dynamics, transitions were found between the classical kink-pair mechanism in quasi-static conditions to a mechanism of profuse kink-pair nucleation leading to dislocation processes typical of shock-loading tests and, further, to mechanical twinning. The boundaries between the different regimes appear to be sensitive to the type of core structure and imposed boundary conditions. Further, the small durations of these simulations entail very large dislocation velocities and high stresses. Periodic boundary conditions along the screw direction also pose problems because they induce spurious kink annihilations. Finally, Domain and Monnet (2005) tentatively corrected MD simulations carried out in constant imposed strain rate for the too high velocities (see also Monnet *et al.* 2010 and Section A.2). Rodney (2007) discussed a method for drawing the stress dependence of the activation enthalpy from MD simulations carried out in dynamic conditions, in consistence with the results of static simulations.

The dislocation glide velocities can also be determined directly with the help of a KMC scheme (Section 1.5) as was shown by Deo *et al.* (2005*b*) for tantalum and tantalum-based dilute alloys. In that case, the activation energy for kink-pair nucleation governs plastic flow and its stress dependence can be fitted to phenomenological laws. The situation is simpler in silicon (Cai *et al.*, 2000), for which use can be made of the kink-diffusion model (Section 3.4.3).

3.2.4.3 Lattice resistance at the mesoscale There is a need for models deriving from atomistic investigations that allow accounting at an upper scale for the energetics of the kink pair mechanism combined with the orientation dependencies of the slip geometry and the yield stress. What is presently available to feed local rules in DD simulations for bcc metals is now discussed. There are alternative rules drawn from experiment or current phenomenology (Sections 4.3.2.2 and 4.3.3.2), but they are far from satisfactory.

Thermally activated velocity of screw dislocations At the mesoscale, the velocity of a screw dislocation moving by the kink-pair mechanism is described by an Arrhenius form involving an activation enthalpy, or a free energy when an entropy contribution has to be taken into account. The pre-exponential factor v_o is derived almost in the same manner as for localized obstacles in materials without a lattice resistance (Section A.1). The attempt frequency of a kink-pair of width w is

$\nu = \nu_D b/w$. Each successful kink-pair nucleation event results in a forward motion of the line by h_o, the periodicity of the Peierls energy barrier, which is usually taken on {110} planes. In addition, there are L/w competing nucleation sites along a screw line of length L limited by fixed obstacles (Fig. 3.12). At the mesoscale, the analysis is usually carried out in terms of effective stress τ^*, as in Appendix A. Hence, the Arrhenius form is written

$$v = h_0 \frac{\nu_D b}{w} \frac{L}{w} \exp - \left(\frac{\Delta H(\tau^*)}{k_B T} \right) \tag{3.2}$$

The stress dependence of w and L is usually considered negligible with respect to the stress dependence of the exponential term. The ratio $C = \Delta H(\tau^*)/k_B T$ is then assumed to be approximately a constant irrespective of temperature (eqn. A.5). Experimental values and numerical estimates yield respectively $C = 23{-}27$ and $C = 25 - 27$ (Tang *et al.*, 1998). A connection can then be established with static atomistic simulations of the kink-pair process, which yield a dependence of the activation enthalpy on the scaled stress τ^*/τ_P (cf. Fig. 3.8).

When temperature decreases to very low values, the phonon spectrum provides less and less activation energy and the quantity C is no longer a constant. This was verified experimentally by Takeuchi *et al.* (1982) in Nb, Fe, Ta and β-CuZn at very low temperatures. The pseudo-constant C starts decreasing progressively below about 50 K and more rapidly below about 10 K. In parallel, the strain rate sensitivity decreases but does not vanish at $T = 0$ K (Section 3.2.3). A phenomenological quantum mechanical rate model was then proposed to replace the usual Arrhenius form at very low temperatures. A full quantum model was eventually proposed by Proville *et al.* (2012). Equation 3.2 also looses its validity at low stresses when the lattice resistance enters in competition with a strong forest strengthening (Section 3.2.5.1 and Fig. 3.2*b*). The factor L/w, which accounts for competing nucleation sites is characteristic of the kink-pair mechanism. It induces a length dependence of the dislocation velocity, which was observed by *in situ* TEM experiments on α-iron (Caillard, 2010*a*), during prismatic slip in hcp metals (Farenc *et al.*, 1993) and in semiconductors (Section 3.4.8).

Temperature and orientation dependence of the yield stress. At present, these two dependencies seem too complex to be fully predicted. A very elegant model was developed by Edagawa *et al.* (1997) for calculating the saddle point configurations of a screw dislocation moving in a two-dimensional Peierls energy barrier with ternary symmetry. The energy barrier is triangular, with an adjustable maximum height in the plane perpendicular to the $\langle 111 \rangle$ screw direction. The formation of kink-pairs under stress is treated using linear elasticity and in the line tension approximation, as in Section 3.2.4.1. A straight screw dislocation jumps by kink-pair nucleation from a stable position to any of the three next stable positions.

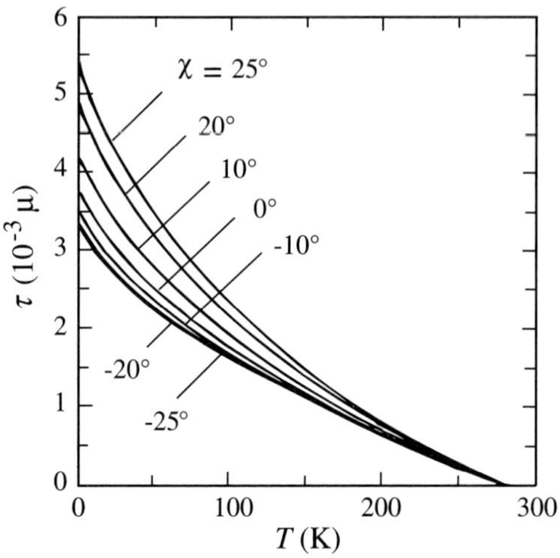

FIG. 3.9. Temperature and orientation dependence of the critical stress for the motion of a screw dislocation in a two-dimensional Peierls potential with threefold symmetry. The angle χ has same definition as in Fig. 3.6. From Edagawa *et al.* 1997, courtesy K. Edagawa.

The velocity of screw dislocations is thermally activated and yield is assumed to occur when it reaches a predefined value. This allows determining the stress and temperature dependence of the total activation enthalpies for slip on the average slip plane and on its three slip components. The results are generic and globally predict rather well the twinning–anti-twinning asymmetry of bcc metals (Fig. 3.9) and the transition from MRSS slip to primary {110} slip at low temperatures. The only feature that cannot be accounted for is the tension–compression asymmetry induced by hydrostatic components of the applied stress tensor. As emphasized by the authors, threefold symmetry is almost sufficient to account for non-Schmid behaviour with only two adjustable parameters, irrespective of the dislocation core structure and the assumed potential.

In a set of three consecutive articles (Gröger *et al.*, 2008*a*; Gröger *et al.*, 2008*b*; Gröger and Vitek, 2008), Gröger and co-workers devised a method for determining non-Schmid behaviour in group-VI molybdenum and tungsten as a function of orientation and temperature. This work expands previous studies where the orientation dependence of the CRSS was investigated at $T = 0\,\text{K}$ by MD simulations, using an analytical yield criterion proposed by Qin and Bassani (1992*a*, 1992*b*). The effective critical resolved shear stress τ_c^* is expressed as a linear combination of stresses on {110} planes in zone with the [111] primary slip direction (see e.g. Gröger *et al.* 2008*b*)

$$\tau_c^* = \tau_S^{(\bar{1}01)} + \alpha_1 \tau_{\parallel}^{(0\bar{1}1)} + \alpha_2 \tau_{\perp}^{(\bar{1}01)} + \alpha_3 \tau_{\perp}^{(0\bar{1}1)} \tag{3.3}$$

The first stress at the right-hand side is simply the resolved shear stress in the primary slip plane, as given by the Schmid law. The other three shear stresses are non-Schmid stresses that affect the dislocation core but produce no work during dislocation glide. The second stress is parallel to [111] in the $(0\bar{1}1)$ plane; it is responsible for the twinning–anti-twinning asymmetry. The last two stresses are perpendicular to [111] and account for the tension–compression asymmetry. The coefficients α_i are temperature dependent; they are determined at $T = 0\,\mathrm{K}$ from atomistic simulations.

For going further, a two-dimensional Peierls barrier is constructed, which depends on the full stress tensor of applied stresses. For this purpose, the method proposed by Edagawa *et al.* (1997) is complemented by phenomenological models for the kink-pair energy and coupled to a rate equation for dislocation velocities. The whole procedure is rather complex and relies on experimental results. It reproduces rather well the dependence of the CRSS for primary slip in molybdenum and tungsten measured by Seeger and Hollang (2000), Hollang *et al.* (2001) and Brunner and Glebowsky (2000), as well as non-Schmid behaviour in molybdenum at low temperatures.

Once the coefficients involved in eqn. 3.3 are determined from atomistic simulations at $T = 0\,\mathrm{K}$, the orientation dependence of the yield stress can be modelled in a reasonably simple manner as a function of temperature (see, for instance, Koester *et al.* 2012). The stress-dependent activation energy for kink-pair formation is adjusted to experiment using the generic form given by eqn. E.13. At present, there is no dislocation-based model that could be used to incorporate the orientation dependence of strain hardening. Therefore, the stress–strain curves can be only obtained by also adjusting current phenomenological models to experiment.

The model by Edagawa *et al.* (1997) outlined above was revisited with the help of DD simulations to examine mesoscopic cross-slip in α-iron (Chaussidon *et al.*, 2008). The temperatures considered were approximately in the lower half of the low temperature range. As was shown in Section 3.2.2, {112} slip is most probably a composition of atomic cross-slip steps on two {110} planes. The primary and cross-slip planes are defined, respectively, as the closest and second closest {110} planes to the MRSS plane. The thermally activated dislocation velocity is derived from experimental results using the procedure described in Section 4.3.2.2. Cross-slip between the two {110} planes is allowed for all orientations of the resolved stress except for anti-twinning orientations. For the latter, only primary slip occurs according to previous MD simulations (Chaussidon *et al.*, 2006). For other stress orientations, the cross-slip probabilities on the slip and cross-slip planes are derived from the corresponding average velocities in a form adapted to mesoscopic time steps. The DD simulations yield a twinning–anti-twinning asymmetry consistent with previous MD simulations and a predominance of primary slip from the twinning orientation to orientations that approach the anti–twinning region as temperature decreases.The potentialities of this type of model are further discussed in Section 4.3.3.2.

3.2.4.4 The locking–unlocking model Anomalies such as knees or humps are frequently observed on the stress–strain curves of bcc metals, hcp metals when they deform by prismatic glide, and $L1_2$ compounds (Caillard and Couret 1996; Caillard and Martin 2003). In bcc metals, the stress anomalies are more or less well-marked depending upon the initial dislocation density (Fig. 3.1) and the impurity content (Section 3.2.5.2). They are usually better seen in the stress dependence of the activation volumes. Several explanations were proposed for their occurrence, for instance a transition in the slip plane where kink-pairs are nucleated (Seeger, 1984) or a transition between bowed and geometric critical kink-pairs (Escaig, 1968*d*). The locking–unlocking model was initially developed to describe *in situ* observations showing that such transitions are associated with jerky prismatic glide of screw dislocations in hcp metals (Section 3.3.3). The transition that was investigated in bcc metals can be seen around 250 K in tantalum (Fig. 3.1*b*). It occurs at the same temperature in α-iron and corresponds to a well-defined maximum in the activation volume (Kuramoto *et al.*, 1979).

In the past years, *in situ* TEM deformation did not allow reaching resolutions better than typically a few tens of nanometres. With only one exception (Kubin and Louchet, 1979), straight screw dislocations were always observed to move smoothly in bcc metals. In the more recent experiments by Caillard (2010*a*, 2010*b*), grains of ultra-pure polycrystalline specimens of α-iron were deformed between 100 K and 300 K with an improved resolution of about 1.5 nm.[7] Local stresses were measured from the curvature radii of non-screw segments, in reasonable agreement with experimental CRSSs. Smooth motion of screw dislocations was recorded between 300 K and about 250 K. Below this temperature, screw dislocation motion was observed to always occur by abrupt jumps in {110} planes, characterized by glide distances λ larger than the maximum height h_o of a single kink-pair (Fig. 3.10*a*). The waiting times between jumps as well as the jump distances were found to be temperature dependent (Fig. 3.10*b*).

All observations strongly suggest that the jump-like behaviour is not influenced by the local environment and is associated with an intrinsic property of the screw dislocation cores. The locking–unlocking model is based on a postulate by Vitek (1966), who assumed that the transformation from a glissile core of high energy to a low energy core may require some thermal activation. If so, a glissile configuration may survive in a metastable manner and overcome several Peierls valleys before being trapped again into its low energy configuration. The locking–unlocking mechanism describes the successive transitions of the screw dislocation

[7] References to video sequences are given in the two supplementary materials.

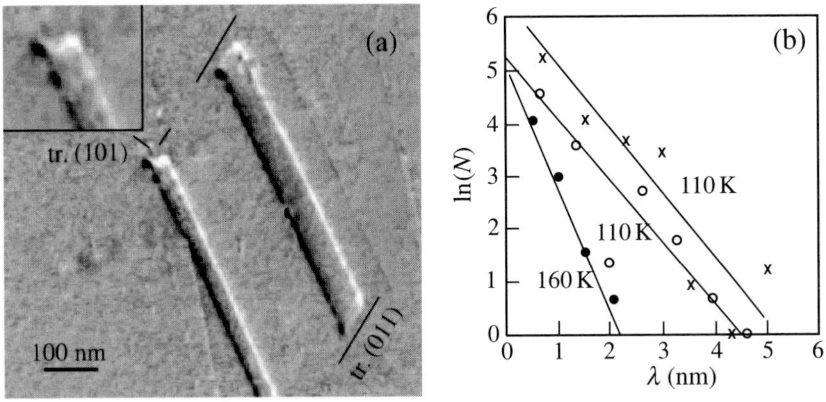

FIG. 3.10. Jerky motion of screw dislocations in α-Fe during TEM *in situ* deformation. (a) Subtracted image contrasts of the initial (black) and final (white) positions of screw lines between two successive frames ($T = 150\,K$). The surface traces correspond to {110} slip planes. The trace of the left dislocation (insert) indicates that it cross-slipped during the jump. (b) Semi-logarithmic graphs of the distributions $N(\lambda)$ of jump distances at two temperatures. After Caillard 2010*b*, courtesy D. Caillard.

cores from glissile to sessile (locking, index ℓ) and sessile to glissile (unlocking, index $u\ell$).

A complete discussion of the locking–unlocking model is given by Caillard and Martin (2003). The numerical estimates are based on phenomenological elastic models for the transformations of screw dislocation cores and are semi-quantitative. All analyses have in common the starting point that follows. The probability for unlocking is written as in in eqn. 3.2, $P_{u\ell} = P_{u\ell o} exp[-\Delta G_{u\ell}(\tau)/k_B T]$, where the activation energy depends on the measured (effective plus athermal) resolved stress τ. The locking of glissile cores is described by a similar form with a stress-independent activation enthalpy $\Delta G_\ell \ll \Delta G_{u\ell}(\tau)$. It follows that the time scale of a jump event is governed by the unlocking probability $1/P_{u\ell}$, whereas the length scale is governed by an average jump distance, $\bar{\lambda}$. The average velocity is $v \approx \bar{\lambda} P_{u\ell}$ and the flight velocity is $v_g = \bar{\lambda} P_\ell$, so that $v = v_g P_{u\ell}/P_\ell$. The general form of the screw dislocation velocity during locking–unlocking processes is then written

$$v = \frac{v_g P_{u\ell o}}{P_{uo}} \exp -\left(\frac{\Delta G_{u\ell}(\tau) - \Delta G_\ell}{k_B T}\right) \quad (3.4)$$

The predicted kink-pair mechanism is based on the interplay between the height of the critical kink-pair configuration, which decreases with increasing

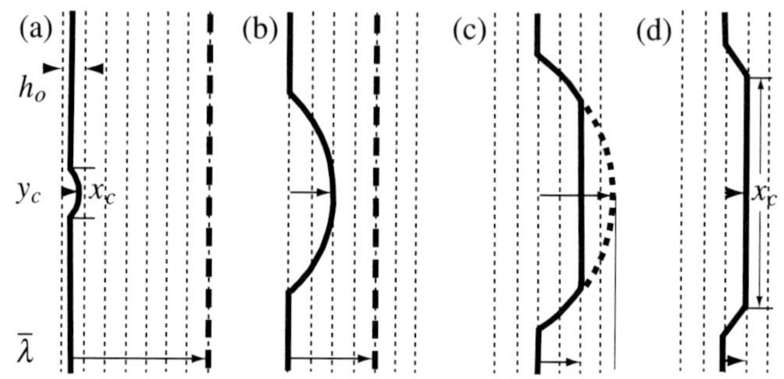

FIG. 3.11. Evolution of the height (y_c) and width (x_c) of the critical configurations and of jump distances ($\bar{\lambda}$) in the locking-unlocking model. The stress decreases from left to right. (a) Small bulge; (b) large bulge; (c) superkink-pair; (d) geometric kink pair. After Caillard 2010*b*.

stresses and decreasing temperatures, and the average jump distance $\bar{\lambda}$, which evolves in the opposite sense (Fig. 3.11).

At low temperatures and large stresses, the critical configuration consists of small bulges of height $y_c < h_o \ll \bar{\lambda}$, with substantial jump distances (Fig. 3.11*a*). At moderate stresses, the critical bulges are larger, but still smaller than the jump distances, $h_o < y_c < \bar{\lambda}$ (Fig. 3.11*b*). At lower stresses in the domain of the anomaly, $h_o < \bar{\lambda} < y_c$. The critical configurations consist of superkink-pairs, that is, truncated bulges with heights multiple of h_o (Fig. 3.11*c*). The transition between bulges and superkink pairs corresponds to a macroscopic anomaly in the mechanical response. The latter arises because y_c and the activation area, as defined from y_c and the width x_c of the critical configuration, are reaching a maximum value. At very low stresses, $\bar{\lambda}$ cannot be smaller than h_o and the usual kink-pair mechanism is recovered (Fig. 3.11*d*).

The anomaly is quantified by estimating the differences in dislocation core energy between the locked and glissile states on both sides of the maximum in activation area (Caillard, 2010*b*). For geometric kink-pairs at very low stresses, the difference is of same order as the *ab initio* value for the total activation energy, $\Delta E \approx 30$ meV/b (Ventelon and Willaime, 2007). For critical bulges at high and moderate stresses, the analysis is inspired from a cross-slip model proposed by Friedel in 1959 to explain prismatic slip in hcp metals (see Caillard and Martin 2003). A fit of this model to experimental results yields $\Delta E \approx 11$ meV/b. This rather small value allows critical bulge-like kink-pairs to develop under reduced critical stresses and reach large critical heights. The proposed description of the anomaly is tentative and may be seen as a way to reconcile elastic models with experiment. Finally, the model may have to be modified at very low temperatures, as the Arrhenius form breaks down (Section 3.2.4.3).

Jerky screw dislocation motion may occur in other bcc metals than α-Fe where the same anomaly was observed. So far, no evidence was found for metastable glissile configurations in atomistic simulations. No alternative explanation to the one provided by elastic models was, however, found for the jerky motion of screw dislocations at low temperatures.

3.2.5 Strengthening and softening in bcc metals

3.2.5.1 Strengthening by forest obstacles Strain hardening is more difficult to model in bcc metals at low temperatures than in fcc metals because forest strengthening and the lattice resistance are not additive mechanisms. The considerations developed in this section apply also to other strong localized obstacles such as precipitates.

The interaction of moving screw dislocations with forest obstacles is sketched in Fig. 3.12 as a function of temperature. The forest density is mostly made up of long screw lines generated during or after the pre-yield stage (Section 3.2.1) by the motion of non-screw segments. Although the intersection processes locally proceed as in fcc metals, the lattice resistance induces the formation of specific critical configurations. The screw dislocations do not bow between forest obstacles under stress but continuously move by kink-pair formation. In consequence, the strengthening constant in eqn. 2.20 becomes temperature-dependent, all the more

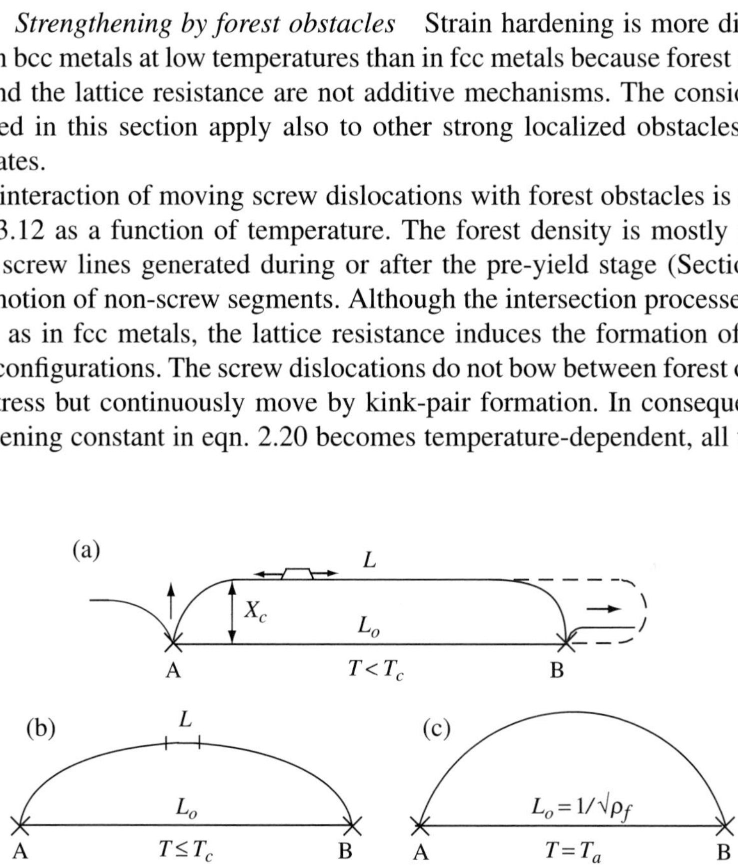

FIG. 3.12. Critical configurations for screw dislocations interacting with forest obstacles (crosses). (a) Low temperature and high stress. The motion of a straight screw line produces kinks that accumulate at obstacles. L_o and L are the initial and critical lengths of the screw segment and X_c is its critical travelled distance. (b) At lower stresses elementary kinks progressively invade the whole line, producing a continuous curvature at a critical temperature T_c. (c) Above the athermal transition temperature, T_a, the mobile line behaves as in fcc metals.

as temperature is low, as was observed for example by Keh and Weissmann (1963) in α-iron.

At low temperatures (Fig. 3.12*a*), the elementary kinks produced by the motion of screw segments accumulate on forest obstacles, where they form small curved non-screw segments. The critical configuration corresponds to a critical angle for the bowing of the non-screw segments and to a critical distance X_c travelled by the screw line. The number of piled-up kinks of height h_o is then X_c/h_o. Depending upon the relative positions of the neighbouring screw segments, forest obstacles can be crossed sideways by a non-screw segment (right part of Fig. 3.12*a*) or in the forward direction (left part of Fig. 3.12*a*). As X_c is proportional to the curvature radius of the non-screw segments, the number of piled-up kinks needed for obstacle crossing is inversely proportional to stress. *In situ* measurements show that X_c is typically a few tens of nanometres for junctions between screw segments at low temperatures and high stresses (Louchet *et al.*, 1979). Assuming that $X_c \ll L_o$ and $L_o \approx L$, the area associated with the critical configuration is $A \approx L_o X_c$. During the steady-state motion of screw dislocations, every time a screw line breaks through an obstacle, it is pinned by another one. Then, A also corresponds to the average area $1/\rho_{fj}$ associated to the density ρ_{fj} of junction-forming screw dislocations.

$$LX_c = \frac{1}{\rho_{fj}} \tag{3.5}$$

This relation was verified by DD simulations (Tang *et al.*, 1999). As X_c is very small at low temperatures, L is very large. For a total forest density $\rho_f \approx 2\rho_{fj} \approx 2\,10^{12}\,\mathrm{m}^{-2}$, and with $X_c \approx 30\,\mathrm{nm}$ in niobium at 110 K (Louchet *et al.*, 1979), $L \approx 15\,\mu\mathrm{m}$. As the non-screw segments are made up of elementary kinks, the time needed for reaching the critical configuration is the total time needed to generate a critical number of kinks. This introduces a coupling between forest hardening and the energetics of the kink-pair mechanism. In consequence, strain hardening takes on a thermally activated character.

With increasing temperature and decreasing stress, the radii of the critical non-screw segments increase and the length L of the critical screw segments decreases. Above a temperature $T = T_c$ that corresponds to rather low stresses, the critical configuration entirely consists of piled-up interacting kinks (Fig. 3.12*b*). The resulting apparent curvature is called a *cold curvature*; it is not an equilibrium configuration and it cannot be described in terms of line tension. This intermediate domain is understood as reflecting an increasing contribution of forest hardening to the flow stress. When the athermal transition temperature is reached, line tension effects predominate and forest obstacle overcoming occurs as in fcc crystals (Fig. 3.12*c*). It follows that the temperature domain between T_c and T_a is associated with the progressive occurrence of stages in the stress–strain curves (Fig. 3.2a, see also Kumagai *et al.* 1990).

Attempts at modelling the flow stress focused on the low temperature domain. The parameters involved in eqn. 3.5 can be determined from line tension or line energy arguments and from the stress–velocity relation for a screw segment of length L_o (eqn. 3.2). This yields, an athermal hardening proportional to the forest density (Rauch, 1994) and a thermal contribution that logarithmically depends on strain rate (Louchet *et al.*, 1979). These two aspects were simultaneously considered by Tang *et al.* (1999), but a global model encompassing the intermediate regime and predicting a smooth transition to fcc-like behaviour does not exist yet.

The stress–strain curves of bcc metals exhibit a variety of shapes, which depend on temperature, strain rate and initial dislocation density. There is a predominance of pseudo-parabolic deformation curves, with an initial domain of steep hardening that extends to larger strains with decreasing homologous temperatures. This last feature is a extension of the pre-yield stage; it arises from the exhaustion of non-screw segments while the screw lines are not sufficiently mobile to accommodate the imposed plastic strain rate.

3.2.5.2 Solid–solution effects As illustrated by Fig. 3.13, solute atoms significantly harden bcc metals at low temperature.

For small solute concentrations, an unexpected reduction of the lattice resistance is often observed in an intermediate domain of temperature. This softening effect typically occurs for solute concentrations of one to a few atomic percent in substitutional alloys and in the range of 100–1000 at. ppm for carbon and nitrogen

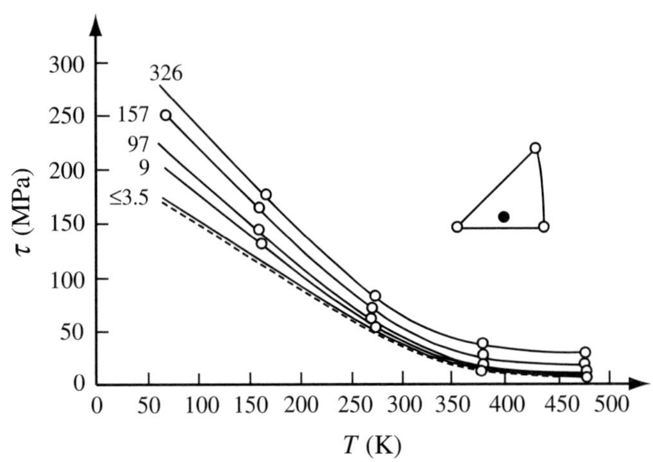

FIG. 3.13. Flow stress–temperature curves of ultra-pure tantalum single crystals containing small controlled amounts of nitrogen. The dashed line is an extrapolation to a nitrogen-free material. From Lachenmann and Schultz 1970.

interstitials. In ultra-pure alloyed materials, strengthening and softening phenomena are two facets of the interaction between solute atoms and screw dislocations cores. Moving kinks also interact with solute atoms, especially with positive concentration fluctuations. For a compilation of experimental aspects and a discussion of past models, the reader is referred to Pink and Arsenault (1980).

In bcc metals of group V (V, Nb, Ta) solute strengthening seems to prevail when the base metal is sufficiently pure (Fig. 3.13). When solute softening is observed, there seems to be evidence that it results from the scavenging of residual interstitial impurities by solute atoms. The recombination of the two types of defects leads to new defects that induce a smaller strengthening. Oxide layers can also induce softening by enhancing dislocation multiplication at free surfaces. In contrast, there are evidences for a single intrinsic mechanism for the strengthening and softening of group-VI metals by substitutional solutes. Ultra-pure α-iron also exhibits intrinsic softening by substitutional and interstitial solutes (cf. Fig. 3.14). This was confirmed by deformation tests carried out on ultra-pure single and polycrystals irradiated at low temperatures, which contained well-defined concentrations of immobile self-interstitials. These experiments were summarized by Kubin (1982).

Interstitial solutes induce in general a substantial strengthening of the CRSS extrapolated to $T = 0\,\text{K}$. When the athermal stress is subtracted from the CRSS, the temperature dependence of the resulting effective stress differs from the one shown in Fig. 3.14. It is not granted, however, that the athermal stress is temperature-independent, even after correction for the temperature dependence of the shear modulus. In binary substitutional alloys, especially α-iron alloys,

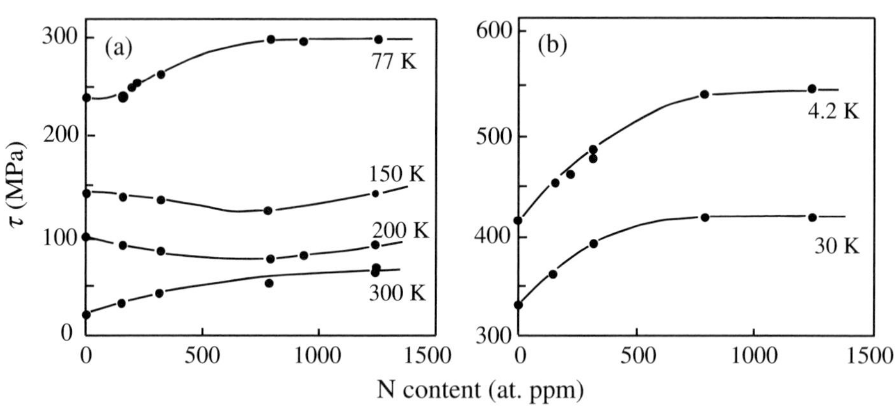

FIG. 3.14. The CRSS of ultra-pure α-Fe single crystals as a function of nitrogen content and temperature. (a) Below room temperature and with increasing solute content, softening is followed by hardening. (b) At very low temperatures, only strengthening is observed. After Aono *et al.* 1980.

attempts were made to establish correlations between the elastic interactions of solute atoms with screw dislocations and the evolution of the CRSS with concentration and temperature (Takeuchi, 1969; Stephens and Witzke, 1976; Chomel and Cottu, 1982; Okazaki, 1996).

Elastic models were proposed to explain intrinsic hardening and softening in bcc metals (Pink and Arsenault, 1980). They estimate in a simplified manner the combined interaction of a perfect screw dislocation with a Peierls energy barrier and misfit centers. Sato and Meshii (1973) investigated the distortions of the core induced by solute atoms that locally modify the atomic radius (size misfit). Solute atoms were found to enhance kink-pair nucleation but impede the sideways motion of kinks. The two processes involve different interaction energies and their interplay results in strengthening or softening depending upon the concentration or temperature. For interstitial solutes, the model yields a reasonably good agreement with experiment. The agreement is less good for substitutional alloys, which is attributed to the clustering of solute atoms. In another model, which is rather complex, both size and modulus effects are investigated (Suzuki, 1979). This model is comprehensively discussed by Argon (2008) and compared to experimental results on substitutional α-Fe alloys. As a whole, correct trends are obtained but the limits of elasticity theory have clearly been reached.

Trinkle and Woodward (2005) performed the first *ab initio* calculations of softening effects, specifically on the interaction between transition substitutional elements with different numbers of *d* electrons and screw dislocation cores in molybdenum. It was found that solutes with a lower number of *d* electrons than Mo, (Hf and Ta) strengthen the Peierls stress, whereas those with a larger number (Re, Os, Ir) soften it, except for Pt. Platinum atoms exhibit, however, a strong and relatively long-ranged attractive interaction with the screw dislocation core that induces softening. Thus, solute atoms globally modify the critical stress for kink-pair nucleation along a screw line, but may also modify it locally through their interaction with the core. In consequence, negative 'Peierls misfits' and attractive interaction energies promote alloy softening at low temperatures. An estimate was also performed for the strengthening induced by the interaction of solute clusters with moving kinks. As a result, it was shown that the predicted strengthening–softening behaviour was in quantitative agreement with the available experimental results on Mo-Re and Mo-Pt alloys.

In a further study (Medvedeva *et al.*, 2007), the effect of solutes introducing an excess ($\delta n > 0$) or deficit ($\delta n < 0$) of valence $s + d$ electrons was investigated in group-V (Nb, Ta) and group-VI (Mo, W) metals. Use was made of an atomistic description of interatomic interactions parametrized with the help of *ab initio* calculations. The solute-dislocation interaction energies were considered as driving forces of electronic origin for kink-pair nucleation and elastic relaxations were neglected. In group-VI metals, softening was predicted for $\delta n > 0$ (Re, Os, Ir, Pt), as in *ab initio* calculations, and strengthening was predicted for $\delta n < 0$. For Nb and

Ta, the reverse behaviour was obtained, with a less pronounced effect of the number of valence electrons. This reinforces the hypothesis of a predominant extrinsic origin of softening in group-V metals.

Two other investigations emphasize the effect of solutes on the slip path of dislocations, in Ta-W (Chen *et al.*, 2008*b*) and W-Ta and W-Re alloys (Li *et al.*, 2012).

In dilute Fe-C alloys, softening occurs in the same temperature range as the transition discussed in Section 3.2.4.4 and seems to remove it (Kuramoto *et al.*, 1979). *In situ* TEM experiments were performed to compare the locking–unlocking mechanism in Fe-C alloys containing 110 at. ppm of carbon (Caillard, 2011) to the one observed in pure Fe (Caillard, 2010*b*). Jerky motion was characterized by more numerous and smaller jumps (less than 2.5 nm) than in pure α-iron (cf. Fig. 3.10*b*). Some observations attested that non-screw segments were interacting with carbon atoms, but no local interaction of screw dislocations with strong obstacles could be detected. This, of course, does not exclude a global weakening of the lattice resistance, like that evidenced by *ab initio* and atomistic calculations in substitutional alloys.

With respect to pure α-iron, the transition between smooth and jerky screw dislocation motion is shifted by more than 100 K towards lower temperatures. As a consequence, the rather sharp transition between truncated large bulges and geometric superkink-pairs (Fig. 3.11) is replaced by the classical transition between elementary bulges and geometric kink-pairs. Thus, a direct correlation is established between alloy softening, the removal of the anomaly in activation area and the shift of jerky screw motion to lower temperatures.

3.3 Prismatic slip in hcp metals

This section discusses first the reason why some hcp metals preferentially glide in the basal plane and others in prismatic planes. The emphasis is here on deformation by prismatic slip, which exhibits the characteristic properties associated with a lattice friction on screw dislocations. Titanium and zirconium are the most investigated hcp transition metals; they preferentially deform by prismatic slip and, as they are located in column IV of the periodic table, they behave in a very similar manner. Their specificity is to exhibit a lattice friction that decreases with decreasing impurity content.

3.3.1 Slip systems and screw dislocation cores

Hexagonal close-packed metals exhibit two predominant slip systems which share a common Burgers vector, $\mathbf{a} = \frac{1}{3}\langle 11\bar{2}0\rangle$. Deformation by glide preferentially occurs either on the basal (0001) plane or the prismatic $\{1\bar{1}00\}$ planes (Fig. 3.15).

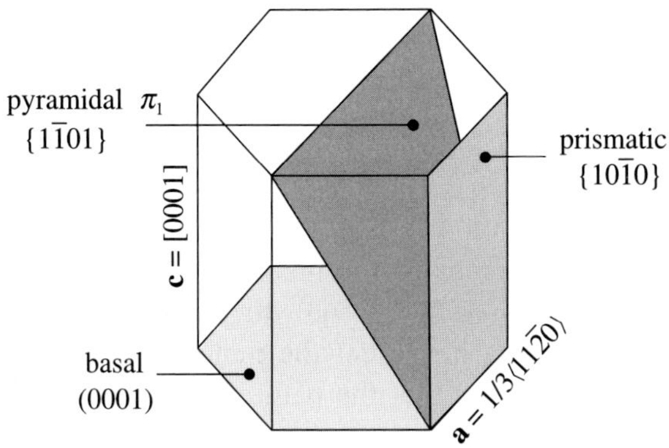

FIG. 3.15. Slip systems with Burgers vector $\mathbf{a} = \frac{1}{3}\langle 11\bar{2}0\rangle$ in hcp crystals.

The preferred secondary slip plane is most of the time the less favoured of the two or, occasionally, a pyramidal slip system with Burgers vector **a**. Pyramidal $\mathbf{c}+\mathbf{a}$ slip and mechanical twinning provide additional deformation modes that accommodate non-basal deformations. The active slip systems in hcp metals were compiled by Legrand (1985) and Bacon and Vitek (2002). More detail on dislocation properties and mechanical behaviour in hcp metals, can be found in the review by Bacon and Vitek (2002) and other articles in the same special issue.

According to a traditional viewpoint, the selection of the preferred slip plane is explained in terms of the Peierls–Nabarro model (Appendix D) and the favoured plane should be the most compact one. Then, basal slip should be preferred for c/a ratios larger than the ideal value $(8/3)^{1/2} = 1.633$ for compact packing, whereas prismatic slip should occur when $c/a < 1.633$. Although this rule is globally valid, it suffers a few notable exceptions. For instance, beryllium has the smallest of all c/a ratios in hcp metals ($c/a = 1.568$) and nevertheless glides preferentially in the basal plane.

In fact, all divalent hcp metals (Be, Mg, Zn, Cd) preferentially glide in the basal plane, whereas transition metals deform either by basal slip (Co, Tc, Re) or by prismatic slip (Y, Ti, Zr, Hf, Ru, Os), depending on the partial filling of their electronic d band. Thus, as was first noticed by Legrand (1984, 1985), the selection of the preferential slip plane is clearly an electronic property.

In divalent hcp metals, the Peierls stress for basal slip is very small and the Schmid law is well verified. Indeed, it was established from experimental results on Zn, Mg and Cd. This led to the assumption of a glissile dissociation of **a**-type dislocations into two Shockley partials according to the reaction

$$\frac{a}{3}[11\bar{2}0] \rightarrow \frac{a}{3}[10\bar{1}0] + \frac{a}{3}[01\bar{1}0] \qquad (3.6)$$

This reaction scheme is confirmed by atomistic calculations of the γ-surface. The SFE in the basal plane is, however, strongly material-dependent (Appendix B) and the dissociation width sometimes reduces to a small core extension by a few Burgers vectors. Shockley dislocations were observed by TEM in cobalt (Korner and Karnthaler, 1983), which has a low SFE.

In hcp transition metals, the $\frac{1}{3}\langle 11\bar{2}0\rangle$ screw dislocation cores have two metastable configurations that are extended in the basal plane or in the prismatic plane. The preferred glide plane is the one where the core has the lowest energy. Using a tight-binding approach to treat the non-central character of electronic bonding, Legrand (1985) found that the nature of the preferred slip plane is an oscillating function of the number of electrons in the *d* band. For Ti and Zr, *ab initio* simulations performed by Domain and Legris (2004) confirmed that the energy of screw dislocation cores is lower in the prismatic planes, in agreement with experimental observations of prismatic slip. The corresponding γ-surfaces exhibit deep minima for a translation $\mathbf{a}/2$, which corresponds to the stable stacking fault in the prismatic planes. The calculated prismatic SFEs are $\gamma = 174\,\mathrm{mJ\,m^{-2}}$ for Ti and $145\,\mathrm{mJm^{-2}}$ for Zr. The configurations of the two screw dislocation cores are reproduced in Fig. 3.16.

Atomistic simulations using a bond-order potential for Ti (Bacon and Vitek, 2002) and a Mendelev–Ackland potential for Zr (Khater and Bacon, 2010) further confirmed these results. In a further DFT study, the secondary spreadings

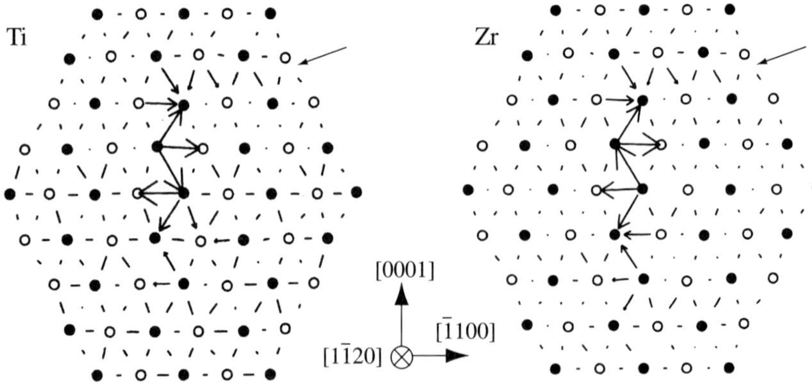

FIG. 3.16. DFT calculation of the screw dislocation core structures in Ti and Zr. The screw components of the relative atomic displacements are represented using the same convention as in Fig. 3.5. The projection plane is normal to the Burgers vector $\langle 11\bar{2}0\rangle$. The traces of the basal and prismatic planes are respectively horizontal and vertical. There is a large prismatic spreading with small secondary edge spreadings in the basal plane (arrows). After Domain 2002.

look different and the SFE is $220\,\mathrm{mJm^{-2}}$ (Ghazisaeidi and Trinkle, 2012). High resolution TEM observations of edge dislocations in Ti (de Crécy *et al.*, 1983) showed that they are dissociated in a glissile manner in the prismatic planes (Fig. 1.2*b*). The SFE is $\gamma = 155 \pm 25\,\mathrm{mJm^{-2}}$, in fair agreement with *ab initio* results for screw dislocations.

3.3.2 The Peierls stress in Ti and Zr

Titanium and zirconium are usually quoted as having a strong lattice resistance at low temperatures owing to their slightly non-planar core structure in prismatic planes. These two elements also have a strong affinity to oxygen, which stabilizes the hcp α phase with respect to a high-temperature bcc β phase. As it is very difficult to control the interstitial concentrations, the total impurity content is often expressed in oxygen equivalent concentration ($\mathrm{O^* = O + 2N + C}$). Figure 3.17*a*

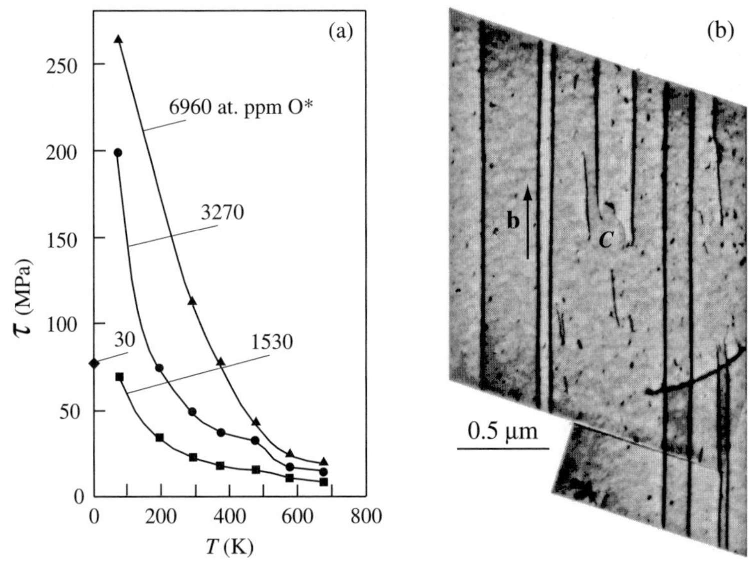

FIG. 3.17. Titanium crystals deformed in single prismatic slip. (a) Temperature dependence of the CRSS for several interstitial impurity contents (O* is the equivalent oxygen concentration). After Naka *et al.* 1988. Courtesy S. Naka. The data point at 0 K is an extrapolated value obtained by Biget and Saada (1989) in rather pure crystals. (b) Straight screw dislocations in a single crystal with 1530 at. ppm O* deformed by a few percent at 77 K. Tilted view of the basal plane. Note the occurrence of a cusp (*C*) on a relaxed edge segment. Courtesy S. Naka.

shows the evolution of the CRSS in single crystals with different purities. For the purest sample, the effective stress extrapolated to 0 K is $\tau_P/\mu \approx 1.3\times10^{-3}$ (Biget and Saada, 1989). This value is much smaller than the Peierls stress of pure bcc metals (Table 3.1) and even less than that of potassium.

The athermal transition temperature of titanium, which is $T_a \approx 550$ K in impure specimens, decreases with decreasing impurity content (Fig. 3.17*a*); it corresponds to measured activation energies in the range of 1–1.5 eV. From similar tests performed in zirconium crystals (Soo and Higgins, 1968; Akhtar and Teghtsoonian, 1971), the effective CRSS extrapolated at 0 K is $\tau_P = 25{-}30\,\text{MPa} \approx 0.7\times10^{-3}\mu$. The static atomistic simulations carried out by Khater and Bacon (2010) yield a similar value, $\tau_P = 22$ MPa for screw dislocations, and 16 MPa for edge dislocations.

The stress–strain curves of titanium crystals with high and intermediate purity show evidence for a significant lattice friction below approximately 470 K (Naka *et al.*, 1988). The CRSS strongly increases at low temperatures (Fig. 3.17*a*) and does not follow the Schmid law. For suitably oriented specimens, the stress–strain curves exhibit two stages at all temperatures, which are related to single prismatic slip followed by double prismatic slip. *In situ* and conventional TEM observations show the formation of long screw segments trailed by edge segments during the pre-yield stage and the predominance of straight screw dislocations at all strains (Fig. 3.17*b*). Non-screw segments sometimes show irregular curved shapes denoting an interaction with unresolved obstacles, possibly impurity clusters. The same trends are observed in zirconium crystals of intermediate purity strained at 77 K (Akhtar and Teghtsoonian, 1971). There are, however, some differences confirming that the lattice friction is smaller than in titanium. For orientations such that twinning does not occur, the CRSS follows the Schmid law. In addition, whereas there is a predominance of straight screw dislocations in stages I and II, dislocation cell structures are observed in a subsequent stage III.

The mechanisms governing the plasticity of titanium and other hcp transition metals such as zirconium were often debated in the past. These discussions, which were summarized by Naka *et al.* (1988), were not revisited since. In hcp metals, there is no elastic interaction between perfect screw dislocations and the tetragonal distortion of the lattice induced by interstitial solutes located in octahedral sites. The latter have a relatively weak interaction with edge dislocations (about 0.3 eV in Ti), which cannot account for the very strong strengthening and the rather high athermal transition temperature (Fig. 3.17*a*). The evidences for a substantial lattice resistance in impure titanium point at an interaction between interstitial solutes and the screw dislocation cores that globally increases the lattice resistance. This chemical interaction should then be relevant to electronic structure calculations.

3.3.3 Locking–unlocking in hcp metals

The occurrence of anomalies in mechanical behaviour associated with prismatic slip in hcp crystals was the subject of many speculations that are reviewed by Caillard and Martin (2003). In titanium, a broad and ill-defined stress anomaly occurs below $T = 500$ K. In Fig. 3.17*a*, it is visible only for the intermediate impurity content. In parallel, the activation area exhibits a well-marked peak around $T = 350$ K and decreases until about $T = 500$ K. *In situ* TEM observations performed by Farenc *et al.* (1993, 1995) between 150 K and 473 K showed the occurrence of jerky screw dislocation motion (Fig. 3.18). The peak in the activation volume was described as a transition between critical large bulges and geometric superkinks.The activation area was drawn from the model by Seeger and Schiller (1962) extended to geometric superkink-pairs and, for large bulges, from the early model by Friedel mentioned in Section 3.2.4.4.

Magnesium and beryllium usually glide in the basal plane but can be forced to glide on a secondary prismatic plane by a careful choice of the loading axis. A stress anomaly is known to occur in such situations. In beryllium, it clearly appears in the stress dependence of the activation volume. Jerky motion of screw dislocations was observed at 150 and 300 K in Mg and between 80 and 480 K in Be. Possible interpretations are discussed by Couret *et al.* (1991) and Caillard and Martin (2003), but uncertainties remain regarding the prevailing mechanisms for the core transformations.

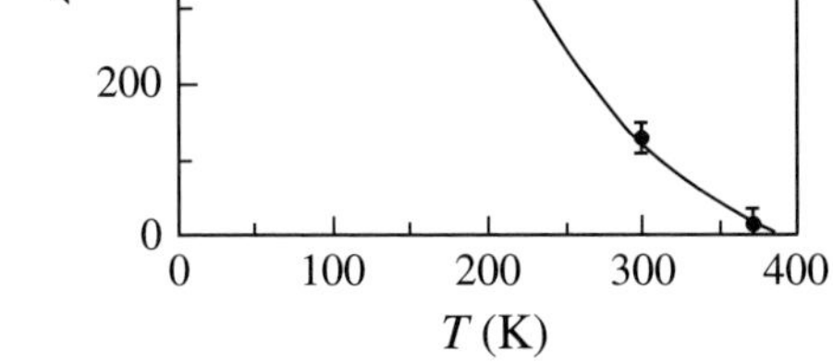

FIG. 3.18. Jerky screw dislocation motion in a prismatic plane of a titanium crystal containing 3270 at. ppm O*. The average jump length, λ_g, continuously decreases with increasing temperature. It reduces to atomic dimensions near the lower end of the stress anomaly shown in Fig. 3.17*a*. Notice that the jump distances are much larger than in Fig. 3.10. From Farenc *et al.* 1995.

3.4 Dislocations in silicon

3.4.1 Introduction

Silicon is often considered as a model material for the study of the high Peierls barriers in dc semiconductors with strong covalent bonding and in the more complex compounds with sphalerite (zinc-blende) structure. It is also a vital material in the domain of optoelectronics, microelectronics and nanotechnologies. With respect to metals, semiconductors exhibit the additional property that dislocations and secondary defects, that is, other defects along the dislocation lines, can be associated with localized electronic states. This induces a coupling between mechanical and electrical properties, which manifests itself in many different manners (Alexander and Teichler, 1991; Labusch and Schröter, 1983). Core excitations by light or electronic illumination usually increase dislocation velocities (Maeda and Takeuchi, 1996); doping with electronically active elements also modify them. The dislocations lines can, in addition, behave as conduction paths that affect the electrical properties of semiconductors. Thus, controlling dislocation generation and the location of generation sites during the processing of micro- and nano-devices becomes all the more important as the size is reduced.

This section is essentially devoted to the energetics of the kink-pair mechanism in silicon. Studies of the core structure of glide dislocations and secondary defects along their lines are first discussed. Within a significant scatter, they lead to estimates for the formation and migration energies of a wide variety of kink configurations. Next, the connection with the mesoscopic scale is examined in relation with current models for the kink-pair mechanism. The comparison between experimental results, models and calculations of defects core structures shows, however, that there are still fundamental open problems. Finally, recent investigations of silicon under high stresses brought answers to long-standing questions about the properties of shuffle dislocations.

The considerable amount of available data on dislocation velocities and activation energies in silicon and other elemental or compound semiconductors is summed up in several articles or chapters, which discuss the evolution of the field at various scales of investigation (Alexander, 1986; George and Rabier, 1987; Rabier and George, 1987; Duesbery and Richardson, 1991; Maeda and Takeuchi, 1996; George and Yip, 2001; Caillard and Martin, 2003; Cai *et al.*, 2004; Spence, 2007). In silicon, the bulk of experimental results was obtained in what is called the central range of stresses and temperatures. The boundaries of the temperature domain accessible to current experiments depend on strain rate and initial dislocation density. They typically extend from the brittle–ductile transition (Section 3.4.9.1), for which a lower bound is about one-half the melting temperature ($T_m = 1690\,\text{K}$ in Si), to the transition towards obstacle-controlled motion around 1300 K ($0.77\,T_m$). In terms of stresses, this domain approximately extends from 1 to 100 MPa, that is, approximately from $1.5\times10^{-5}\mu$ to $1.5\times10^{-3}\mu$ ($\mu_{Si} = 64.9$ GPa).

3.4.2 Dislocations in the diamond cubic lattice

The dc lattice of elemental semiconductors (Fig. 3.19*a*) is a non-primitive fcc lattice containing a base of two atoms located at (0, 0, 0) and (1/4, 1/4, 1/4). Each atom is connected to four nearest-neighbours by tetrahedrally directed covalent bonds. Because of the high energy associated with the breaking of a covalent bond, about 2.2 eV, the lattice and its defects can to some extent be visualized using simple ball-and-spoke models. When viewed along a $\langle 110\rangle$ direction, the structure appears as a stacking of two sets of $\{111\}$ planes (Fig. 3.19*b*). If the regular stacking of the fcc structure is denoted *ABCA*, the duplication of the compact planes in the dc structures leads to a stacking of the type *AaBbCcAa*, with narrowly spaced planes of the *aB* type and widely spaced planes of the *Aa* type.

As in the fcc lattice, the perfect dislocations have $\frac{1}{2}\langle 110\rangle$ Burgers vectors and glide in the close-packed $\{111\}$ planes. Thus, two sets of dislocations can be distinguished. Those which produce a shear between the narrowly spaced planes can further lower their energy in full analogy with fcc metals through a dissociation into glissile Shockley partials with Burgers vectors $\frac{1}{6}\langle 112\rangle$. For this reason, these dislocations are said to belong to the glide set. In silicon, the intrinsic SFE is $\gamma = 65\,\mathrm{mJm^{-2}}$ and the dissociation width in equilibrium is about 3.2 nm (Appendix B).

The dislocations producing a shear between the widely spaced planes *aA* define what is called the shuffle set. As a shear between planes like *Aa* produces a fault

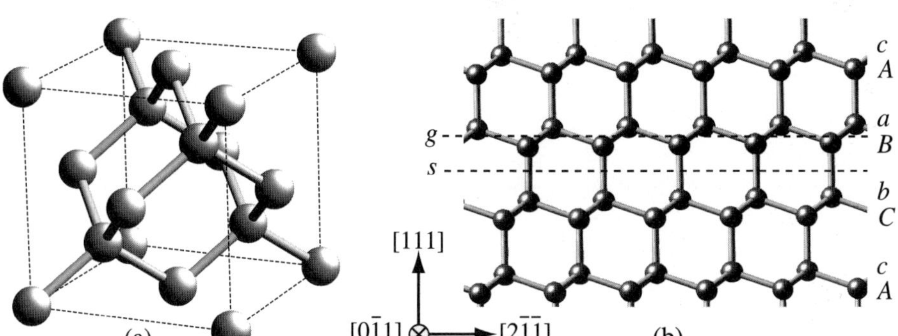

FIG. 3.19. (a) The unit cell of the diamond cubic lattice. (b) A $\langle 110\rangle$ view showing the double stacking of $\{111\}$ planes, *ABC* and *abc*, in the two interpenetrated lattices. The planes of the glide and shuffle set are denoted *g* and *s*, respectively.

of very high energy, there is no possible glide dissociation in the shuffle set. From geometric considerations, Hornstra (1958) and other authors deduced several possible types of shuffle dislocation cores, of which only few were further found to be stable (Section 3.4.9.2). For instance, Hirth and Lothe (1992) describe an extended dislocation core that can only move by non-conservative atomic rearrangements (reshufflings), hence the denomination of 'shuffle' generally attributed to dislocations that shear the widely spaced {111} planes.

Owing to the reasonably large dissociation width, weak-beam TEM observations show that, in equilibrium or after deformation in the central range of stresses, dislocations are always dissociated in a glissile manner (Fig. 3.20*a*, see also Fig. 1.2*a*). The dislocations lines tend to be aligned along the dense ⟨110⟩ rows of the {111} planes. In these orientations, the core structures of partial dislocations can reach minimum energy configurations via a reconstruction process that is discussed in the next section. As dislocation loops achieve pseudo-hexagonal shapes with screw and ±60° orientations (Fig. 3.20*a*), there are only two types of partial dislocations, a 30° partial and an edge 90° partial (Fig. 3.20*b*).

It was believed for some time that dislocations were not moving in the glide set but in the shuffle set. Indeed, if one ignores possible core reconstructions, glide dislocations break three covalent bonds per atomic step of motion, whereas perfect shuffle dislocations break only one (Fig. 3.19*b*). Duesbery and Joós (1996) proposed an explanation for the prominence of glide dislocations at low stresses. The shuffle dislocations were assumed to be non-dissociated and located in shuffle

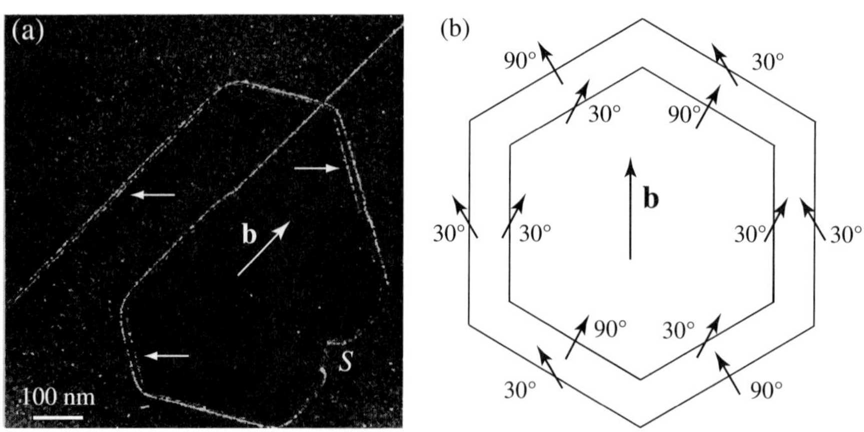

FIG. 3.20. (a) A dissociated glide dislocation loop in silicon (weak-beam dark field). The two branches of the loop lie in neighbouring {111} slip planes and are connected by the inclined segment *S*. From Wessel and Alexander 1977. (b) Dissociation of a hexagonal dislocation loop of the glide set in the dc structure.

planes. The glide dislocations do have a larger Peierls stress than the shuffle ones, but they are more sensitive to thermal activation because the activation energy for dislocation motion scales like μb^3. As the Shockley partials have a smaller Burgers vector than perfect shuffle dislocations, their motion requires a smaller activation energy. Hence, a crossover should occur between slip by shuffle dislocations at high stresses and low temperatures and slip by glide dislocations at smaller stresses and higher temperatures. Substantial progress was achieved in the resolution of the so-called glide-shuffle controversy thanks to recent experiments and simulations that are discussed in Section 3.4.9.

3.4.3 Dislocation cores in the glide set

The core structure of partial glide dislocations and of secondary defects on their lines was investigated for many years using interatomic potentials adapted to the fourfold coordination and strong resistance to bond-bending in the dc structure. Tight-binding and *ab initio* methods are now also employed. A comprehensive review of this field is given in a chapter by Cai *et al.* (2004).

Figure 3.21 shows a (111) projection of the core of a dissociated half-dislocation loop with its two partials at 30° and 90°. There are rows of unpaired orbitals in the core of the partials, which are represented as lines of dangling bonds. The dislocation cores can further lower their energy by the pairing of dangling bonds, which restores the coordination number at the expense of some local distortion. This process, which is called *reconstruction*, has been investigated since

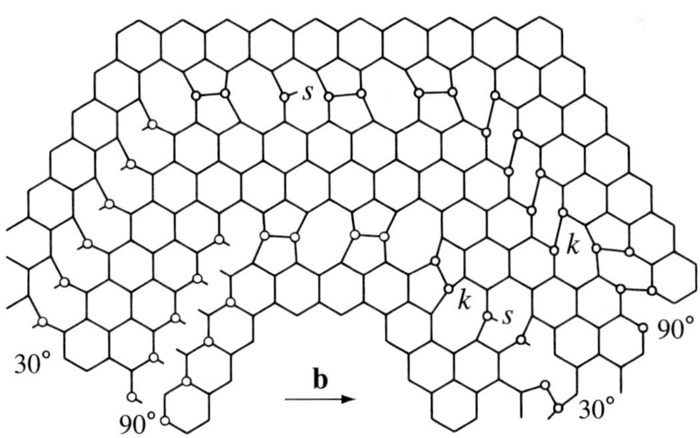

FIG. 3.21. A dissociated half-dislocation loop in silicon. The cores of the 30° and 90° partials of the mixed 60° dislocations are unreconstructed at left and reconstructed at right. On the screw dislocation, the cores of the 30° partials are reconstructed. The defects denoted *k* and *s* are respectively kinks and solitons. After George 1997.

the late 1970s. Their occurrence is confirmed by the measurement of electrical properties, which shows that the density of deep electronic states in the gap corresponds to a low density of unpaired orbitals. A possible reconstruction mode consists in the pairing of unsaturated orbitals that preserves the periodicity parallel to the dislocation line. This is called a single period (SP) reconstruction. There is a second type of core reconstruction, which involves a doubling of the period (DP reconstruction). This last core configuration is energetically favoured in the 30° partial (Fig. 3.21), with a substantial energy gain of the order of 1 eV per dimer. In the 90° partial, the SP and DP reconstructed lines have almost same energy and may coexist (Bennetto *et al.*, 1997; Lehto and Öberg, 1998; Valladares *et al.*, 1998). Both types of reconstructions break the mirror symmetry with respect to the ⟨011⟩ plane normal to the dislocation line. In addition, the translational symmetry is broken in DP cores, which results in a variety of possible different core configurations. When the latter occur along the same line they are separated by reconstruction defects, which are called antiphase defects or solitons.

Owing to the energetic cost of bond distortions, the kinks are geometrical and have narrow reconstructed cores. Thus, they have to break covalent bonds during their migration. In consequence, they have a significant migration energy, which is at the origin of the high secondary Peierls stress that characterizes dc semiconductors. For the most probable configurations, the migration energy is smaller than that corresponding to the breaking of a reconstructed bond, which is around 1.6 eV. It is also smaller than the value of 2.2 eV required for breaking a non-distorted covalent bond. This is due to bond switching effects as the weakening of one bond is compensated for by the simultaneous formation of another one.

In 30° reconstructed cores, the breaking of the mirror symmetry with respect to the ⟨011⟩ plane perpendicular to the lines implies that the two kinks that form a kink-pair have different core structures. One then has to make a distinction between configurations that are denoted 'left' and 'right' kinks. Two additional kink configurations can be identified in the DP reconstructed cores. In 90° DP reconstructed cores, the breaking of translational symmetry induces an additional degeneracy of the kink configurations. The kinks can also bind to reconstruction defects, forming complexes that may have lower formation and migration energies. Taking into account the possible types of dislocation cores, reconstruction defects and kink configurations, one obtains an extremely complex pattern of configurations, which was rationalized by symmetry considerations (Bulatov *et al.*, 1997, 2001*a*). For instance, there are no less than eight topologically different kink families for the DP-reconstructed 90° partial dislocation (Bennetto *et al.*, 1997). Some of these configurations are, however, unstable or marginally stable (Cai *et al.*, 2004; Valladares and Sutton, 2007).

Compiled values of the formation and migration energies for the various types of kinks are given in Tables 2 and 3 of the chapter by Cai *et al.* (2004). The results obtained from atomistic simulations performed with various potentials generally exhibit a wide scatter. Quantum-based simulations yield in principle more accurate results but they are restricted to a small number of atoms so that the treatment of boundary conditions is critical. In the simple example of the two elementary left and right kinks in 30° partials, the formation energy of the left kink is found between 0.35 and 1.2 eV and that of the right kink between 0.39 and 1.24 eV. The corresponding migration energies are distinctly higher; they range respectively from 0.82 to 1.53 eV and from 0.74 to 2.1 eV.

In spite of these numerical uncertainties, an interesting attempt was made by Cai *et al.* (2000) to derive dislocation velocities from the available results. Experimental measurements, which are also not exempt of uncertainties, do not differentiate the different types of kinks. To establish a micro-meso connection, a KMC scheme (Section 1.5) was applied to a simplified catalogue of kink formation and activation energies in the relatively simple 30° partial. The data were taken from an atomistic simulation and a tight-binding scheme. An entropy contribution of $S = 3k_B$ was included to account for non-zero temperatures. The two sets of results yielded velocities that were respectively two orders of magnitude larger and smaller than experimental values from George (1979) and Imai and Sumino (1983). The total activation energy found for the tight-binding set of results was nevertheless identical to the experimental value ($\approx$ 2.2 eV). Discrepancies between estimated and predicted dislocation velocities are in fact arising at all scales. The origin of this recurrent problem may be related to the presence of undefined defects along the lines and their possible influence on dislocation velocity. A discussion of this question is postponed to Section 3.4.8, in order to include additional information drawn from experiment and continuum modelling of the kink-pair mechanism in dc semiconductors.

3.4.4 Experimental methods

The kink properties can be derived as a function of stress and temperature from direct measurements of the dislocation velocities. A review chapter by Spence (2007) provides an assessment of these measurements and discusses possibilities for further investigations on defect cores.

The main methods for measuring dislocation velocities are X-ray topography, TEM and etch-pit techniques. Each method has its limitations and may be submitted to artefacts, as recalled by Alexander (1986), Duesbery and Richardson (1991) and Cai *et al.* (2004). Internal friction was employed too (see Gottschalk *et al.* 1993), but the interpretation of results proves to be difficult. X-ray topography observations on bright synchrotron radiation sources allow conveniently performing direct measurements of dislocation velocities. They also show the occurrence of a number of dislocation mechanisms such as generation at free surfaces, multiplication and cross-slip in the bulk or at free surfaces and intersections (see for example Vallino *et al.* 2000, 2001). In the double etch-pit technique, the position of a dislocation segment emerging at a free surface is revealed before and after a load pulse. In the intermittent loading technique developed by Nikitenko and co-workers (1985, 1993), periodic stress pulses are applied with controlled height, duration and period.

Post mortem TEM observations on thin foils extracted from specimens deformed at high temperatures allow examining unrelaxed deformation microstructures. The specimens are cooled down to room temperature under load, to freeze in the dislocations under the increasing lattice resistance. The possibility of visualizing partial dislocations under load led to some ingenuous methods for estimating the velocities of the partials (Wessel and Alexander, 1977) and their migration energies (Gottschalk *et al.*, 1993). These studies are summarized by Alexander and Teichler (1991). Dislocation motion can also be investigated by *in situ* TEM (see Caillard and Martin 2003). Finally, at a scale closer to atomic scale, high-resolution TEM using forbidden Bragg reflections from the stacking faults allows visualizing kinked partial dislocations (Kolar *et al.*, 1996; Spence, 1999), as illustrated by Fig. 3.22. The migration energy of kinks, W_m, is deduced from their motion during *in situ* annealing and their formation energy, F_k, is approximately estimated from the equilibrium width of kink pairs in relaxed configurations under stress. For the 90° partial, $W_m \approx 1.24\,\text{eV}$ ($\tau = 275\,\text{MPa}$, $T = 400\,\text{K}$) and $F_k \approx 0.73\,\text{eV}$ ($\tau = 250\,\text{MPa}$, $T = 700\,\text{K}$).

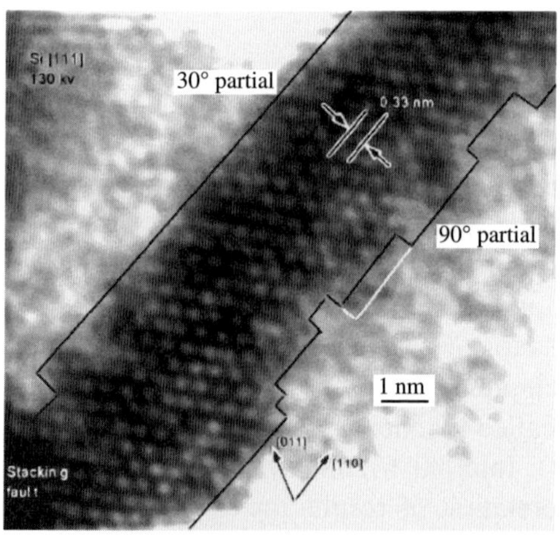

FIG. 3.22. High-resolution TEM micrograph of a dissociated 60° dislocation in a {111} silicon film after relaxation. The light circles in the area bounded by the kinked partials are a lattice image of the double layer of atoms forming the stacking fault. After Kolar *et al.* 1996.

As discussed by Spence and Koch (2001) and Spence (2007), the forbidden-reflection method and electron microdiffraction potentially allow going further towards the determination of core reconstructions and, possibly, defect core structures.

3.4.5 The multiplication yield point of silicon

Information about the energetics of the kink pair mechanism can be drawn from deformation tests by a classical thermal activation analysis, or by a mesoscopic model involving a phenomenological stress versus velocity law. Estimates of the dislocation densities under stress can also be obtained. In the central range of stresses, the stress–strain curves of silicon single crystals are characterized by the occurrence of a yield point, of which the amplitude increases with decreasing temperature (Fig. 3.23). This yield point appears at the athermal transition temperature ($T_a \approx 1320$ K) and was observed down to $T \approx 700$ K under a confining pressure of 1.5 GPa (Castaing *et al.*, 1981; Rabier and Demenet, 2000).

As the upper yield point is sensitive to the initial density, thermal activation analyses are carried out at the lower yield point where the microstructure can be considered as approximately constant (Rabier and George 1987; Omri *et al.* 1987). For crystal orientations favouring easy glide, the stress–strain curves exhibit deformation stages, especially at the highest temperatures (Fig. 3.23). The dislocation microstructures are uniform at small strains. Although there are very few observations at large strains, dislocation patterning is observed, as in bcc metals (Fig. 3.2) and for the same reason. Very interesting wall and channel patterns are

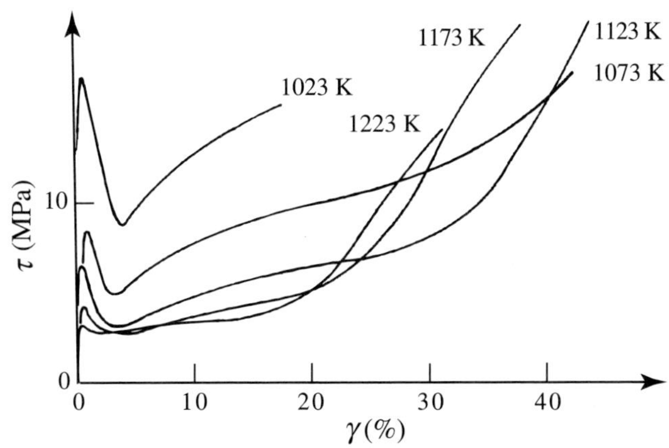

FIG. 3.23. Resolved stress–strain curves of dislocation-free silicon single crystals oriented for easy glide and deformed in compression at various temperatures. The imposed strain rate is 2×10^{-5} s^{-1}. From Omri *et al.* 1987.

also obtained in cyclic deformation when the lattice resistance is small (Legros *et al.*, 2004; George *et al.*, 2007). These microstructures are akin to the persistent slip bands found in fcc metals (Section 2.6.6).

Multiplication yield points are a common feature in the plasticity of dc semiconductors. They appear when two conditions are simultaneously met.

- The initial density of mobile dislocations must be low. Then, extensive multiplication occurs at yield in order to accommodate the imposed strain rate. A high temperatures prestrain reduces the amplitude of the yield point and sometimes suppresses it.
- The sensitivity of the dislocation velocity to the applied stress has to be small. This means in practice that the velocity should be typically a linear or quadratic function of stress. As all the segments emitted by a dislocation source are submitted to a lattice resistance, the rate of operation of sources and the multiplication processes are governed by the dislocation velocities and are also weakly stress-sensitive. The plastic strain rate produced by the mobile dislocations increases upon loading; when it matches the imposed strain rate, the rate at which it increases cannot slow down instantaneously. This results in a rather smooth stress relaxation that accommodates the excess plastic strain rate. Stable flow conditions are eventually reached at the lower yield point.

The determination of the activation energy for plastic flow can be carried out by fitting the stress–strain curves to a simple and efficient phenomenological model for the yield point behaviour that was proposed by Alexander and Haasen (1968). This model also applies to creep deformation where the excess multiplication of dislocations leads to sigmoidal creep curves. In tensile deformation with a constant applied strain rate, the resolved applied shear rate in single slip, $\dot{\gamma}_a$, is split into its elastic and plastic components

$$\dot{\gamma}_a = \dot{\tau}_a/M + \rho_m b v(\tau^*) \tag{3.7}$$

The first term at the right-hand side represents the elastic strain rate obtained under the resolved applied stress rate $\dot{\tau}_a$. M is a combined stiffness of the specimen and the deformation apparatus. The second term expresses the plastic strain rate in terms of the Orowan law, with a velocity that depends on the effective stress τ^*. The applied stress is the sum of this effective stress and of an average athermal interaction stress between mobile dislocations, which is taken in the form of a square root relation

$$\tau_a = \tau^* + A\mu b_m^{1/2} \tag{3.8}$$

It is, thus, assumed that the total density is essentially composed of mobile dislocations in density ρ_m. The coefficient A is given by $A \approx \mu b/2\pi(1-\nu)$. The dislocation velocity is expressed via the empirical relation

$$v = v_o \left(\frac{\tau^*}{\tau_o}\right)^m \exp - \left(\frac{Q}{k_B T}\right), \tag{3.9}$$

where v_o and τ_o are constant reference values for velocity and stress and Q is a constant total activation energy. The stress exponent m is usually found between 1 and 2. This is, indeed, a very low value in comparison to the ones found in pure fcc metals, typically $m \approx 100$. Equation 3.9 is frequently used because it is easily fitted to experimental data in the central range of stresses and temperatures. The set of equations is closed by a multiplication law for the mobile density

$$\dot{\rho}_m = k\rho_m v(\tau^*)\tau^*, \tag{3.10}$$

where k is a constant. This empirical relation is deduced from density measurements performed by the double etch-pit technique (Alexander and Haasen, 1968). It is specific to dc semiconductors, as one usually assumes that the multiplication rate is simply proportional to the area swept per unit time, $\rho_m v$ (Kocks *et al.*, 1975).

Omri *et al.* (1987) measured apparent activation volumes at the lower yield point of the stress–strain curves shown in Fig. 3.23. A classical thermal activation analysis yielded a total activation free energy $\Delta G_o = 2.6\,\text{eV}$. DD simulations of the yield point properties in silicon were carried out by Moulin *et al.* (1999*a*) using a length-independent velocity law with a total free energy of activation $\Delta G = 2.2\,\text{eV}$ that included an entropy contribution of $S = 5k_B$. A good agreement was obtained with the predictions of the Alexander–Haasen model for the stress and temperature dependencies of the upper and lower yield points. Some doubts were nevertheless raised about the meaning and validity of the multiplication law for the mobile density (eqn. 3.10), for which a modified form was proposed.

3.4.6 Velocities in the kink-diffusion model

The kink-diffusion model developed by Hirth and Lothe (1992) considers the steady-state motion of a dislocation overcoming the lattice resistance. Its basic equations are derived in Section E.4. The model predicts the existence of two velocity regimes, according to whether the mean free path of kinks is governed by the annihilation of kinks of opposite sign or by their storage at fixed obstacles. These obstacles can be forest dislocations, nodes or jogs, dislocation corners on pseudo-hexagonal loops and intrinsic or extrinsic point defects.

For short segments or small kink densities, the mean free path of kinks is governed by the distance L between fixed obstacles. The velocity of a segment is thus proportional to the number of competing sites for kink-pair nucleation along the line, as in traditional kink-pair models. In this *length-dependent regime*, the dislocation velocity is given by eqn. E.29, which is written here in developed form

$$v_d = \frac{\tau b^2 h_o^2}{k_B T} \frac{L}{b} \nu_D \exp - \left(\frac{2F_k + W_m}{k_B T} \right) \tag{3.11}$$

In this regime, the total activation free energy is the sum of the migration and formation energies of a kink-pair. All other terms have same meaning as in eqn. 3.2. In this model for single dislocation behaviour, no distinction is made between applied and effective stresses; both are denoted τ.

For long segments or large kink densities along the lines, the free-flight distance of kinks is governed by their mutual annihilations. In this *kink-collision regime*, the dislocation velocity is no longer length-dependent. It is obtained from eqn. E.30 in the form

$$v_d = \frac{2\tau abh_o^2}{k_B T} \nu_D \exp - \left(\frac{F_k + W_m}{k_B T} \right), \tag{3.12}$$

where a is the reconstructed period along the dislocation line.

A correction to these velocities can be made by introducing entropy terms. An approximate estimate performed by Marklund (1985) for the 90° partial suggests $S \approx 0.5k_B$ for the formation energy and $S \approx 5k_B$ for the migration energy. At high stresses, the formation energy of kink-pairs can also be corrected to take into account their geometric nature (see Section E.4.2 and Alexander *et al.* 1999). This is performed with the help of the model by Seeger and Schiller (1962), which is discussed in Section E.3. The length L^* of segments at the transition between the length-dependent and length-independent regimes is given by eqn. E.31; its estimated values are of the order of $L^* \approx 1\ \mu\text{m}$.

3.4.7 Dislocation velocities and activation energies

In this section, we consider first the effect of impurity atoms and next that of dislocation splitting on the velocity of perfect dislocations.

At low stresses, typically below 10 MPa, the stress–velocity relation is often found to become non-linear and the activation energy strongly increases with decreasing stress. In addition, there is a threshold stress below which no dislocation motion occurs. This behaviour is not observed in ultra-pure silicon (Sumino and Imai, 1983; Imai and Sumino, 1983), as illustrated by Fig. 3.24*a*. Thus, impurities atoms, and especially oxygen, are most probably responsible for this effect. Departures from the linear stress–velocity law found in silicon at high temperatures are also attributed to the presence of point defects or impurities.

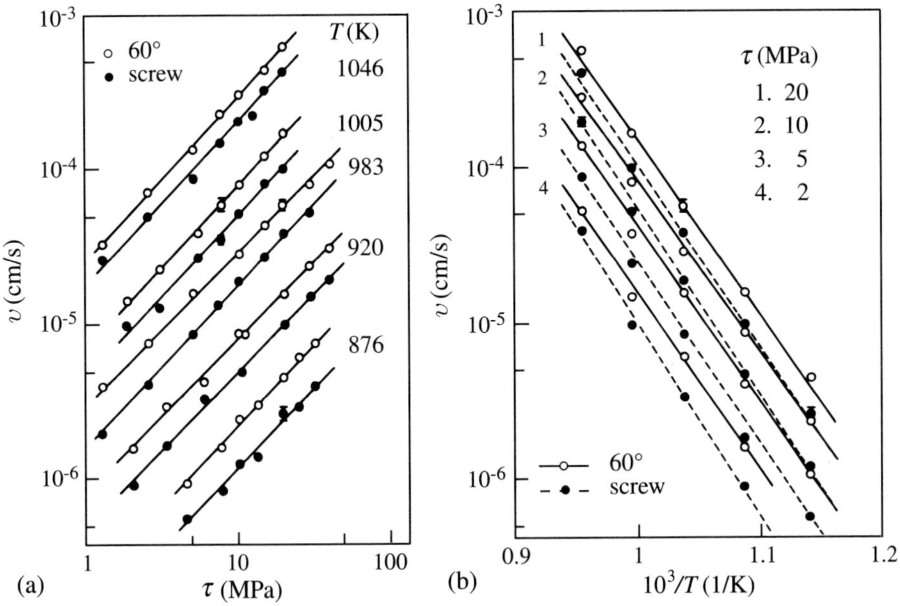

FIG. 3.24. Dislocation velocities in high-purity silicon as measured by X-ray topography. (a) Linear stress–velocity dependencies at different temperatures. Screw dislocations are slightly less mobile than 60° dislocations. (b) Arrhenius plot showing that the total activation energy is a constant for a given dislocation character. The 60° dislocations have a slightly smaller activation energy than the screw dislocations, hence a higher velocity. From Imai and Sumino 1983.

The formation of a kink-pair on a trailing partial favours the formation of a kink-pair ahead of it on the leading partial, via its repulsive field. Möller (1978) investigated this correlated formation of kink-pairs on partial dislocations at stresses smaller than 10 MPa. Working out the kinetics of this process leads, however, to the conclusion that it has no significant consequence (Caillard and Martin, 2003). The correlated motion of partials across a periodic lattice resistance was also discussed by several authors (see Spence 2007). When the dissociation width under stress is an integral multiple of the repeat distance, the two partials simultaneously move upwards in the potential, which hinders the nucleation of correlated kink-pairs. For particular stress values, the dissociation width can also be a half-integral multiple of the repeat distance. One partial can then move upwards in the potential profile while the other one moves downwards, which enhances the rate of kink-pair formation. Therefore, the velocity of a perfect dislocation should be an oscillatory function of stress when the motion of partials is correlated. This behaviour was convincingly reproduced by KMC simulations of single dislocation motion (Cai *et al.*, 2000).

For stresses above 10 MPa, the motion of partial dislocations is not strongly correlated. Nevertheless, the velocity of the whole dislocation line depends on the individual velocities of the two partials at 90° and 30°. As the latter have different core structures, their velocities have no reason to be identical. This was confirmed by TEM measurements of dissociation widths in microstructures quenched under stress (Wessel and Alexander, 1977). The observed widths and their distributions can be rationalized in terms of three factors. One is, indeed, the intrinsic difference in the velocity of partial dislocations, which does not depend upon whether they are leading or trailing (Vanderschaeve and Caillard, 2007). The second one is the orientation-dependent Escaig effect (Section 2.4.4). The third one arises from the fact that dislocations with small dissociations widths are more likely to reach their equilibrium distance under stress than those with large dissociation widths (Caillard and Martin, 2003). The reason is that strong interactions between the partials are needed to overcome the primary Peierls barrier and relax to equilibrium. Taking these effects into account, one reaches the conclusion that under a given stress, 90° partials move faster than 30° partials. This difference simply reflects the more efficient core reconstruction of the 30° partials. Hence, the screw dislocation, which is dissociated into two 30° partials, should exhibit a lower velocity than the 60° dislocation, which is dissociated into a 90° partial and a 30° partial (Fig. 3.20*b*).

The bulk of experimental results shows that the velocities linearly depend on stress up to about 30–50 MPa, whereas the total activation energy remains constant (Fig. 3.24). The activation energies drawn from Fig. 3.24*b* are $Q = 2.20$ and 2.35 eV for screw and 30° dislocations respectively. Under higher stresses, the velocities progressively tend to exhibit a quadratic stress dependence. In parallel, the activation energies decrease to typically $Q = 1.8$ eV for $\tau = 300$ MPa. These evolutions are thought to reflect a stress dependence of the kink-pair formation energy, a feature that is not included in the classical kink-diffusion model. It is accounted for by introducing the Seeger–Schiller correction mentioned in Section 3.4.6. Then, the total activation energy is found to be a constant.

A critical discussion of kink migration and formation energies measured by different methods on 60° dislocations in silicon and germanium is given by Spence (2007). For silicon, W_m ranges between 1.2 and 1.8 eV and F_k ranges between 0.4 and 1 eV. Thus, kink migration appears again as the rate-controlling step for dislocation velocities, as was also deduced from atomic-scale simulations. The scatter largely depends on whether the corrections mentioned in Section 3.4.6 were made or not and on how experimental observations were analysed. Nevertheless, direct measurements of the total activation energy lead to a consensual value of about 2.2 eV. As expected from the kink-diffusion model, a length-independent (or kink-collision) regime is identified for long dislocation segments and a length-dependent regime is observed for short segments in thin TEM foils. However, as discussed in the next section, analyses carried out in the kink-collision regime reveal some unexpected features.

3.4.8 The length-independent regime

Several experimental results suggest revisiting the transition between the two velocity regimes predicted by the kink-diffusion model. A few of them are reported in this section; the reader is referred to Maeda and Yamashita (1993), Maeda and Takeuchi (1996) and Spence (2007) for a more complete account.

Louchet (1981) and Hirsch *et al.* (1981) analysed by *in situ* TEM experiments the motion of short dislocation segments under high stresses in silicon. A transition from length-independent to length-dependent dislocation velocities was found for segments of rather small lengths, respectively smaller than ≈ 0.4 and ≈ 0.2 µm. In both cases, the nature of the pinning obstacles at the origin of the length-dependence could not be resolved.

The intermittent loading technique yielded results incompatible with the kink-collision model, which are discussed by Farber *et al.* (1986), Nikitenko *et al.* (1993) and Iunin and Nikitenko (2001). Although the results are difficult to interpret (Alexander and Teichler, 1991), these observations lead to the suspicion that the experimentally observed length-independent regime is not a kink-collision regime but a regime governed by point defects on the dislocation lines.

It was noticed by Louchet and George (1983) that in the kink-collision regime, the dislocation velocities measured in silicon and other dc semiconductors are systematically larger by several orders of magnitude than the predicted ones. A similar result was also obtained in the KMC simulations by Cai *et al.* (2000) that are discussed in Section 3.4.3. The contribution from entropy terms to the total free energy of activation was invoked to explain these discrepancies. However, for the velocities determined by Imai and Sumino (1983), which are too large by two to three orders of magnitude, this contribution should be $S = 7k_B$ (Cai *et al.*, 2004). This value looks a bit high with respect to usual estimates (Section 3.4.6).

These uncertainties suggest that the kink-collision regime and the transition to the length-independent regime should be better characterized. For this purpose, Yamashita *et al.* (1993) and Maeda and Yamashita (1993) employed an experimental setting that allows following accurately the velocity of a mobile dislocation segment as a function of its length under a known resolved stress. The moving segments are threading dislocations generated by indentation in epitaxial layers of $Si_{1-x}Ge_x$ on a silicon substrate. The layer thicknesses are larger than the critical value at which the threading dislocations start relaxing the lattice mismatch. The dislocations then move with a constant velocity which is measured by a double-etching technique. The experimental results are illustrated by Fig. 3.25.

The length-dependence of the dislocation velocity saturates for $L^* \approx 2$ µm. According to eqn. E.31, the activation energy at this transition is expected

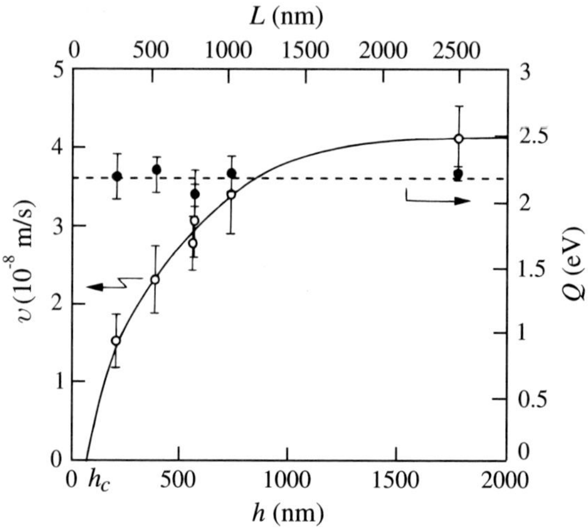

FIG. 3.25. Dislocation velocity, v, and activation energy, Q, of dislocation segments of length L in $Si_{0.9}Ge_{0.1}$ heteroepitaxial layers with various thicknesses h. The length of the moving segments is $L = \sqrt{2}h$ and dislocation motion occurs above a critical film thickness h_c. From Yamashita *et al.* 1993.

to be $F_k \approx 0.5\,\text{eV}$. Nevertheless, the total activation energy does not change and saturates at the maximum value reached in the length-dependent regime, $Q = 2F_k + W_m \approx 2.2\,\text{eV}$. Thus, at saturation, the mean free path of kinks is determined by a mechanism other than kink collisions.

The interpretation suggested by Yamashita *et al.* (1993) and Maeda and Takeuchi (1996) involves the presence of obstacles along the lines with a spacing L^* that limits the mean free path of kinks before the kink-collision regime is reached. Although these obstacles may block kink motion, a blocked kink may be annihilated by a kink of opposite sign arriving at the same obstacle from the other side. Above saturation, the dislocation segments are then still in the length-dependent regime, but their length is a constant, $L = L^*$. If so, a good quantitative agreement is obtained without any additional assumption between experimental and predicted dislocation velocities in the whole range of investigated stresses and temperatures.

The nature of the obstacles defining the saturated length L^* and the way they interact with the moving kinks is for the moment largely speculative. Jones (2000) and Jones and Blumenau (2000) suggested several candidate obstacles such as impurity atoms and intrinsic point defects, especially interstitials or interstitial clusters trapped in the dislocation cores. The creation of atomic-scale defects on the lines during dislocation motion may also be considered. It could be responsible for such effects as the non-reversibility of the dislocation velocity upon a stress reversal, which is sometimes observed by X-ray topography (Jacques and George, 2001). As yet, the debate between the obstacle model and the kink-diffusion model is not settled. Thus, the formation energy F_k involved in the length-independent regime is only known within a factor of two, of same order as the scatter in experimental and computed values. As suggested by Spence (2007), it would be helpful to further characterize the saturation length and its temperature dependence in pure silicon or germanium, in order to remove possible alloying effects.

3.4.9 Dislocations at high stress

The suggestion by Duesbery and Joós (1996) that shuffle dislocations may become more mobile than glide dislocations at high stresses (Section 3.4.2) motivated a number of experimental and simulation studies. This section summarizes a body of investigations on silicon that is reviewed in a chapter by Rabier *et al.* (2010) and an article by Pizzagalli *et al.* (2011), to which the reader is referred for more detail and additional references.

3.4.9.1 Deformation tests and microstructures The brittle–ductile transition in silicon crystals arose some interest because of its original properties (Roberts *et al.*, 1994; Michot *et al.*, 2000; Gally and Argon, 2001). It is thermally activated and structure-sensitive; it is also remarkably abrupt in dislocation-free crystals and smoother in prestrained crystals. Thus, the temperature of the brittle-ductile transition depends on deformation conditions and material history. In virgin pure crystals, its values are found in the range of 850–1200 K. There are, nevertheless, several ways to obtain plastic flow at low temperatures and high stresses. For example, it was recently discovered that nanowires and nanopillars of silicon and other materials reputed for their brittleness become ductile when their diameter decreases below about 100 nm (Gerberich *et al.*, 2009). In such conditions, dislocation-mediated plasticity occurs under high stresses determined by size, in addition to temperature and initial dislocation density. So far, only preliminary TEM observations of the dislocation microstructures were performed (see e.g., Korte *et al.* 2011). High-stress plasticity can also be obtained in small volumes under the tip of a nanoindenter.

Large volumes of silicon crystals were strained in deformation machines where compressive stresses are applied to specimens that are confined by a hydrostatic stress. The latter reduces crack propagation and shifts the brittle–ductile transition to high stresses and low temperatures. In the so-called Griggs machine, the confining stress allows obtaining plastic flow up to about 1.5 GPa. In such conditions, plasticity was obtained under resolved shear stresses up to 1 GPa at temperatures as low as 400–500 °C. The microstructures of ⟨100⟩ and ⟨123⟩ crystals only contained dislocations of the glide set (Castaing *et al.*, 1981; Rabier and Demenet, 2000). In a prestrained ⟨100⟩ crystal, where the Escaig stresses tend to expand the stacking fault ribbon in the slip plane (Section 2.4.4), the glide dislocations exhibited very large dissociation widths.

Geophysicists use multi-anvil apparatuses to deform specimens under higher confining pressures, up to 25 GPa. A modified version of such a device was designed for dislocation studies in plastically deformed materials (Cordier and Rubie, 2000). The uniaxal compression stress cannot be determined, but the sample volumes (about 1–12 mm^3) still allow performing *post mortem* TEM observations. Under a confining pressure of 5 GPa, at room temperature and 150 °C, a new type of microstructure was found in non-prestrained ⟨123⟩ and ⟨100⟩ crystals (Rabier *et al.*, 2000; Rabier *et al.*, 2001). In the ⟨123⟩ crystals, it consists of perfect $\frac{1}{2}\langle 110\rangle\{111\}$ dislocations exhibiting preferential line orientations along the screw direction, the ⟨112⟩ direction at 30° and the ⟨123⟩ 41° direction (Fig. 3.26*a*). A similar microstructure of non-dissociated dislocations was also found in the ⟨100⟩ crystals; no trace was found of dissociated dislocations.

Other TEM evidences for the occurrence of perfect dislocations at very high stresses are reported in the chapter by Rabier *et al.* (2010). They include deformation tests performed at room temperature under a confining pressure of 15 GPa. Although the specimens underwent several phase transformations under stress, they mostly relaxed to their dc structure after unloading.

Figure 3.26*b* shows a compilation of data on resolved yield stresses versus temperature, as obtained by conventional tests and tests under hydrostatic pressure. The inflection towards a less steep temperature dependence seen around 300 °C for resolved stresses above 1 GPa is assimilated to a transition towards

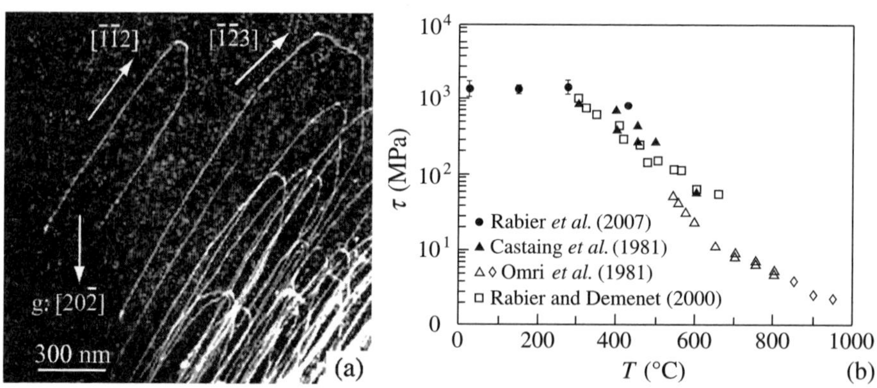

FIG. 3.26. (a) Perfect $\frac{1}{2}[10\bar{1}](111)$ dislocations in a ⟨123⟩ silicon crystal after deformation at room temperature under a confining pressure of 5 GPa (weak-beam TEM dark field). Two preferential line directions are indicated. From Rabier *et al.* 2004. (b) Resolved yield stress versus temperature data for silicon crystals showing a transition near 300 °C. After Rabier *et al.* 2007.

plastic flow mediated by perfect shuffle dislocations. The highest stresses were determined by different methods. They were measured under very high pressures by Rabier *et al.* (2007) with the help of another modified multi-anvil apparatus, the D-DIA. which allows recording *in situ* stress–strain curves in a synchrotron beamline (Wang *et al.*, 2003*b*). In frozen-in microstructures of perfect dislocations, the change in direction between straight lines is accommodated by small curved segments. Rough estimates of average local stresses were deduced from TEM measurements of the small curvature radii (Rabier and Demenet, 2005). As these local values are smaller than the applied stresses at yield, the combination of all sets of results indicates that the resolved stress to nucleate perfect dislocations is about 1.1–1.5 GPa (2.2–$3\times10^{-3}\mu$) under conventional strain rates. Thus, the transition discussed by Duesbery and Joós (1996) is experimentally observed, but the crossover occurs at stresses larger than the expected ones (about $10^{-3}\mu$).

It may be that perfect and dissociated dislocations transform into each other by cross-slip or climb upon a change in deformation conditions. To check this possibility, microstructures of each type were produced in the conditions where they were previously observed and further submitted to deformation conditions typical of the other type. TEM observations performed by several authors (see Rabier *et al.* 2010) showed that there is no massive transformation from one microstructure to the other. Whereas dissociated glide dislocations do not transform into perfect dislocations, a limited and localized transformation from perfect to dissociated dislocations was occasionally observed. Perfect and dissociated dislocations seem to nucleate and move independently of each other, in different conditions of stress and temperature. The results concerning the transformations between the two types of cores are, however, not clear-cut. Saka *et al.* (2006) observed a transformation from perfect to dissociated dislocations by heating a sample containing a saturated concentration of interstitials at 400 °C in the TEM. A similar result was obtained by Izumi *et al.* (2010) in a SiN film implanted with phosphorus atoms at low temperature and further submitted to severe thermal processing.

3.4.9.2 Core structure of shuffle dislocations Geometric analyses of the location and core structure of perfect dislocations by Hornstra (1958) and related results, including the above-mentioned ones, were revisited by Rabier (2007). In particular, the occurrence of preferential $\langle 112\rangle$ 30° and $\langle 123\rangle$ 41° low-velocity directions (Fig. 3.26*a*) was rationalized by considering the density of dangling bonds in the cores and their possible reconstructions. The core structures of the perfect dislocations were directly computed on screws and 60° dislocations. The 30° and 41° orientations were examined as yet since as they involve too long periodicities along the dislocation lines. The simulations were carried out using various atomistic potentials, *ab initio* tight-binding and DFT methods, as well as tight-binding approaches, which provide a good compromise between reliability and computing efficiency.

Of the two possible non-dissociated configurations for screw dislocations, only one was found to be stable by DFT first-principles calculations (Pizzagalli *et al.*, 2011), a perfect shuffle dislocation denoted A. Its energy was compared to that of a stable perfect screw dislocation of the glide set, with a DP-reconstructed core (C2). The same initial configuration with an SP-reconstructed core of high energy (C1) was found to be unstable (Wang *et al.*, 2006*a*). Two recent sets of tight-binding and DFT studies (Wang *et al.*, 2006*a*; Pizzagalli *et al.*, 2009*a*) show that the C2 glide structure is the most stable. Thus, the effect of core reconstruction in the glide screw dislocation prevails over the low bond density in the shuffle core. The Peierls stresses of the A and C2 cores, as yielded by DFT calculations, are approximatively 4 GPa (Pizzagalli and Beauchamp, 2006) and 6 GPa (Wang *et al.*, 2006*a*), suggesting a higher velocity of the A core. As this velocity is likely to be thermally activated, the formation and migration energies of kinks were recently computed using the NEB method and either an atomistic potential or a DFT *ab initio* method for computing the images. As the A core is not reconstructed, the kinks are only doubly-degenerate; their formation energy is about 1 eV and their migration energy is quite low, less than 50 meV (Pizzagalli *et al.*, 2008). Related results are reported by Rabier *et al.* (2010), and Pizzagalli *et al.* (2011), where an updated discussion can be found. The conclusions are that there is no possible stable transformation between A and C2 cores and the A core is likely to be the preferred shuffle configuration at low temperatures. The effect of a hydrostatic pressure was also examined; it enhances the stability of the A core and its velocity (Pizzagalli *et al.*, 2009*a*).

The 60° shuffle dislocation seems to be very mobile, as TEM observations do not show straight segments having this orientation. Nevertheless, it is obtained in simulations of dislocation nucleation at surface steps (Godet *et al.*, 2004, 2006, 2009) or at sharp corners (Izumi and Yip, 2008). In simulations like that illustrated by Fig. 3.27, which were carried out with different loading axes, only 60° shuffle dislocations are nucleated.

Two stable core configurations were found and investigated for this 60° dislocation (Pizzagalli *et al.*, 2009*b*). One is a perfect shuffle core S1 proposed by Hornstra (1958), which is centred in a shuffle plane, and a perfect glide core called G, which is located in a glide plane. Several atomistic and DFT-based simulations showed that the G-core, which is reconstructed, is highly stable and sessile, as its Peierls stress is extremely high. The S1 core exhibits a low Peierls stress and should be mobile. Unexpectedly, however, it was found to be unstable and to transform into sessile configurations of lower energy, S2, and further to another sessile configuration S3. This seems to contradict the result shown in Fig. 3.27, where the freshly nucleated S1 core glides in a shuffle plane. However, it turned out that the S1 core can be displaced when it is submitted to stresses of 1–2 GPa, which are presumably above its Peierls stress. Upon moving, its structure is approximately preserved and is not submitted to a thermally activated lattice resistance.

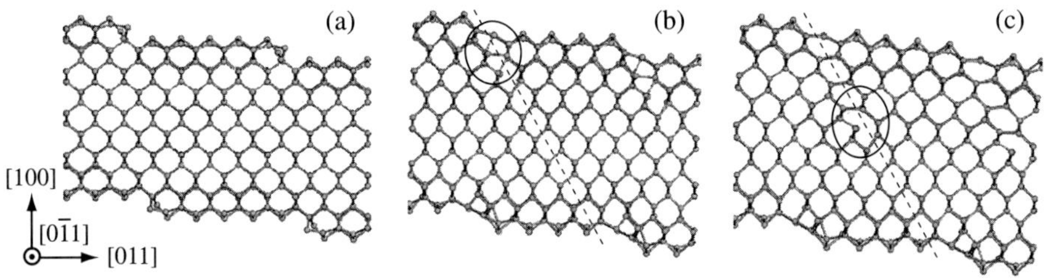

FIG. 3.27. *Ab initio* simulation of the generation and propagation of a shuffle 60° S1 dislocation in compression. (a) Initial configuration (duplicated for clarity) containing an atomic step at the reconstructed surfaces. Under a compressive resolved stress of about $0.16\mu \approx 10$ GPa, the dislocation is nucleated in the dashed shuffle plane (b) and propagates (c). From Godet *et al.* 2006, courtesy J. Godet.

Thus, the freshly nucleated 60° core shown in Fig. 3.27 is the only one that can be glissile, at least in a transient manner, when it is loaded above a certain threshold. When, for any reason, the stress decreases below its Peierls stress, it stops moving and irreversibly transforms into a sessile core. This result can have significant consequences regarding the nucleation of dislocations in silicon nanostructures.

Simulations of dislocation nucleation at surface defects are sensitive to many geometric, structural or chemical parameters. In particular, the number of competing nucleation sites along a surface step is considerably smaller in simulated than bulk crystals, which may lead to a size effect. Figure 3.28 shows that MD simulations performed for a single surface step as a function of temperature yield a glide–shuffle transition similar to the experimentally observed one. A temperature gap is observed, during which complex atomic rearrangements occur but no glide event is recorded (Godet *et al.*, 2009). The nucleation stresses are higher

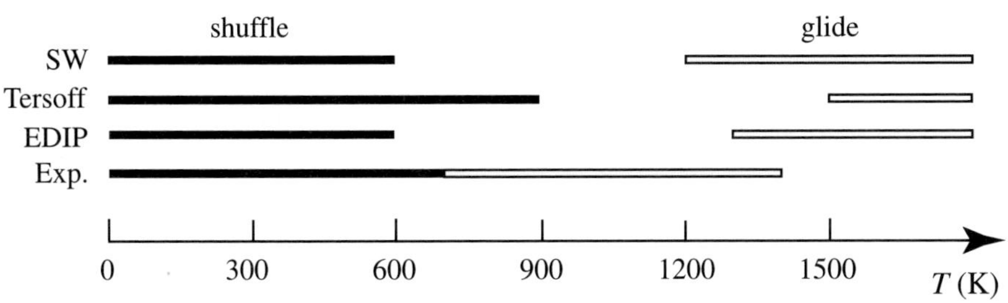

FIG. 3.28. Temperature dependence of the generation of shuffle and glide dislocations in silicon as obtained experimentally and by MD simulations with three different potentials. The experimental observations of shuffle dislocations were carried out on deformed virgin samples (Rabier *et al.* 2010, see Fig. 3.26*b*). Courtesy S. Brochard.

than the experimental ones, as is also the case for the *ab initio* simulation of the 60° core (Fig. 3.27). As discussed by Godet *et al.* (2009), one cannot expect dislocation generation and multiplication processes to be identical in nanovolumes and bulk specimens. In addition, the very high strain rates of MD simulations affect thermally activated processes (Section A.2). These results, nevertheless confirm that glide and shuffle dislocations nucleate in two different regimes. Further confirmation was brought by the reaction pathway analysis for nucleation at a sharp corner performed by Shima *et al.* (2010).

In other semiconducting materials, the perfect dislocations globally behave as in silicon, with a few specific differences that are attributed to the degree of ionicity or structural differences (see Rabier *et al.* 2010). As far as silicon is concerned, the major conclusions are as follows.

- The perfect dislocations with A screw core and S1 60° core can be called shuffle dislocations, according to the usual terminology, even though they move in a conservative manner. Simulating the cores of the 30° and 41° dislocations would bring a more complete insight into mobile shuffle dislocations.
- There is no massive transition from glide to shuffle dislocations or from shuffle to glide dislocations in non-irradiated specimens. The transition observed at high temperatures in irradiated specimens is still an open issue.
- Further experimental progress may be achieved by directly imaging the dislocation cores of shuffle dislocations by high resolution methods, as was suggested by Spence and Koch (2001). Use could also be made of current optical and electronic methods for characterizing the properties of dislocation cores. Finally, it would be interesting to examine whether shuffle dislocations play any role in the brittle–ductile transition.

4

A GUIDE TO 3D DD SIMULATIONS

4.1 Introduction

The input of a DD simulation contains a set of parameters specifying crystallographic, elastic and core-related properties. Temperature is also specified when necessary. Plastic deformation is often carried out in strain control, although other options are usually available. For simulations carried out in the quasi-static regime, the applied strain rates are taken higher than those conventionally used in laboratory tests in order to reduce the computing time. There is, however, an upper limit that depends on the simulation settings and should not be exceeded (Section 1.4.3). Another important problem to care of is the implementation of an initial dislocation configuration that does not introduce spurious effects. For instance, dislocation multiplication is not driving plastic flow in bulk fcc crystals and should be allowed to proceed in a natural manner. Examples of initial microstructures prepared for bulk and small-scale specimens can be found in Section 5.5.5.3. Chapter 5 also provides many examples of the wealth of information that can be extracted from the output of model, small- or large-scale simulations.

Large-scale simulations must capture the statistically representative volume element of a microstructure. In conditions of obstacle-controlled plastic flow, this condition is particularly critical because dislocation pattern formation is governed by collective phenomena. The representative elements are then the areas swept by large dislocation avalanches (Section 2.7.2). Size effects occur when a too small cell size limits the area swept by a significant fraction of the avalanche size distribution (Section 5.5.5). Since the upper limit for size effects is well documented in fcc and bcc crystals, it provides guidelines for defining a simulation cell size adapted to a given range of dislocation densities. For instance, Fig. 5.19 shows that the linear dimension of a simulation cell should be about 10–20 μm in fcc crystals.

During easy glide, however, the glide paths of dislocations are so large (Section 2.6.2) that size effects cannot be avoided. In bulk bcc crystals deformed in the regime of high lattice resistance, there are no avalanches and microstructures are rather uniform. The mean free paths of non-screw dislocations can be also very large during the preyield stage; they determine the initial density of screw dislocations at yield and subsequent forest strengthening (Section 3.2.5.1). Within some limits, extended glide paths can be implemented with the help of periodic boundary conditions (Section 4.4.1).

Obtaining meaningful mesoscopic data for multiscale analyses requires extensive calculations. This explains why optimizing the efficiency of DD codes is a permanent concern. In many cases, however, there is no need to attempt reaching ultimate accuracy. Indeed, simulations are mostly carried out on perfect dislocations and in isotropic elasticity, whereas the treatment of core mechanisms and properties is very approximate (Sections 4.2.4 and 4.3).

This chapter presents basic elements that are common to all 3D DD codes and emphasizes issues, solved or in progress, which are critical for obtaining reliable results. References describing the computational procedures of existing codes can be found in Section 4.5. In addition, the book by Bulatov and Cai (2006) and the related websites provide an insight into a wealth of computing recipes. The topics discussed in the next three sections expand and apply basic notions that were introduced in previous chapters. The last section provides references for further reading.

- Section 4.2 is devoted to the implementation of the elastic properties of dislocations, which constitutes the backbone of all DD simulations. The discretization of the dislocation lines is a convenient way to obtain tractable analytical forms for the dislocation fields. The final objective is to compute the net forces governing the evolution of a dislocation microstructure. The major issues are the treatment of the self-stress and the search for a reasonable compromise between computing efficiency and accuracy.
- Section 4.3 discusses local rules for mechanisms that cannot be treated by the elastic theory of dislocations. The implementation of dislocation velocities, or mobilities, and cross-slip is based on models that are detailed in Chapter 2 for fcc crystals and in Chapter 3 for bcc crystals.
- Two types of boundary conditions are presented in Section 4.4. Periodic boundary conditions are employed for large-scale simulations. The treatment of boundary conditions in finite volumes is often carried out with the help of the superposition method. Other methods are also utilized.
- Section 4.5 gathers references discussing the technical bases of the available simulation codes.[8]

4.2 Elastic properties

4.2.1 Outline

This section discusses the steps involved in the implementation of elastic properties in a DD simulation, more or less in the same order as in specialized

[8] In what follows, simulations are often referred to by their code names as given in Section 4.5.

publications. The description of continuously curved lines as a succession of segments (Section 4.2.2) and of their evolution during a simulation (Section 4.2.3) is followed by a short assessment of the approximation made upon neglecting core fields (Section 4.2.4). The regularization of the dislocation core, for which various models were proposed (Section 4.2.5), removes the main obstacle to a calculation of Peach–Koehler forces or effective stresses (Section 4.2.6). If nothing is done to reduce it, the treatment of elastic dislocation properties contributes to an excessively large computing cost. Optimizing this cost is performed with the help of local (Section 4.2.3) or global (Section 4.2.7) procedures. The implementation of anisotropic elasticity in DD simulations has achieved substantial progress in the last years (Section 4.2.8) and simulations of dissociated dislocations are briefly discussed in Section 4.2.9.

4.2.2 Discretization of dislocation lines

In discretized dislocation microstructures, the number of degrees of freedom is determined by the number of nodes, or segments, and their velocities. The discretization can be performed in two different manners, which leads to two families of DD simulations. There were rather sterile debates about the respective merits and drawbacks of the different types of discretization in terms of accuracy, computing efficiency, simplicity or flexibility. In practice, each simulation has its strengths and weaknesses and what matters is the physical relevance and originality of the results that are drawn from it.

In *nodal simulations* (Section 4.5) a set of nodes is positioned on the dislocation lines and the consecutive nodes are most of the time connected by straight segments of lines. The position of the nodes and the orientation of segments can vary in a continuous manner in the simulated volume (Fig. 4.1*a*). The segments are only submitted to the condition of conservation of the total Burgers vector at the

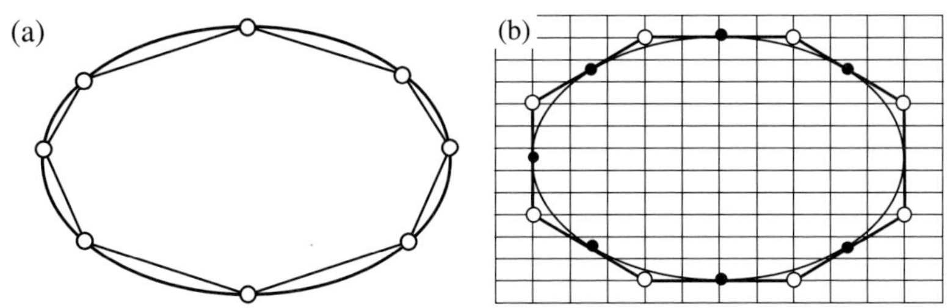

FIG. 4.1. Discretization of a dislocation loop by straight segments. (a) Nodal simulation. The nodes (open circles) are located at corners between adjacent segments. (b) Lattice-based simulation. The nodes are also located at corners, but on a 3D lattice. The forces are computed in the middle of the segments (filled circles).

nodes. The forces are in general computed at the nodes, that is, at corners between two adjacent segments with different orientations. The spacing of crystallographic slip planes is defined as a multiple of the physical one. As recalled by Schwarz (2007) in a brief historical section (see also Section 1.6.1) the first 2D in-plane simulations were also performed using nodal codes.

In the *PDD* code, the local shape of the lines is refined through a discretization into cubic splines. The latter are parametrized in generalized coordinates, that is, coordinates common to all dislocation lines in the investigated configuration. This leads to a technical formulation which is, as a whole, mathematically complex. However, the number of degrees of freedom is in principle less that the one required for a description by straight segments.

All *lattice-based* simulations derive from the same initial code (Section 4.5). The curved lines are discretized into a succession of straight segments, which are positioned on preset directions of an underlying cubic lattice. The length of a segment is a multiple of the lattice translation parallel to its line direction (Fig. 4.1*b*) and its direction of motion is normal to it. The lattice parameter is a multiple of the atomic one and constitutes a scaling length for all dimensions and distances. The precision required for a given problem is obtained by adjusting the minimum length of a segment and the spacing of the slip planes. Force calculations are performed in the middle of the segments. The latter glide in a continuous manner, but Peach–Koehler forces are computed on the closest parallel crystallographic direction of the underlying lattice. The present version of *microMegas* utilizes a base of eight line vectors per slip system, with screw, edge and two mixed orientations, for describing curved lines in all cubic or nearly cubic crystals. In *Tridis*, there are four line vectors per slip system, with screw and edge orientations, like in the initial lattice-based code. Lattice-based simulations are particularly suited for enforcing the rigid motion of straight segments with crystallographic line directions in materials exhibiting a strong lattice resistance.

4.2.3 Local procedures and optimization

Local procedures have to be introduced for treating in an optimal manner the evolution of discretized microstructures. The discretization lengths are such that they minimize the gradients of effective stress along the segments, for instance in regions of high curvature or high local interaction stresses. Thus, these lengths vary between prescribed minimum and maximum values. Small average values improve the spatial resolution of a simulation, whereas larger values improve the computing efficiency. In consequence, the domain of variation of the discretization length necessarily results from an optimization process in each specific case. Adaptative rediscretizations have to be performed to account for the increase in dislocation density and the evolution of the local stress gradients with increasing plastic strain.

The kinetics of segments is governed by two related algorithmic parameters, the time per step and the travelled distance per step, which also have to be optimized in the same spirit as above. In addition, the ratio of these two parameters defines a free-flight velocity, which is governed by the imposed deformation conditions. Depending upon the simulation code and the problem considered, the average discretization lengths and the time per step may vary in very broad ranges. In *microMegas*, typical values are, respectively, between 10 nm and 1 μm and between 10^{-14} and 10^{-3} s.

Implementing modifications of the discretization in evolving microstructures is in general a rather complicated exercise because of the multiplicity of specific topological changes that are required. At each time step of a simulation, the length and orientation of segments describing a local curvature may be modified. In addition, the detection of discrete short-range events has to be carried out and updated. As a consequence, various procedures have to be implemented to modify the connectivity of the nodes by insertion, merging or removal. Typical discrete events are intersections of repulsive dislocations, junction zipping and unzipping, direct dislocation annihilations or interactions with a surface or an interface. In simulations with finite sizes, images forces at free surfaces are computed through the boundary conditions or directly in the DD simulations (Sections 4.4.2 and 4.4.3).

Each simulation has it own local procedures, which are sometimes complex. The references given in Section 4.5 may be consulted for more detail. All procedures for elastic processes are validated against known analytical solutions, for instance for the critical stress of a Frank–Read source. The global outputs of large-scale simulations, typically the well-known value of the strain hardening rate in stage II of fcc crystals (Section 2.6.3), provide more demanding validation tests.

4.2.4 Core fields

The elastic theory of dislocations does not take into account the core fields. These fields are characterized by a volume expansion and shear components; they are more extended in the edge direction than in the screw direction. In atomistic or DFT-based simulations, they are obtained as the difference between the total field and the Volterra field. Core fields decay like $1/r^2$, that is, faster than Volterra fields. They can be described in isotropic or anisotropic elasticity by a distribution of line force couples or dislocation dipoles located near the core (Gehlen *et al.*, 1972; Henager Jr. and Hoagland, 2004; Henager Jr. and Hoagland, 2005; Yang *et al.*, 2010; Clouet, 2011*a*; Clouet, 2011*b*).

As far as DD simulations are concerned, what essentially matters is the equilibrium shapes of the dislocation lines under stress. In a DFT-based study of the screw dislocation core in α-iron, Clouet (2011*b*) characterizes in anisotropic elasticity all

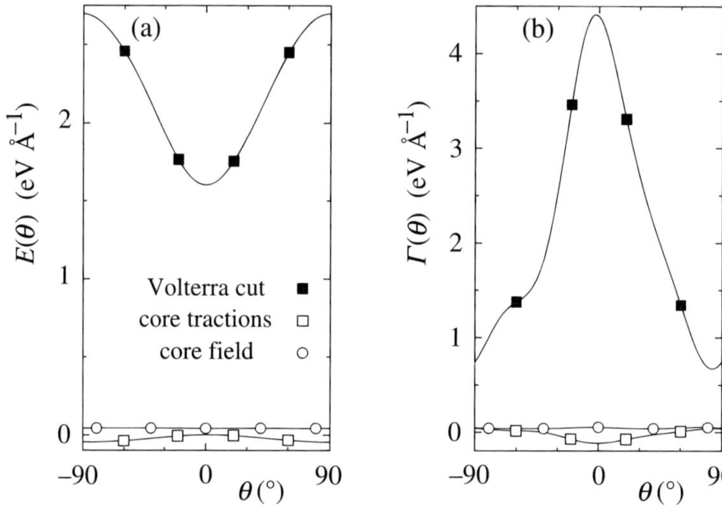

FIG. 4.2. Elastic and core contributions to (a) the total elastic energy $E(\theta)$ and (b) the line tension $\Gamma(\theta)$ of a $\frac{1}{2}\langle 111\rangle$ dislocation in its primary $\{110\}$ plane in α-Fe. The angle θ varies from 0 for a screw line to $\pm\pi/2$ for an edge line. The contribution from the Volterra cut is by far the dominant one (see text). After Clouet 2011*b*, courtesy E. Clouet.

contributions to the line energy (Fig. 4.2*a*), taking cut-off radii of $r_o = 1.2b$ and $R_c = 10^4 b$. The contributions to an effective line tension are shown in Fig. 4.2*b*. They are estimated from the orientation-dependent eqn. 1.13, as given by de Wit and Koehler (1959). The line tension due to the Volterra cut is by far the leading term. The so-called core tractions yield a small elastic contribution. This term arises from the tractions exerted by the Volterra elastic field on the cylindrical surface defining the core radius. The core field also contributes to two small terms, one associated with its elastic field and the other deriving from the core energy. The three elastic contributions to the total energy and the line tension are depicted in Fig. 4.2 as a function of the dislocation character. The core energies are only known for the screw direction; in these estimates, their orientation dependencies arise from anisotropic elasticity.

In the elastic calculations that follow, the core radius is unknown. Most of the time, its value is set to $r_o = b$ and assumed to be small enough to compensate for the absence of core energy. Schwarz (1999, 2007) estimates the resulting systematic uncertainty on the self-stress to about $\pm 10\%$ at micron scales and $\pm 20\%$ at nanometer scales. In addition, the detailed atomistic studies by Henager Jr. and Hoagland (2005) show that the effect of the core fields may significantly influence dislocation–dislocation interactions at distances below about 50*b*.

4.2.5 The self-stress

4.2.5.1 Brown's model and sequels The interaction stress of a line element with itself logarithmically diverges. Commonly employed solutions for regularizing this self-interaction derive from a heuristic method that was proposed by Brown (1964) in an article where the term *self-stress* appeared for the first time. A dislocation loop is in equilibrium under its self-stress and a shear stress τ resolved in its plane. The application of the method to a field, or integration, point P_0, where the Peach–Koehler force is computed, is illustrated by Fig. 4.3. A core radius r_c is defined. Two points P_0^+ and P_0^- are constructed at distances $\pm r_c$ along the normal to the dislocation line at P_0. Equilibrium considerations suggest taking the local self-stress at P_0 as the half-sum of the stresses due to the initial loop on P_0^+ and P_0^-. As these stresses are smooth, the singularity at P_0 is removed.

The whole self-stress in the loop plane is obtained by adding interactions with other line elements. The calculations are performed on simple configurations, for which exact analytical solutions are known, and for large curvature radii. The self-stresses obtained for prismatic and shear loops with radius $R \gg r_c$ reproduce the exact results within a small additional term of unknown origin. In the case of a small bulge of length $2\lambda \gg r_c$ on an infinite straight line, the leading term of the self-stress is similar to de Wit and Koehler's line tension (eqn. 1.14), except that the outer cut-off radius is replaced by λ.

In nodal 2D simulations, where the stress diverges at the nodes between adjacent segments, the application of Brown's method requires that the discretized

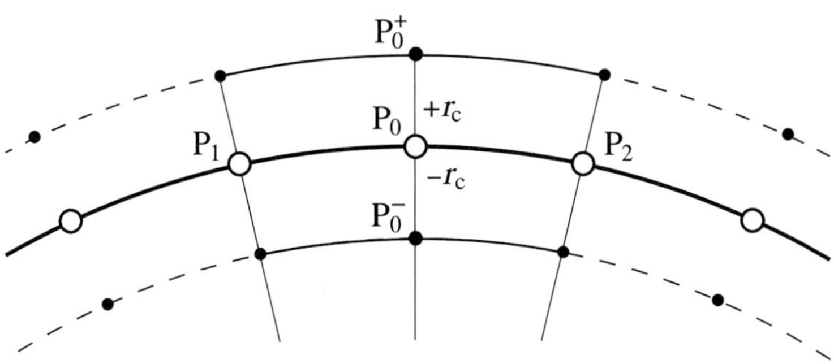

FIG. 4.3. Core splitting methods for regularizing the self-stress (see text). Brown's (1964) method applies to a field point P_0 of a curved segment, and two points displaced upwards (P_0^+) and downwards (P_0^-) by $\pm r_c$. Foreman (1967) and Bacon (1967) consider the self-stress of a circular arc passing through points P_1, P_0 and P_2 of a line discretized into straight segments. Schwarz (1999) extends this last method by splitting the core over larger, adjustable distances (dashed lines).

shapes be locally replaced by continuously curved arcs. In his simulations of segments bowing between two pinning points, Foreman (1967) applies Brown's method to a circular arc of line limited by two nodes, P_1 and P_2 and centred on the field point P_0 (Fig. 4.3). The self-stress of this arc is directed towards the centre of curvature. It is given in analytical form by[9]

$$\tau_{self}(P_0) = K\frac{b}{R}\left\{\left(1 + \nu - 3\nu\, sin^2\gamma\right) \ell n\left(\frac{2\lambda}{r_c}\right) + 2\nu\, cos2\gamma\right\}, \tag{4.1}$$

where $K = \mu b/4\pi(1-\nu)$, R is the curvature radius, 2λ is the length of the curved arc, r_c is a core radius and γ is the angle between the Burgers vector and the curved segment at P_0. Equation 4.1 is obtained by assuming that $r_c \ll \lambda$ and $\lambda \ll R$. The contributions from remote segments to the self-stress are calculated numerically, except for the fields of two semi-infinite side-arms that extend the curved segment beyond its pinning points. As mentioned in Section 1.3.3, this allowed estimating curved dislocation shapes under large stresses, in rather good agreement with the line tension model by de Wit and Koehler (1959). Foreman's method is employed in *Tridis*.

Bacon (1967) introduced several improvements to Foreman's method in order to treat high local curvatures and configurations more complex than single bowing segments. The density of nodes is increased when necessary to ensure that the assumption of large curvature radii remains valid. Remote segments are taken as straight lines and their stress fields are expressed using an analytical form given by Li (1964). Thus, no recourse is made to numerical calculations. The calculated equilibrium shapes of a Frank–Read source yield critical stress values slightly more accurate (by 5 to 10%) than the expression given by Foreman (eqn. 1.15).

Brown's approach was eventually formalized in a self-consistent manner by Gavazza and Barnett (1976). All calculations are carried out in anisotropic elasticity. A dislocation loop is considered and a tube with arbitrary radius r_c is defined around it. Account is taken that the displacement field undergoes a discontinuity given by the Burgers vector b upon crossing the tube surface. The self-energy of the loop is defined as the elastic energy external to the tube; it contains two contributions, one from the Volterra cut and the other from the tube energy, which takes into account the core tractions (Section 4.2.4). The contributions to the self-force are determined from variations of the corresponding self-energies. Hence, all calculations are performed in consistence with the initial

[9] Equation 4.1 includes a correction made by Bacon (1967).

definition of the core radius. For a planar dislocation loop, the leading term of the in-plane self-stress is given by

$$\tau_{self} = \frac{1}{R}\left[E(\mathbf{t}) + \frac{\partial^2 E(\mathbf{t})}{\partial^2 \alpha^2}\right] ln\left(\frac{8R}{r_c}\right), \tag{4.2}$$

where **t** is the unit tangent to the line at the considered point and E is the pre-logarithmic energy factor of a straight, infinite dislocation. R and r_c are outer and inner cut-off radii with $R \gg r_c$. As in previous results, the leading term has same form as an orientation-dependent line tension. The only difference is found in the argument of the logarithmic term. The spurious terms obtained by Brown no longer appear; they arose because the arc length of a line element changes when it is displaced (Fig. 4.3). It follows that Brown's results can be made more exact by simple corrections to the displaced stress fields (Schwarz, 2003*b*).

Another way to treat the self-stress was proposed by Duesbery *et al.* (1992*b*). In his 2D simulations, use is made of a linear stress versus velocity law and the dislocation shapes are no longer obtained from a succession of equilibrium states as in most previous simulations. A discretization into polynomial spline elements, which prefigures the one used in the *PDD* code, is introduced to replace the self-stress calculations based on circular arcs. It is exposed and justified in great detail. Several applications are presented, of which some involve dissociated dislocations (Fig. 1.7*a*).

4.2.5.2 Implementation in DD codes In nodal simulations it happens that curvatures are calculated at nodal corners where two segments meet and the stress diverges. Regularization by a simple cut-off procedure induces numerical instabilities for very small segment lengths in the 10 nm range. When this problem occurs, a circular arc can is defined, as in Foreman's method, and the self-stress is obtained from an analytical solution (Scattergood and Bacon, 1975).

Such instabilities also occur in Brown's treatment and similar approaches where a circular arc is passed through only two segments adjoining the considered point. The reason is that these segments are not regularized, so that their contribution to the self-stress diverges when their length becomes very small. A solution to this artefact based on a refinement of Brown's method was proposed by Schwarz (1999). It consists in splitting the core and describing it by circular arcs over an extended neighbourhood (Fig. 4.3). Numerical stability requires that the length of this arc be at least one order of magnitude larger than the splitting width r_c. This solution is implemented in *PARANOID*, where it allows developing applications down to the nanometre scale. It is also implemented in the Weygand code and as an option in *microMegas*.

Mohles proposed a new approach to the treatment of the self-stress, which is similar to the one proposed by Bacon (1967) but avoids making use of Brown's splitting procedure (see Section 4.5 for references). In fact, the removal of an element from the line may result in uncontrolled errors since the integration of the total self-stress is no longer carried out over a closed path. Mohles' method is based on a double discretization of the curved lines into main segments and connecting secondary segments containing no nodes. This particular discretization allows computing the self-stress in the middle points of three main segments. It takes advantage of the fact that the self-stress of a straight segment vanishes in its mid-point by reason of symmetry (the interaction forces with the two parts of the segment have same magnitude but opposite sign). Thus, it is not necessary to make use of Brown's core splitting method. The nodal points are no more material points of the loop. An adaptation of Bacon's method is used for regularizing the local self-stresses at these points and introducing a cut-off radius. Although the loop is planar, the total self-stress is calculated in three dimensions. This solution is also implemented as an option in *microMegas*, together with that of Bacon and the line tension approximation by de Wit and Koehler (1959).

The self-stress of a dislocation with arbitrary shape given by Gavazza and Barnett (1976) requires too intensive computations to be of practical use in DD simulations. In the *PDD* code, use is made of a simplified formulation (eqn. 2.17 of Ghoniem *et al.* 2000). Only the leading line tension term is retained in the *PARANOID* code for evaluating the self-stresses of extended arcs.

An alternative method for deriving the self-stress was devised by Cai *et al.* (2006) and is implemented in *ParaDiS*. In its principle, the method is reminiscent of the Peierls–Nabarro model, in which the Burgers vector is spread out in a continuous manner between two atomic planes to account for their relative displacement (Appendix D). The force on a dislocation segment can be obtained from the negative gradient of the total elastic energy (eqn. 1.6). The latter is classically given by a double integration performed on a closed path. In substance, the self-energy diverges when the approach distance $R = |\mathbf{r} - \mathbf{r}'|$ between two points $\mathbf{r}$ and $\mathbf{r}'$ on this path approaches zero. The singularity can be removed by finding a non-singular distribution function f of the Burgers vector around the dislocation lines, which transforms $R = \sqrt{R^2}$ into $R_a = \sqrt{R^2 + a^2}$. The regularization parameter a is a small distance that plays the role of a core width or cut-off radius.

The function f has to fulfill several requirements. It must lead to simple analytical solutions for straight or curved segments without exhibiting the defects of Brown's cut-off procedure. It also has to be consistent in that sense that the force on a dislocation segment computed from the gradient of elastic energy should be identical to the force obtained by integrating the Peach–Koehler relation. The mathematical solution proposed for the core spreading is axisymmetrical and the fields derived from it are three-dimensional. They reproduce classical solutions for an infinite dislocation line or a dislocation loop within a second-order term in

$a^2/\mathcal{R}^2$, where $\mathcal{R}$ is an outer cut-off radius. The value of the regularization parameter a is arbitrary; it is suggested that a safe value should be of the order of a real core radius, that is, a few Burgers vectors.

4.2.6 From self-stress to effective stress

Once the self-stress at a node is estimated, there are basically two methods for determining the effective stress or driving force acting on it.

The first method consists in summing up one by one the missing elastic contributions at the nodes. In a single crystal, this implies, in particular, summing up at each node individual long-range interactions stresses with remote segments over the whole microstructure. There are in the literature several analytical expressions for the stress field of a straight segment. They are drawn from classical integrals, where the integrand differs by terms that vanish upon integration over a closed path. Others are imperfect because they are not determined from summations over a closed path. More detail on the available solutions, including those drawn from Brown's formula (Brown, 1967), can be found in the book by Hirth and Lothe (1992) and in Section 4.2.3. of the review article by Bacon *et al.* (1979). An article by Yin *et al.* (2010) discusses solutions in anisotropic elastic media.

Three equivalent formulations for the stress field of a straight segment are utilized in DD codes. The stress fields of the source segments are expressed in generalized coordinates, which facilitates their summations at nodes. The solution given by Hirth and Lothe (1992) is used in various forms in the *MDDP* code and its earlier versions. Another form given by Devincre and Condat (1992) is employed in lattice-based simulations (a full demonstration is given by Devincre 1995). It is based on an earlier work by de Wit (1967). A last equivalent expression was derived by Schwarz (1999) and is implemented in the *PARANOID* code.

Alternatively, the driving force on a node can be derived from the change in negative gradient of the total energy induced by small displacements of the nodes in the considered microstructure (eqn. 1.6). In *ParaDiS*, an analytical expression is established, which yields the total force on a given node as a closed integral of the Peach–Koehler forces over all individual line elements. For straight segments, the change in shape induced by the displacement of a single node is only accommodated by the segments connected to it and the neighbouring nodes are not displaced. Analytical expressions are available for computing the total force on a single node; they are implemented in MATLAB (Section 4.5).

In the *PDD* code, the discretization into flexible segments makes the calculation of the total energy change over the whole microstructure more complex. Use is made of the principle of virtual work, which expresses that during a time step the work of the effective forces should balance the dissipated energy. The energy dissipated by a moving line element is proportional to its velocity, which, at this step, has to be specified by a local rule. The energy balance is eventually carried out over all the loops defining the microstructure and redistributed to the nodes. Wang *et al.* (2006*b*) provide a concise description of the related procedures. The principle of virtual work is also utilized in the Weygand code.

Within a few unavoidable approximations, the procedures based on stress calculations and energy variations should lead to identical results.

4.2.7 Further optimization

Large-scale simulations of plastic flow require large simulation cells (Section 4.1), in which the number of segments may increase up to relatively large values. The local procedures for optimizing a code, which are discussed in Section 4.2.3, have to be complemented by more global procedures. The reason is that the computational cost of long-range interaction stresses between segments constitutes the most CPU time-consuming part of a DD simulation. It scales as $O(N^2)$, where N is the number of interacting segments. Three methods can significantly improve the efficiency of a code, parallel computing, the fast multipole method and under-integration. The objective is to reach within a reasonable time sufficiently large strains for obtaining valuable insights into microstructures and the mechanical response.

4.2.7.1 Parallelization DD simulations have taken advantage of the growing computing power provided by clusters of computers and multi-processor supercomputers with parallel architecture. Hence, DD simulation codes are written, or were rewritten, for parallel computing. There are several parallelization schemes and algorithms, of which the efficiency depends on the problem considered and the computing environment. In every case, the communication time between processors has to be minimized and the processor efficiencies have to be optimized by balancing the computing load assigned to each of them. Several additional features, which are specific to DD simulations, have to be cared of. The dislocation density increases with time, the continuity of the dislocation lines has to be preserved, the microstructure often evolves by localized strain bursts and it tends to form spatially heterogeneous patterns.

The most commonly employed parallelization scheme, to which an introduction is provided by Bulatov and Cai (2006), is called the domain decomposition method. The simulation space is first partitioned into domains of equal size. This size decreases with time, and the number of domains increases, to better accommodate the increasing computational load. The dislocation lines contained in one or several domains are then assigned to a given processor. The strength of this approach is that it is well established since it is utilized in many areas of physics. Its well-known weakness arises from the fact that its efficiency sensitively depends on the number of available processors. Most DD simulations (*Tridis*, *MDDP*, *PDD*, the Weygand code and *ParaDiS)* utilize this method; references to its implementation are given in Section 4.5. *PARANOID* is the first code that was parallelized but, seemingly, no technical detail was published.

Another method is implemented in *microMegas*. It aims at optimizing the load balance and adopts for this purpose a domain decomposition devised for the fast multipole method (see next section). An evaluation of the computing costs of long- and short-range interactions is performed for each segment. Each processor is then assigned a number of segments representing a computing cost adapted to its speed. This solution presents the advantage of being independent of the number of processors. It also allows taking easily into account spatial heterogeneities in the dislocation microstructure. On the other hand, it is only focused on interactions and cannot accommodate possible load unbalances caused by contact reactions between dislocations.

4.2.7.2 Fast multipole method As defined by Greengard and Rokhlin (1987), the N-body problem is concerned with the evolution of potential and force fields in many-particle systems interacting via long-range gravitational or Coulomb forces. This definition can obviously be extended to long-range dislocation interactions. Beyond $N = 2$, general solutions of the N-body problem have to be computed. A fast algorithm for reducing the asymptotic scaling from $O(N^2)$ to $O(N)$ was proposed by Greengard (1987) and Greengard and Rokhlin (1987). It is known as the fast multipole method (FMM). Many variants, which yield scalings as $O(N \log N)$ and $O(N)$, were further developed (see Greengard 1997). Like for parallelization, the efficiency of the various algorithms depends on the problem considered and the computing environment. The basic strategy of the FMM is rather simple and can be applied in a controlled manner.

In the implementation to DD simulations the simulation cell is decomposed into a more or less complex hierarchical structure of sub-cells, as schematically illustrated by Fig. 4.4. A distinction is made between long-range and short-range interactions. The latter are computed explicitly at every time step in a given cell and its nearest-neighbouring cells. The remaining interactions are considered as long-ranged; they are small and evolve slowly in time and space.

The interaction fields of segments in remote cells are lumped together and computed at the centre of the considered cell by means of a multipole expansion up to a few moments. As they evolve slowly in space, these fields are applied to all the segments contained in the cell. In addition, these interaction fields are updated using a time step that is a multiple of the simulation time step. Both simple and quite sophisticated solutions were developed for DD simulations. The method becomes all the more complex as the order of expansion and the level of cell subdivisions increase. In all simulations, the efficiency of the FMM method is increased by parallelization.

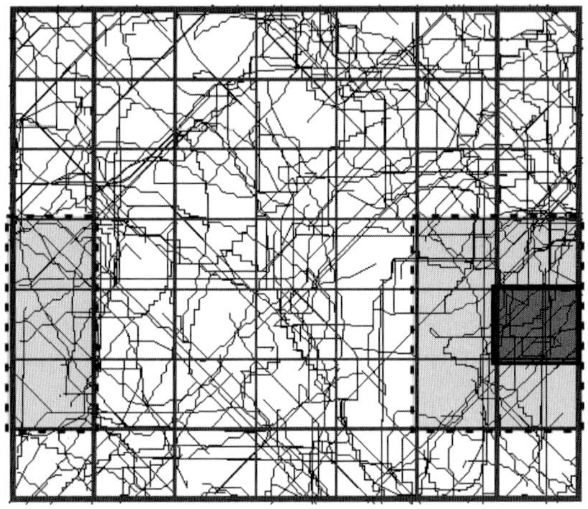

FIG. 4.4. Domain decomposition in a 3D DD simulation with periodic boundary conditions (see Section 4.4.1). The elementary cell is shown in a ⟨001⟩ view; it is divided into 7^3 sub-cells of average linear dimension about 1.4 μm. For dislocations located in the dark sub-cell, all interactions with other dislocations in the same sub-cell and in the neighbouring (light grey) cells bounded by dashed lines are explicitly calculated. The contributions from dislocations in more distant (white) sub-cells and in image volumes (not shown) are calculated using a fast multipole scheme. This procedure is repeated on each sub-cell of the primary cell. Courtesy R. Madec.

The implementation of the FMM requires lot of care when high performance is required. In its simplest variant, which is discussed by Shin *et al.* (2006), the simulation cell is decomposed into a single set of sub-cells (Fig. 4.4) and the remote fields are computed explicitly without expansion. The optimization procedure is rather simple and the acceleration is attractive, typically between 15 and 50 for about 2×10^4 segments. This solution constitutes a good compromise between efficiency, accuracy and simplicity. It is implemented in *Tridis* and *microMegas*, which are both parallelized.

The FMM can be extended to higher orders for treating several hundreds of thousands segments. Its optimization requires solving issues concerning the numerical convergence of the calculations and their accuracy, as well as other problems mentioned above in the context of parallelization (Zbib *et al.*, 1998; LeSar and Rickman, 2002). The benefit is that an acceleration by two orders of magnitude can be obtained in some cases (Wang *et al.*, 2004*a*). In the last version of the *PDD* code (Wang *et al.*, 2006*b*), an $O(N)$ algorithm akin to the FMM is naturally embedded into the parallel data structure. The fast O(N) FMM implemented in the highly parallel *ParaDiS* code is detailed by Arsenlis *et al.* (2007);

it extends previous works by LeSar and Rickman (2002) and Wang *et al.* (2004*a*). Further progress was obtained by applying for the first time to dislocations a new FMM that was proposed by Greengard and Rokhlin (1997). Zhao *et al.* (2010) demonstrated on simple tests the efficiency of this new version.

4.2.7.3 Under-integration This method also makes a distinction between slow and fast-evolving dislocation events, but at a more local scale. It applies in particular to situations where a large fraction of the dislocation density is immobilized at strong obstacles, for instance forest obstacles or grain boundaries. In such conditions, most of the dislocation microstructure is immobilized or evolves slowly during plastic flow. As small fraction of the lines undergoes much faster local events, typically by up two orders of magnitude larger. Such events are caused by fast dislocation motion past a critical configuration or abrupt topological changes induced by cross-slip or dislocation reactions.

Under-integration schemes are based on the idea that the two types of events can be cared of by a two-level time step procedure. The latter is implemented in DD simulations by defining a global time step Δt, at the end of which the position of all segments is updated and all related changes in the distribution of nodes along the lines are performed. This time step is essentially dealing with the treatment of events that evolve under the effect of long-range interactions. It incorporates much smaller sub-steps, δt, which are exclusively devoted to the treatment of fast events. Criteria and testing procedures have to be defined for characterizing individually each type of event. When they are carefully implemented, a gain in computing efficiency is obtained without loss of accuracy. Under-integration is implemented in *microMegas* and in the Weygand code.

The combination of the three methods presented in this section can accelerate DD simulations by a factor well over two orders of magnitude.

4.2.8 Elastic anisotropy

Early attempts at implementing anisotropic elasticity in DD simulations showed that it induces a prohibitive increase in computing load. The latter can be somehow alleviated by making use of tabulated values and interpolations (Rhee *et al.*, 2001). Simulations were carried out on model dislocation configurations, with the objective of comparing isotropic and anisotropic shapes (Han *et al.*, 2003). Reasonably accurate shapes are obtained with the Reuss average elastic constants or, better, those of Scattergood and Bacon (1975, 1982, see Section B.2). Some local discrepancies may also average out in large-scale simulations performed in isotropic elasticity. In addition, it may sometimes be useful to introduce an effective, orientation-dependent, Young's modulus to account for the elastic response.

In a more recent DD simulation study (Yin *et al.*, 2010), alternative analytical solutions for the anisotropic interactions and self-stresses of straight segments are compared to isotropic solutions in terms of accuracy and computing efficiency. The results obtained on Frank–Read sources confirm that the isotropic elastic constants proposed by Scattergood and Bacon lead to the best accuracy. Furthermore, the new anisotropic solutions are found to be much more computationally efficient than those tested earlier. The amount of CPU time is increased by only a factor of five for elementary configurations. In another study (Yin *et al.*, 2012), an expansion of the fast multipole method (Section 4.2.7.2) to anisotropic elasticity was also developed. It was tested on single dislocation loops in molybdenum and sodium, which are respectively weakly and highly anisotropic materials. With respect to isotropic calculations, the increase in computing cost for more complex configurations may be within one order of magnitude.

A strong motivation is needed for performing anisotropic simulations. One was recently found in some properties of defects in α-iron irradiated at high temperatures (about 550 °C). The anomalous occurrence of square edge loops with a $\langle 001 \rangle$ Burgers vector is a long-standing unresolved problem. Indeed, isotropic elasticity predicts that the Burgers vector $\frac{1}{2}\langle 111 \rangle$, of smaller magnitude, is always the most stable in the bcc structure. The explanation proposed by Dudarev *et al.* (2008, 2009) establishes a link between several properties that are specific to α-iron. Its magnetic properties are responsible for the occurrence of an α–γ transition at 912 °C and for the softening of the stiffness constant $c' = (c_{11} - c_{12}/2)$ at high temperatures. This softening results in an increased elastic anisotropy. The value of the stiffness constant c' significantly decreases when temperature increases from room temperature, at which α-iron is already highly anisotropic, to high temperatures. Isotropic elasticity is unable to reproduce this behaviour as it cannot account for the softening of the phonon mode in question.

Estimates for the self-energy of dislocation lines and small loops were carried out using anisotropic elasticity (Dudarev *et al.*, 2008) and quantum mechanical calculations based on a magnetic tight-binding Hamiltonian (Dudarev *et al.*, 2008). They suggest that $\langle 001 \rangle$ prismatic loops are the most stable loops at elevated temperatures provided that they are small enough. No change in slip geometry is, however, expected. More results may be obtained by taking advantage of potentials that were recently developed for atomistic simulations of magnetic iron (Section 3.2.2).

The impact of elastic anisotropy on the temperature dependence of the yield stresses of α-Fe and ferritic-martensitic steels is now becoming an interesting problem. For studying it, the effect of the high elastic anisotropy of bcc iron on the equilibrium shapes of Frank–Read source segments was investigated (Fitzgerald, 2010; Fitzgerald and Aubry, 2010). DD simulations of dislocation reactions in anisotropic elasticity may, however, yield a more direct insight into yield and flow stresses. The shapes of several types of loops were also investigated up to temperatures close to the α–γ transition by 3D DD simulations (Aubry *et al.*, 2011). At high temperatures, the loops with $\langle 001 \rangle$ Burgers vectors exhibited

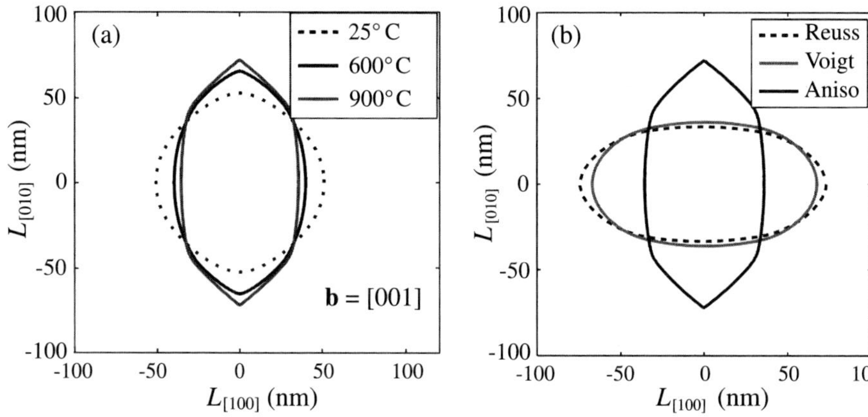

FIG. 4.5. Simulations of a [100](001) dislocation loop in α-iron. The equilibrium shapes are computed at constant loop area. (a) Dependence of shape on temperature in anisotropic elasticity. (b) Comparison of anisotropic shapes and isotropic shapes yielded by the Reuss and Voigt averages at 900 °C. From Aubry *et al.* 2011.

cusps reflecting a range of unstable line orientations, which cannot be predicted by isotropic elasticity (Fig. 4.5). In contrast, the shapes of $\frac{1}{2}\langle 111\rangle\{011\}$ loops were not much affected by temperature and remained smooth.

4.2.9 Dissociated dislocations

The dislocation fields considered till now in this part were those of perfect dislocations. Several DD simulations were adapted to deal with planar slip of dissociated dislocations in fcc crystals. The topics investigated were the interactions of dislocations with two-dimensional precipitates (Duesbery *et al.*, 1992*b*), with three-dimensional distributions of precipitates in long-range ordered alloys (Mohles, 2004*a*) and the properties of dislocation junctions in 3D (Shenoy *et al.*, 2000). A 3D simulation of planar slip was also performed on dc silicon (Moulin *et al.*, 1999*b*). More recently a 3D simulation of dissociated dislocations in fcc crystals was derived from *ParaDiS* (Martinez *et al.*, 2008).

The SFEs and dissociation widths under zero stress are reasonably well known in fcc crystals (Appendix B). The short-range interactions of partial dislocations with other dislocations were revisited by Bacon (1978), who proposed to a correction taking into account the effect of dissociation on the energy of a dislocation line. Under stress, the dissociation widths depend on the Schmid factors of the edge components of the partials (Fig. 2.9), hence on the orientation of an uniaxial applied stress. The so-called Escaig stresses on the Shockley partials give rise to the Escaig effect, which is experimentally found to be rather moderate (Section 2.4.4). The cross-slip mechanism and its orientation dependence can be implemented with the help of local rules deriving from those discussed in Section 4.3.3.

4.3 Local rules

4.3.1 Outline

Local rules are implemented in DD simulations to account for dislocation core mechanisms and properties that cannot be fully modelled by the elastic theory. Although these rules are often imperfect by lack of atomistic input, the predictive ability of a simulation depends on the care taken to establish and implement them. This section discusses successively the free-flight mobility of dislocations in materials where plastic flow is governed by athermal obstacle strengthening, dislocation velocities in materials with high lattice resistance, in particular bcc metals, local rules for dislocation cross-slip and other local rules. The basic content of this section is for a large part borrowed from other chapters.

4.3.2 Dislocation mobility and velocity

When the relation between effective stress and velocity is linear, one can equivalently use mobility laws or velocity laws. For thermally activated mechanisms, which introduce temperature in a simulation, the velocity law is non-linear and is preferably treated as such.

4.3.2.1 Obstacle-controlled mobility In fcc metals and other materials above their athermal transition temperature, the lattice resistance is negligible and it is necessary to keep track of dislocations during their free-flight motion between obstacles. For this purpose, use is made of the results obtained in Section 2.2.2.3 on free-flight motion governed by phonon damping. In the regime of quasi-static deformation, inertia can be neglected. The critical stress for overcoming obstacles is then always much larger than the stress opposing free-flight motion between obstacles. For the sake of simplicity, it is usually assumed that dislocation motion is always overdamped. In other words, dislocations are assumed to move with their steady-sate velocity. A small number of very fast events may occur in DD simulations, and it is common practice to limit the dislocation velocities to some value below one-half of the sound velocity.

The linear relation between applied stress and velocity given by eqn. 2.8 only accounts for the Peach–Koelher force induced by externally applied stresses on straight dislocations. Thus, it has to be modified. In DD simulations, the effective shear stress τ^* on a node or a line element is given by a summation over all stress contributions, including the self-stress (eqn. 1.18). In some simulations, it is convenient to use a reversed relation between the velocity, v, and the driving glide force $F_g = \tau^* b$ along the slip direction. In such cases, and by analogy with the physical terminology, the mobility M of a dislocation is defined as the ratio of driving force to velocity. The local rule for free-flight dislocation motion in fcc crystals can then be written in the two following manners:

$$\tau^* b = Bv; \quad v = \frac{\tau^* b}{B} = MF_g, \tag{4.3}$$

The damping coefficient B exhibits a temperature dependence, even above the Debye temperature (Fig. 2.1). Typical values for fcc metals are discussed in Section 2.2.2.2. They are not all accurately known, but seem to be between 10^{-5} and 10^{-4} Pa s^{-1}. Taking excessively high values of B has the same effect as increasing the velocity or the plastic strain rate. At some point, the small serrations induced by dislocation avalanches on simulated stress–strain curves disappear (see Fig. 2 of Devincre and Kubin 2010). This indicates that plastic flow is no longer obstacle-controlled.

A small lattice resistance is sometimes subtracted from the effective stress. It is meaningless, unless simulations are carried out at cryogenic temperatures (Section 2.2.1). The glide mobility is sometimes made anisotropic to account for the higher mobility of screw segments with respect to edge ones. For this purpose two distinct drag constants or mobilities, M_e and M_s, are introduced for screw and edges. A smooth function varying between the assumed two extreme values allows treating all dislocation characters. This refinement is usually ignored since the values of the two free-flight mobilities are not known. The anisotropic equilibrium shapes of curved segments and dislocation loops are, of course, incorporated into the self-stress.

In the DD simulations that were carried out so far at high strain rates, (Section 5.4.5), the damping constant has same value as in the quasi-static regime, inertia is taken into account and use can be made of the rest mass of dislocations, m_o up to velocities of at least 10^6 m s^{-1}. This is justified because at imposed strain rates of 10^6 m s^{-1}, most dislocations move at velocities much smaller than one-half the velocity of transverse shear waves in DD simulations (Shehadeh *et al.*, 2005*a*; Wang *et al.*, 2007*a*). The equation of motion implemented in the *MDDP* code is similar to eqn. 2.5; it includes the relativistic mass m and is written

$$m\frac{dv}{dt} + Bv = F_g \tag{4.4}$$

A climb mobility was sometimes introduced in early simulations, which is equivalent to treating climb as a very slow glide process. The *PDD* code is the only one where a model based on point defect diffusion diffusion was implemented.

Models for climb processes in DD simulations, coupled or not to glide processes, have notably progressed in the recent years (see Section 5.3.3).

The lattice resistance on screw dislocations in bcc metals was, and is, also sometimes modelled as a constant mobility (Ghoniem *et al.*, 2000; Zbib *et al.*, 2000; Bulatov and Cai, 2006; Wang and Beyerlein, 2011). The reason is that some discretization schemes make it very difficult to extract the driving force from the principle of virtual work unless the mobility is a constant (Arsenlis *et al.*, 2007). A smooth treatment of the strong contrast between screw and non-screw mobilities, which introduces an empirical orientation dependence, is then performed. If the lattice friction was governed by the kink-diffusion model, it would be posssible to assume a linear relation between effective stress and velocity (eqns. 3.11 and 3.12). This is not the case, however, in bcc metals. In simulations performed at a given temperature, the corresponding mobility has to be specified but the strain rate dependencies of the yield and flow stresses are lost. In addition, if the velocity does not depend linearly on the length of screw segments, the kinetics of dislocation–obstacle interactions is also lost (Section 3.2.5.1).

4.3.2.2 Lattice resistance This section is focused on screw dislocations in bcc metals, where the lack of atomic-scale input is compensated by an extended experimental literature. It is not so for prismatic slip in hcp transition metals, where screw dislocation motion is also governed by the kink-pair mechanism. The kink-diffusion model was implemented in DD simulations of silicon by Moulin *et al.* (1997, 1999*a*, 1999*b*). In more recent simulations, however, silicon is treated as an fcc-like material since the emphasis is on the high temperature regime at which micro- and nano-devices are processed (Section 5.5.3).

Models for the kink-pair mechanism in bcc metals are discussed in Section 3.2.4. As there is no secondary Peierls stress, the kinks are extremely mobile. All non-screw segments, behave like dislocations in fcc metals and their velocity is governed by phonon damping (eqn. 4.3). The interactions of kinks, non-screw segments and screw lines with forest obstacles are schematically depicted in Fig. 3.12. At low temperatures and high stresses, the non-screw segments move so fast between obstacles that extremely small times steps would be required to track their position. As what matters is only the strain they produce, two procedures were developed to simplify their treatment. A simple method consists in setting the time scale on screw dislocations and allowing non-screw segments to move until they are blocked between two steps of screw motion (Tang *et al.*, 1998; Chaussidon *et al.*, 2008). The other method consists in attributing to non-screw segments a velocity rule proportional to that of screw segments. The proportionality coefficient, typically 10^2, is determined from the condition that it does not affect the stress–strain response (Monnet *et al.*, 2004).

At the mesoscale, the velocity of screw segments is given by an Arrhenius form, which is discussed in Appendix A and Section 3.2.4.3. Since both experiments and atomic-scale simulations yield activation enthalpies (Appendix A), the local rule is written

$$v = (A/L_o)L \exp -\left(\frac{\Delta H(\tau^*)}{k_B T}\right) \tag{4.5}$$

A correction for entropy terms (Schoeck, 1965) can be added if necessary. The velocity is proportional to L, the length of a screw segment between obstacles, and the coefficient A/L_o depends on the critical length of a kink-pair, the Debye frequency and other quantities (eqn. 3.2). The stress-dependent activation enthalpy should in principle be obtained from atomistic simulations. When no such data are available, use is made of a two-parameter generic form for $\Delta H(\tau^*)$ proposed by Kocks *et al.* (1975), which is given by eqn. E.13. Simple manipulations, which are described by Tang *et al.* (1998), allow obtaining a parametric temperature dependence of the lattice resistance, which is adjusted to an experimental stress versus temperature curve. The obtained values for the back-extrapolated Peierls stress and the total activation energies must be consistent with known values, like those given in Table 3.1.

To bypass difficulties associated with non-Schmid behaviour, the orientations selected are in the centre of the standard triangle, where the primary slip system is active. Once the velocity rule is implemented in a DD simulation, a simple check consists in verifying that the simulated temperature dependence of the yield stress is consistent with the data used for the fit. This verification was successfully performed in tantalum (Tang *et al.*, 1998) and α-Fe (Naamane *et al.*, 2010), and for prismatic slip in zirconium (Monnet *et al.*, 2004). Another, more demanding, check consists in verifying the geometric relation 3.5.

The predictive ability of DD simulations can be improved in a rather simple manner. Figure 3.8 shows that the atomistic stress dependence of the activation enthalpy written in dimensionless form,

$$\frac{\Delta H}{\Delta H_o} = f\left(\frac{\tau^*}{\tau_P}\right), \tag{4.6}$$

is in quite reasonable agreement with the empirical dependencies drawn from experiment. In the spirit of multiscale modelling, it would more natural to draw from atomistic simulations numerical data for the energetics of the lattice resistance. The total activation enthalpies ΔH_o are usually in good agreement with measured values and the only quantity that requires an adjustment is the too high Peierls stress yielded by atomistic simulations (Section 3.2.3). Finally, Yang *et al.* (2010) provide sets of parameter values for vanadium, tantalum and molybdenum to be inserted into the two-parameter eqn. E.13.

4.3.3 Dislocation cross-slip

This section discusses successively cross-slip in fcc crystals, for which a semi-quantitative rule is usually adopted, and in bcc crystals for which there is no established local rule.

4.3.3.1 Cross-slip in fcc crystals Among the various types of cross-slip mechanisms listed in Section 2.4.8, the most important one is involving the interaction of a screw dislocation with a nearby dislocation. This process includes the annihilation of screw dislocations by cross-slip, which occurs at a stress τ_{III} and marks the onset of stage III in the stress–strain curves of fcc crystals (Section 2.6.4). The Escaig stresses are often assumed to drive the occurrence of cross-slip, but they are much too high for doing so (Section 2.4.4). Thus, the Escaig model incorporates a rigid barrier to the motion of a screw dislocation in its slip plane, which is intended to reproduce in a generic manner the additional interaction of a screw dislocation with its immediate environment. The obtained numerical predictions are sometimes approximate and, globally, in semi-quantitative agreement with experiment. Atomistic simulations have, however, not brought any additional information about the stress τ_{III} and its dependencies on SFE and temperature. For this reason, the local rule for cross-slip implemented in DD simulations is inspired from Escaig's model (Kubin *et al.*, 1992). It is written in the form of an activation probability, which is an impressionistic transcription of eqn. 2.17, where local stresses are replaced by macroscopic stresses.

$$P(\ell) = \beta \frac{\ell}{\ell_o} \frac{\delta t}{\delta t_o} exp\left(\frac{\Delta H_o}{k_B T}\right) exp\left(\frac{|\tau_g| - \tau_{III}}{S}\right) \tag{4.7}$$

In eqn. 4.7, an activation enthalpy is substituted to the traditional notation (E) adopted in Chapter 2. In the pre-exponential factor, β is a constant, ℓ is the length of the cross-slipping segment and δt is the time per simulation step. These two quantities are scaled by a reference length ℓ_o and a reference time δt_o. The probability is, therefore, expressed per simulation time step. In Escaig's model the activation volume V is a constant in the low stress regime. The latter actually extends to reasonably high resolved stresses according to the criterion given by the model (cf. Section 2.4.3). The activation energy is then of the form $\Delta H = \Delta H_o - V\tau_g$, where ΔH_o is the total activation energy and τ_g is the resolved shear stress on the perfect screw dislocation in its primary glide plane. The total activation energy of the screw dislocation is written as a separate term, which can be lumped into the constant β at a given temperature. The effective stress is taken to be $\tau_g^* = \tau_g - \tau_{\text{III}}$ in tension, where the stress τ_{III}, is considered as representing the Escaig barrier. Finally, the effective stress is scaled by a factor $V/k_B T = 1/S$, where S is the strain rate sensitivity (Section A.1).

The numerical adjustment of eqn. 4.7 is often performed on experimental values and at $T = 300\,\text{K}$. The activation volumes for copper and aluminium were measured (Section 2.4.5); they follow the scaling law predicted by Escaig (eqn. 2.19). Values for τ_{III} are given as a function of temperature by Mitchell (1964) for several fcc metals deformed in a broad range of low-symmetry orientations. At and close to high-symmetry orientations, the values are substantially different (Kubin *et al.*, 2009). The three constants β, ℓ_o and δt_o are lumped together and adjusted as follows. The typical length of a cross-slipping segment is taken to be one micron. Then, in a model DD simulation, the critical annihilation distance of two attractive screw segments must be about 50 nm under a resolved stress of 30 MPa at $T = 300$ K (Essmann and Mughrabi, 1979). For other temperatures, the cross-slip probability has to be rescaled. ΔH_o can be drawn from experiment, eqn. 2.16 or reliable atomistic simulations.

The cross-slip probability is implemented in DD simulations through a classical Monte Carlo scheme. A preliminary test on effective stresses in the primary and cross-slip planes may help eliminating cross-slipped segments which cannot expand in their new plane. In Escaig's model, no distinction is made between effective and applied stresses since screw dislocations are supposed to be straight or weakly curved. This raises several problems. First, it is necessary to define a screw length on a curved segment. It is usually taken as the length of an arc that deviates by a few degrees from the screw orientation. When the length of a screw segment is smaller than the activation length, approximately V/b^2, cross-slip cannot be activated. Under the very high stresses reached in nanomaterials, the dissociated ribbon can be extended or contracted depending on the Schmid factor on the Shockley partials (Fig. 2.11*a*). Escaig stresses may then significantly contribute to the cross-slip probability. In that case, and probably in several other ones, the local rule has to be modified in a specific manner. For instance, when screw dislocations pile-up at a grain boundary in a fine-grained polycrystal, the barrier opposing dislocation glide is complex and the local rule has again to be modified. In the procedure that was proposed by Hiritani and Zbib (2003), virtual Shockley partials were associated with screw dislocations and were allowed to move to their equilibrium positions under stress.

The cross-slip rule given above allow obtaining annihilations, as well as repulsive and double cross-slip events. In small-scale pillars, it helps stabilizing pinning configurations for spiral sources (see e.g., Motz *et al.* 2009 and Fig. 5.22*a*). In large-scale simulations of bulk materials, it leads to the formation of incipient dislocation cells (see Madec *et al.* 2002*b* and Fig. 1.9). Traces of cross-slip events are also observed in simulated dislocation avalanches (Fig. 5.11). One has, however, to keep in mind the weaknesses and the semi-phenomenological character of this local rule. On the other hand, a total absence of cross-slip or cross-slip without energy barrier may result in unrealistic configurations.

The annihilations of screw dislocations leave pairs of superjogs in the cross-slip planes, which produce massive annihilations of dislocation with a broad range of characters (Fig. 5.2). Thus, an efficient way to examine microstructure formation without implementing a cross-slip rule consists in introducing pairs of superjogs in the initial configuration (Figs. 2.19*b* and 5.7).

The rule based on Escaig's model is implemented in nodal simulations, in the Weygand code and in the recent versions of the *PDD* code (Wang *et al.*, 2004*b*; Wang *et al.*, 2007*b*). In the *MDDP* code, the formulation is different and based on an elastic model for a critical pair of perfect kinks in the slip and cross-slip planes. Cross-slip occurs when the energy of this configuration is smaller in the cross-slip plane than in the slip plane. A similar rule is implemented in *ParaDiS*. Cross-slip is not implemented in *PARANOID*.

4.3.3.2 The case of bcc crystals When the transition to athermal plastic flow is approached from low to high temperatures, the shape of the stress–strain curves of bcc crystals progressively changes. Pseudo-parabolic hardening evolves towards the classical three-stage hardening that characterizes forest strengthening (Fig. 3.2*a*) and dislocation cell structures are observed after some strain (Fig. 3.2*b*). Above the transition temperature, DD simulations of bcc metals can be carried out using local rules for fcc metals, save for {112} slip that may require further elaboration. In α-iron at 293 K, that is, below the transition temperature of 340 K (Table 3.1), and up to a strains of 15%, the slip geometry is typically the one described in what follows for moderate temperatures. Nevertheless, the stress–strain curves have already started exhibiting three stages (Spitzig and Keh, 1970). This complex regime of mixed high and low temperature behaviour is not discussed here.

Many features of the behaviour of bcc crystals discussed in Chapter 3 can be summarized as follows. Atomistic and DFT calculations indicate that slip occurs almost always by atomic steps of motion on two {110} planes sharing the same Burgers vector (Section 3.2.2). Cross-slip events between two such planes can generate composite slip on any slip plane containing the Burgers vector, depending on the respective activation energies for kink-pair formation. Thus, cross-slip and slip geometry are intimately related, at least at the atomic scale. This relation is not simple because the critical stress for kink-pair formation is governed by non-Schmid effects (Section 3.2.4.3). At this step, it is convenient to split the low temperature domain into two domains corresponding to a very low and a moderate temperature range.

In the lower half of the low temperature range, slip trace observations yield almost the same slip geometry in all transition bcc metals (Section 3.2.1). Crystallographic slip on the primary {110} system predominates in a range of orientations that widens with decreasing temperature. Macroscopic slip on {112} planes is observed, preferentially in the twinning orientation, when these planes bear the maximum resolved shear stress. Non-Schmid effects are all the more pronounced as temperature is low.

The model implemented in small-scale DD simulations of α-iron by Chaussidon *et al.* (2008) below 200 K is outlined in Section 3.2.4.3. It actually constitutes a rule for atomic-scale cross-slip extended to the mesoscopic scale, which predicts both the orientation dependence of the slip geometry and the twinning–anti-twinning asymmetry. This type of model is potentially interesting since it may also bring useful informations about slip geometry at the mesoscopic scale. In addition, it only requires adjusting the stress dependence of the activation enthalpy on experimental data. As indicated in Section 4.3.2.2, this adjustment can be avoided when atomistic data are available. It may also be possible to include the tension-compression asymmetry by adjusting to atomistic data at 0 K two of the coefficients involved in eqn. 3.3 (see Section 3.2.4.3 and Koester *et al.* 2012). In such conditions, no free parameter would remain and a full connection with the atomic scale would be established for the yield stress.

At medium temperatures, primary slip is still observed in the centre of the standard stereographic triangle, but in a range of orientations that shrinks with increasing temperatures. For other orientations, slip traces tend to follow the maximum resolved shear stress (MRSS) plane and {112} slip is observed in same conditions as at lower temperatures. In situ TEM observations show that the elementary slip events still occur in {110} planes, in spite of the high cross-slip probability (Caillard, 2010*a*). In parallel, non-Schmid effects decrease and vanish at the transition to athermal plastic flow. The model by Edagawa *et al.* (1997) is outlined in Section 3.2.4.3. It actually predicts in a generic manner the slip geometry and the twinning–anti-twinning asymmetry over the whole range of temperatures (Fig. 3.9). The model by Chaussidon *et al.* (2008) should yield somehow better predictions, but it has not been tested in this temperature range.

In summary, it seems that no mesoscopic rule would be needed for cross-slip in DD simulations of bcc metals at low temperatures if a microscopic rule could be devised for {110} composite slip.

In the past and present experimental literature on the medium temperature domain, one finds mentions of the so-called pencil glide, which was first observed by Taylor and Elam (1926) and Taylor (1928). This term describes the diffuse and wavy character of MRSS slip in bcc metals and especially α-iron (Section 3.2.1). Pencil glide is often mentioned in the context of bcc steels at large strains, where it seems to be associated with a lack of planar slip traces (Canadinc *et al.*, 2008). This is perhaps not surprising considering the strong non-uniformity of local stresses in such materials. Another example of pencil glide is provided by Fig. 7 of Caillard (2011). The slip traces left by screw dislocations expanding from a crack tip at 96 K in a Fe-C alloy exhibit a spectacular fan shape. All these examples may indicate that mesoscopic cross-slip easily occurs at moderate temperatures under the effect of stress gradients.

The fact that cross-sip is quite easy at moderate stresses does not mean at all that there is no longer an energy barrier to overcome. Indeed, there are still straight screw dislocations and the flow stress is still temperature dependent (see e.g., Caillard 2010*a*). This deformation mode persists till forest strengthening begins to substantially contribute to plastic flow close to the transition to the athermal regime. In *ParaDiS*, it is assumed by default that in the domain of moderate stresses, screw dislocations can be treated through their mobility as perfect dislocations cross-slipping without barrier into any slip plane, crystallographic or not. A probably safer procedure may consist in introducing no cross-slip rule and focusing on an orientation in the centre of the standard triangle, where primary slip predominates.

4.3.4 Other local rules

Local rules have to be devised when there is little, or no, input on a given mechanism. The most important one is nucleation at free surfaces in small-scale materials (Section 5.5.5.4) or in silicon under extreme deformation conditions (Section 3.4.9.2). The state of the art has, however, notably progressed in this domain (Weinberger *et al.*, 2012) and it is possible to adjust local rules to MD simulations in the case of nanopillars (Zhu *et al.*, 2008). Phenomenological rules are available for dislocation–grain boundary interactions in thin metallic films; they apply to blocking, transmission and nucleation of dislocations (Section 5.5.4). In bulk polycrystals, it is reasonable to assume that grain boundaries are impenetrable to dislocations.

The impact of atomic jogs on plastic flow of fcc crystals does not seem to be very significant. For a long time, it was thought that jogs could pin moving screw dislocations. Actually, it seems that these defects are extremely mobile since the activation energy for their conservative motion along a screw dislocation line is 15–20 meV in copper (Vegge and Jacobsen, 2002). Thus, in pure fc crystals, jog pinning and dragging occur only under high strain rates and dislocation velocities. As discussed in Section 2.5.3, the formation energy of jogs upon repulsive dislocation intersections induces a moderate strain hardening at low temperatures. A local rule could be drawn from the available data, but this was not deemed necessary till now.

Only simplified phenomenological rules can be employed for treating the various interactions of dislocations with grain boundaries. In large grain-sized polycrystals, the assumption that grain boundaries are impenetrable to dislocations may not hold in alloys exhibiting planar slip (Section 2.7.2.1). Then, devising local rules for dislocation emission and absorption by grain boundaries is simply a matter of ingenuity (see Section 5.5.4).

4.4 Boundary conditions

This section discusses successively the method employed for implementing periodic boundary conditions and the most commonly employed method for implementing boundary conditions in materials with finite sizes. Other methods were also developed for this last purpose.

4.4.1 Periodic boundary conditions

Periodic boundary conditions are employed for investigating representative volume elements embedded into large crystals or grains of polycrystals. The traditional Born–von Karman periodic boundary conditions were implemented in the first 2D in-plane simulations (Lépinoux and Kubin, 1987; Ghoniem and Amodeo, 1988). An example drawn from the study of vortex lines in ^{4}He shows that, although they present some artificial features, these conditions can be extended to three dimensional lines (Schwarz, 1988). This fact was ignored at the time of the first 3D DD simulations, in which dislocations were left to freely escape at external surfaces. Alternatively, methods were devised to deal with the significant loss in density, but they were rather unsatisfactory. Periodic boundary conditions were considered as inapplicable to three-dimensional dislocation microstructures because it was thought that they would not preserve the continuity of the lines. It became clear that this was not a problem when Bulatov *et al.* (2001*b*) proposed a method for implementing them in three dimensions.

This method is not always compatible with mechanical boundary conditions solved with the help of FE codes, for instance in the case of samples that should be infinite in one direction and not in the others. This problem arises because periodic boundary conditions require periodicity everywhere, whereas in some FE codes periodicity is only imposed at mesh boundaries.

The basic principle of the method consists in allowing dislocations to glide in a periodic supercell formed by a primary simulation cell and its images. Every time a line element crosses a boundary between two cells, which it does simultaneously in all cells, an image re-enters the opposite boundary at an equivalent position. The example given in Fig. 4.6 shows that the continuity of lines is not destroyed. For clarity, this figure is drawn in two dimensions, but the considerations developed here apply to three dimensions.

We consider first the central cell drawn in thick black lines in Fig. 4.6 and its periodic images. The translation vectors that generate the images of cells and loops are called modulo translations. They are of the form

$$\mathbf{T} = uL\mathbf{x} + vL\mathbf{y}\ (+wL\mathbf{z}), \tag{4.8}$$

where the coefficients u, v and w are integers. Since the origin of the infinite supercell is arbitrary, we consider a random cell represented by dashed lines in Fig. 4.6. This cell and its images obviously generate the same array of loops as the central cell, although the configurations are different in each cell. One can observe that the thick dislocation lines inside the random cell are simply a folded picture of a whole loop. In addition, one can verify that the incoming and outgoing dislocation fluxes are balanced since each portion of line having left the cell is exactly compensated

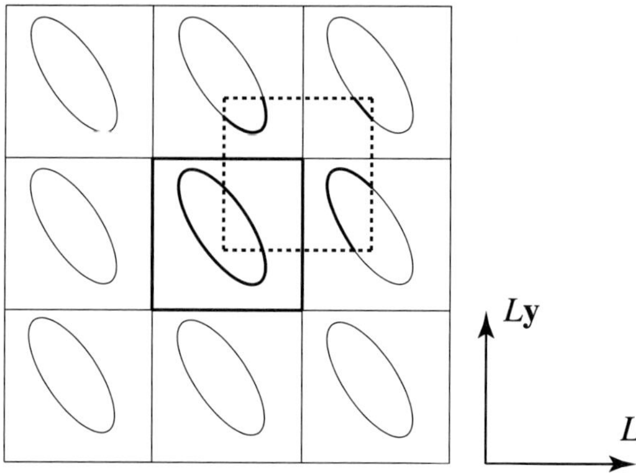

FIG. 4.6. A two-dimensional periodic array of square cells with linear dimension L containing each a single loop. All cells and all loops are displaced with respect to each other by translations which are integer multiples of the elementary translations $L\mathbf{x}$ and $L\mathbf{y}$. The dashed cell is placed at an arbitrary position and contains a folded image of a whole loop (see text).

by an identical image that entered it on the other side. As these lines continuously cross the dashed boundaries, the latter no longer play any role in the whole supercell. Thus, there is no surface effect.

The dislocation configurations found in the dashed cell or in any other arbitrarily positioned cell can be used as initial configurations. As discussed by Bulatov *et al.* (2001*b*), geometrically necessary dislocations induce curvatures that are incompatible with translational periodicity and break mechanical equilibrium. As a consequence, initial configurations should only contain configurations with equilibrated Burgers vectors like glide or prismatic loops and infinite dipoles. In practice, local statistical unbalances in the sign of the Burgers vectors are tolerated.

To estimate the interactions of dislocations with their images, the supercell is centred on the point where all image fields have to be summed. The interactions cannot be computed one by one, as this would entail a prohibitive number of arithmetic operations. The method developed for ParaDiS by Cai *et al.* (2003) applies to DD as well as atomistic simulations. In both cases, the summation of an infinite array of periodic image fields is only conditionally convergent. In other words, the obtained result depends on the way the terms are grouped to calculate the sum. A solution to this problem is derived from the observation that conditional convergence is related to periodicity breaking. For dislocations, the total image stresses on infinitesimal line elements of the primary cell are calculated exactly and the values are incorporated in tabulated form. In the *MDDP* and *PDD* codes, the fast multipole method is employed to estimate the long-range interactions, this time on a finite number of images.

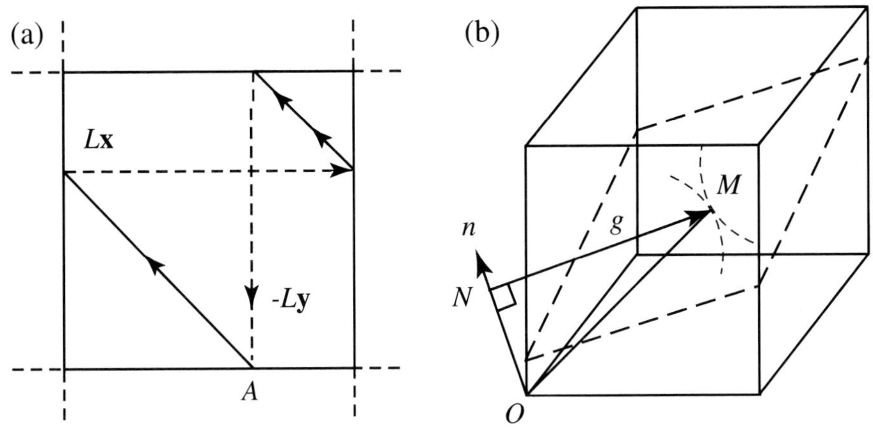

FIG. 4.7. (a) Example of a short glide path along a high symmetry direction parallel to the second diagonal of a square cell. The folded path undergoes two modulo translations, $L\mathbf{x}$ and $-L\mathbf{y}$, and closes up on itself. (b) Self-annihilation in three dimensions. The cell can be orthorhombic or cubic (see text). After Madec *et al.* 2004.

Periodic boundary conditions induce a well-known artefact, which consists in spurious annihilations that limit the glide path of dislocations. As the array of cells is periodic, the dislocation microstructure is also periodic in a single slip plane, albeit with a different period. Thus, along its glide path, an element of an expanding loop necessarily encounters an image element of the same loop with opposite sign and the two mutually annihilate. Fig. 4.7*a* illustrates this undesirable phenomenon in two dimensions. A folded glide path parallel to the second diagonal starts in *A*, crosses two boundaries and re-enters twice in the cell, at which point is has returned to its initial position. Since there is another, symmetrical, closed path going in the opposite sense, an annihilation must necessarily occur.

In the same way, one can easily verify that a dislocation loop or dipole expanding in a {001} face of a cubic cell is self-annihilated after one boundary crossing. In three dimensions premature annihilations are found in low-index planes, which are also glide planes (see Fig. 4.8). Hence, it turns out that periodic boundary conditions are not so well adapted to planar dislocation glide.

At this point there are two options. Bulatov *et al.* (2001*b*) suggest taking primary cells as large as possible (20 μm) as well as high initial densities (10^{12} m^{-2}), which statistically reduce the normal glide paths with respect to annihilation glide paths. Cross-slip may help extending the glide paths of screw dislocations, especially at temperatures where it is no longer thermally activated.

The second option is implemented in *microMegas* and *MobiDiC*. It consists in modifying the shape of the simulation cell to change the modulo translations and shift re-entries into the initial slip plane to a larger number of boundary

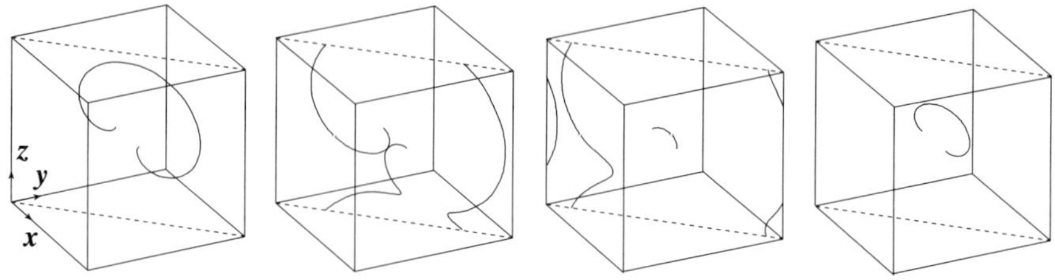

FIG. 4.8. A Frank–Read source expands from left to right in a bcc crystal. The Burgers vector of the source segment is $\frac{1}{2}$ $[1\bar{1}1]$ and its slip plane is (110). After one rotation, all emitted lines are annihilated by images of opposite sign. Courtesy R. Madec.

crossings. A simple method consists in using orthorhombic cells with mutually incommensurate lengths along the [001] axes. The optimal solution is selected from an infinite set of discrete solutions for three integer numbers (see Madec *et al.* 2004).

The starting point is as follows. Figure 4.7*b* shows a point M of an expanding loop in a slip plane with normal **n**. The origin of the vector **OM** can be defined in an arbitrary manner and is taken at the origin of the cell. This vector has a component **g** in the glide plane and a component **ON** along its normal. An index i allows reckoning the height of the slip planes with respect to the origin. It is defined by $i = \mathbf{OM.n}$ and we assume $i \neq 0$, that is, there is no annihilation after a single boundary crossing. When self-annihilation occurs at M, the line elements underwent the same total modulo translation **T**, which is given by eqn. 4.8. As the two elements are in the same slip plane, they must have same index. In consequence, a necessary condition for obtaining annihilations is $\mathbf{T.n} = i$ or, in developed form

$$huL_x + kvL_y + lwL_z = i, \tag{4.9}$$

where (L_x,L_y,L_z) are the cell dimensions, which can differ from each other, (u,v,w) are the numbers of boundary crossings and (h,k,l) are the components of **n**. From this condition, one can derive relations giving the length of the path for first annihilation and the corresponding cell dimensions.

Another artefact arises with too long annihilation paths. In substance, a large number of boundary crossings means that a large number of closely spaced parallel planes are visited. This enhances the probably for artificially forming narrow dipoles. In optimal large cells, the three dimensions differ by about 10%; this can be seen in Fig. 4.4, where the projected cells are clearly rectangular. To obtain realistic microstructures, the primary cell must also fulfill the conditions suggested by Bulatov *et al.* (2001*b*). In such conditions, and for a reasonably isotropic microstructure, the mean travelled path in the unfolded plane can be more than one magnitude larger than the average linear dimension of the primary orthorhombic cell. An example describing how to calculate an optimal solution for prismatic slip in zirconium crystals is given by Monnet *et al.* (2004). In that case, the optimization is performed for both slowly moving screw segments and fast-moving non-screw segments.

For particular slip geometries, the modulo translations are in the slip planes ($i = 0$) and spurious annihilations occur after the first boundary crossing. This occurs not only in {001} planes, but also, as illustrated by Fig. 4.8, in {011} slip planes in bcc crystals.

In bcc metals, the most convenient solution for avoiding annihilations on the six {110} planes consists in rotating in an adequate manner the faces common to two adjacent cells around their normal (Madec *et al.*, 2004). This procedure can also be optimized.

This extension of periodic boundary conditions allows obtaining, in addition to microstructures in the primary cell, microstructures with large periodicities in the unfolded planes. It is implemented in *microMegas* and *MobiDiC*. Examples are given in Figs. 1.9, 2.15, 5.5*b* and 5.11.

4.4.2 Finite boundary conditions

DD simulations performed on volumes of finite size require the implementation of mechanical equilibrium in the volume for given conditions applied at the external surfaces. A method based on the superposition principle was developed by Van der Giessen and Needleman (1995) from an earlier work by Lubarda *et al.* (1993). The solutions are three-dimensional and were applied to 2D simulations. The bases of the method are presented in this section; the calculations that go along with it are not reproduced here. The extension to three dimensions is outlined next.

A simulated volume of finite dimensions is assumed to contain dislocations, as well as elastic phases like precipitates, fibres and other types of reinforcements. This volume is not in mechanical equilibrium since the long-range fields of dislocations extent to the surface. The boundary value problem is solved in linear elasticity, in the usual approximation of small strains (Section C.2.2). The basic idea consists in determining a correction field that once superposed to the initial field establishes the conditions for mechanical equilibrium while preserving the continuity of internal boundaries.

The current state of the body under stress and without correction is depicted in Fig. 4.9*a*. It is submitted at its external surface to time-dependent tractions and displacements, respectively $\mathbf{T} = \mathbf{T}_o(t)$ and $\mathbf{u} = \mathbf{u}_o(t)$. In the classical continuum formulation, these two conditions have to be applied to distinct surface elements, respectively S_f and S_u. The stress, strain and displacement fields are denoted $\boldsymbol{\sigma}$, $\boldsymbol{\epsilon}$ and $\mathbf{u}$. With the help of the superposition principle, they are decomposed as a sum of two fields (Fig. 4.9).

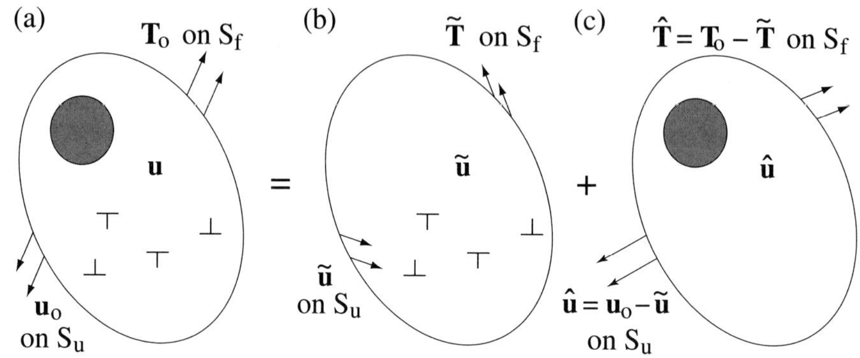

FIG. 4.9. The superposition principle applied to finite boundary conditions (see text for notations). The body shown in (a) contains dislocations and a precipitate (in grey); it is not in mechanical equilibrium. Its fields are decomposed into a sum of (b) the dislocation (˜) fields in an infinite solid, and (c) the complementary (ˆ) fields that enforce boundary conditions for dislocations in the presence of precipitates. Adapted from Van der Giessen and Needleman 1995, courtesy of E. Van der Giessen.

The solution searched for is, therefore, given by

$$\mathbf{u} = \tilde{\mathbf{u}} + \hat{\mathbf{u}}; \quad \boldsymbol{\epsilon} = \tilde{\boldsymbol{\epsilon}} + \hat{\boldsymbol{\epsilon}}; \quad \boldsymbol{\sigma} = \tilde{\boldsymbol{\sigma}} + \hat{\boldsymbol{\sigma}} \quad \text{in } V \tag{4.10}$$

The (˜) and (ˆ) fields at the right-hand sides of eqns. 4.10 are obtained from the conditions of mechanical equilibrium with adequate boundary conditions. The (˜) fields are the sum of the fields of all dislocation lines in the matrix. They have to be non-singular in the volume and must be known in analytical form; these two conditions are met in DD simulations. The (ˆ) fields are derived for a linear isotropic solid. The whole problem is in general solved with the help of FE techniques.

The application to 3D DD simulations is much less straightforward. The shapes of dislocations lines close to external boundaries have to be treated as accurately as possible in order to compute the (ˆ) fields. The major issue consists, however, in treating the singular field of a dislocation line intersecting the surface. A first solution to this problem was given by Weygand *et al.* (2002). Surface sliding nodes are introduced on lines intersecting the surface. A portion of loop that remains inside the volume after the sliding node has emerged at a surface is connected by two new surface nodes to a virtual portion of loop located outside the volume. This procedure brings again the loop in closed form. The shape of the virtual line is such that it approximates the mirror image of the real part of the loop.

Obtaining a good accuracy requires a complex fine-tuning of the FE code and the nodal description to avoid overlap of nodes and integration points of the FE mesh. Validation tests on dislocation loops and Frank–Read sources are presented.

This solution is also implemented in the *Tridis*, *MDDP* and *PDD* codes. In the *PDD* code, the boundary value problem is solved by a boundary element, or integral, method (Beer *et al.*, 2008). This numerical technique solves linear partial differential equations which have been transformed into integral equations applicable over the boundary. It is suited for treating fields in linear homogeneous media and is often more efficient than FE methods when the ratio of surface to volume is small. Chapter 5 presents several examples of application of the superposition principle to finite boundary conditions.

A hybrid DD and FE solution based on the superposition principle was presented by Tang *et al.* (2006). It is based on a known analytical solution for the stress field of a semi-infinite segment. When such a segment emerges at a surface, an image segment is constructed and the surface becomes traction-free. The resulting stress field is analytical. Emerging segments are in general not straight and the semi-infinite segment is positioned in such a way as to have same tangent at the surface as the considered curved segment. Use is made of the superposition principle to obtain the final solution. The traction-free stress field for the straight segment and its image is calculated by the DD code. A correcting stress field accounts for the difference between this stress field and that of the curved segment. This second field is smooth and located beneath the boundary; it is computed by a FE code. As the latter does not have to deal with a singularity, the convergence of the solutions is obtained faster than in the previous solution and coarser meshes can be used. A extension to multiple free surfaces is provided and validations are performed on edge dislocations in a free-standing film and on a half-loop below the free surface of a semi-infinite half-space. This method is implemented in *ParaDiS*.

4.4.3 Other methods for finite sizes

The article by Tang *et al.* (2006) contains a short historical review providing references on various methods utilized to implement boundary conditions in solids with finite sizes. Other references are given in Section 4.5.

Several methods were initially developed for *Tridis* (Verdier *et al.*, 1998). They are replaced by the superposition method for studies of microindentation (Chang *et al.*, 2010; Fivel *et al.*, 2011). A method for calculating images stresses of three-dimensional segments without recourse to the FE method was proposed by Khraishi *et al.* (2001) for implementation in the *MDDP* code. It combines the image stress of the segments, to eliminate shear stresses, with a distribution a small prismatic loops to eliminate normal stresses. In *microMegas*, the boundary value problem is solved using the revisited Continuous–Discrete model (Section 4.5). In this hybrid model a DD simulation is coupled to a FE code. The evolution of the dislocation microstructure is computed by the DD simulation. The long-range mechanical fields, which take boundary conditions into account, are computed by a FE code. The coupling method is based on the eigenstrain theory.

A calculation of images forces performed by Liu and Schwarz (2005) showed that they essentially affect the dislocation shapes at short distances. Advantage was taken of this result to devise a local procedure that reproduces the dislocation shapes near a planar free surface with a very good accuracy (in the range of a few percent). This procedure is implemented in *PARANOID* for studies on thin films and in *microMegas* with or without finite boundary conditions. An efficient algorithm for computing image forces in free-standing films is also discussed by Weinberger *et al.* (2009). It can be implemented in any DD simulation. A good agreement is obtained with the predictions of dynamic and static atomistic simulations, as well as with those of a simpler line tension model.

4.5 Current 3D DD simulations

This section provides references to publications and internet sites where information can be found about the most widespread 3D DD simulation codes and a highly sophisticated 2D simulation.

All lattice-based simulations derive from the same initial code where dislocation lines were discretized into screw and edge segments (Kubin *et al.*, 1992).

- *Tridis*, for 3D Discrete Dislocation Dynamics, is an updated version of the initial edge-screw simulation, which was developed by Verdier *et al.* (1998) and further parallelized (Shin *et al.*, 2006). The boundary conditions are based on the superposition principle (Shin *et al.*, 2005). *Tridis* is dedicated to studies of the mechanical response of fcc and bcc metals and alloys. Downloadable videos and a full list of references on the simulation and its applications can be found in the *Tridis* site (Fivel, 2012). This site is also intended to provide information about a new nodal code, *NuMoDis*.
- *microMegas*, after Voltaire's philosophic tale (1752), is a free software under the terms of the GNU License. It can be downloaded from a site containing other items (Devincre, 2012*b*). Periodic boundary conditions and the treatment of their artefacts (Madec *et al.*, 2004) are included by default. Finite boundary conditions are implemented by a model coupling DD simulations to a FE code, the Discrete–Continuous model (Lemarchand *et al.*, 2001; Vattré *et al.*, 2012). The website contains, in particular, an article describing the simulation (Devincre *et al.*, 2011) and a few other recent publications. Videos of elementary dislocation processes and applications of *microMegas* can be downloaded from another link (Devincre, 2012*a*). The present version of the code treats materials with cubic symmetry, like fcc and bcc metals, plus other structures like hcp (Monnet *et al.*, 2004) and a few model minerals of the earth mantle (Durinck *et al.*, 2007; Amodeo *et al.*, 2011).

- *MobiDiC* is a code derived from *microMegas*, which is dedicated to bcc metals in the athermal temperature regime (R. Madec *et al.*, to be published). It incorporates {110} and {112} slip systems.

Nodal simulations are listed in chronological order of the initial 3D code.

- *MDDP* (Multiscale Dislocation Dynamics Plasticity) is a hybrid code combining a FE code with a DD simulation named *micro3D*, which is based on a previously developed code (Yasin *et al.*, 2001) and on the original one (Zbib *et al.*, 1998; Rhee *et al.*, 1998). The FE code solves the boundary conditions at external and internal interfaces, by means of the superposition principle (Yasin *et al.*, 2001; Zbib and Khraischi, 2008). The FE code also solves the equation of motion of the dislocations, which incorporates inertia and a heat conduction equation taking into account the energy dissipated during plastic flow. *MDDP* was adapted to investigations of shock loading (Section 5.4.5). Other examples of applications are discussed by Yasin *et al.* (2001) and Zbib and Khraischi (2008). Video clips, written information and instructions for downloading codes are provided on a website (Zbib, 2012).
- *PARANOID* (PARAnoid NOdal IBM Dislocation code) is a code dedicated to simulations of dislocations in confined semi-conducting and metallic materials. The initial version was developed by Schwarz (1999) and is complemented by two updates. One describes local rules for dealing with dislocation junctions and multi-junctions (Schwarz, 2003*b*) and the other one presents a treatment of image forces on dislocations intersecting a free surface (Liu and Schwarz, 2005). Cross-slip is not included. A chapter reviewing the simulation method and its applications is also available (Schwarz, 2007).
- *PDD* (Parametric Dislocation Dynamics). The peculiarity of this simulation code is to discretize dislocation loops or curved segments into a succession of parametrized cubic splines (Ghoniem *et al.*, 2000). A diffusion model for dislocation climb is implemented. Further developments include an anisotropic treatment of dislocation fields (Han *et al.*, 2003), the parallelization of the code (Wang *et al.*, 2006*b*), the implementation of boundary conditions based on the superposition principle, which are solved by a boundary element method (El-Awady *et al.*, 2008; Takahashi and Ghoniem, 2008), the implementation of cross-slip (Wang *et al.*, 2007*b*) and an adaptation to high strain rate deformation (Wang *et al.*, 2007*a*). An advanced simulation code named *MODEL* is in preparation; it is intended to be open source and provides a library of available building blocks (see the site Ghoniem 2012).
- The simulation code developed by Weygand *et al.* (2002) incorporates the first 3D extension of the 2D superposition principle. Further development involves the treatment of junctions (Weygand and Gumbsch, 2005) and

the parallelization of the code (Weygand *et al.*, 2009). This simulation is dedicated to the plasticity of small-scale materials and related size effects.

- *ParaDiS* (Parallel Dislocation Simulator), is a dislocation dynamics code designed for running on massively parallel computing systems. The parallelization method is detailed by Bulatov *et al.* 2004. The objective is to treat collective dislocation processes involving large numbers of discretized segments under stress. The simulation code is described in a book by Bulatov and Cai (2006) in this series and by Arsenlis *et al.* (2007) for an evolved version. Further publications discuss a new solution for implementing traction-free surfaces in large-scale systems (Tang *et al.*, 2006) and in free-standing films (Weinberger *et al.*, 2009). A public version of the code, which can be set to simulate bcc and fcc crystals, is available on the *ParaDiS* site (Arsenlis *et al.*, 2012). It includes linear mobility laws for glide and climb, periodic boundary conditions without correction of artefacts, or free surfaces, and a simple rule for cross-slip without barrier in bcc crystals. Users are free to implement their own mobility laws.

  Another site (Cai, 2012) provides case studies and proposes exercise problems to be solved by *ParaDiS* or *DDLab*. *DDLab*, which is available on this site, is a collection of MATLAB scripts that serves as a much simplified version of *ParaDiS* able to treat elementary configurations of moving and interacting dislocations.

  A simulation code deriving from *ParaDiS* incorporates dislocation dissociation in fcc metals, using SFEs fitted to atomistic data (Martinez *et al.*, 2008). The local rule for cross-slip at 0 K is based on an atomistic calculation of Escaig stresses (Section 2.4.4).

The 2D simulation code developed by Mohles is dedicated to the study of precipitate strengthening and makes use of an accurate 3D treatment of the self-stresses. The simulation method is described in two publications (Mohles 2001, 2004*b*) and a few representative applications are discussed in Section 5.3.4. A website (Mohles 2012) provides more detail on the method and additional references to its applications, and displays downloadable video clips of the simulations.

5

APPLICATIONS OF DD SIMULATIONS

5.1 Outline

The objective of this chapter is to provide an overview of major problems that are examined by DD simulations and to place the results in perspective with respect to current issues and debates. Owing to the wealth of available literature, the focus is on investigations which lead to improved understanding, are considered of fundamental importance or have a potential for further development. In these studies, use is made of large-scale or model simulations with various boundary conditions. As a rule, the simulations discussed here are three-dimensional, except when explicitly mentioned otherwise. The contents of this chapter are classified into four topics.

- Dislocation intersections, mainly in fcc metals, in relation with dislocation strengthening. This section is essentially involving elastic processes.
- Strengthening by solutes, radiation defects and precipitates, in model materials and some industrial alloys. These typical multiscale problems involve connections with atomic-scale mechanisms and the properties of bulk materials.
- Collective dislocation processes, that is, dislocation avalanches and their consequences regarding dislocation-based modelling, dislocation patterns in monotonic deformation and potentially important studies on cyclic deformation and on deformation at high strain rates.
- Size effects in small-scale materials are first briefly reviewed and illustrated by a few examples of extrinsic effects. This section details the size effects in thins films and small-scale pillars, which are the subject of most DD simulations on intrinsic size effects.

5.2 Dislocation intersections

When dislocation interactions govern plastic flow, the average stresses for unzipping junctions constitute the largest contribution to the flow stress. The strength of dislocation interactions is usually described by an average coefficient α (eqn. 2.20). The interaction coefficients entering the generalized relation between flow stress and dislocation densities (eqn. 2.23) are more useful. They pave the way for generalized dislocation density-based models at the scale of slip systems, like the storage–recovery models (Sections 2.6.7 and 5.4.2).

In what follows, use is made of large-scale simulations and several types of model simulations on fcc and bcc metals. The first sections are devoted to investigations of individual intersection processes. This rather simple exercise shows that, in addition to junctions, three other types of configurations contribute to a smaller extent to the mechanical response. The following sections are dealing with the respective strengths of the interactions between non-coplanar slip systems and their consequences.

5.2.1 Intersections and reactions

5.2.1.1 Orientation mappings Depending upon their initial orientation, two interacting dislocation lines that glide in non-coplanar slip systems may repel or attract each other. In the last case, they may or may not react upon crossing each other. DD simulations allow predicting the full orientation dependence of the final configurations that are obtained in every case. For this purpose, the simplest initial configurations consist of two straight, finite or infinite, dislocation lines positioned at a very short slip distance of their potential intersection point. The lines are then allowed to relax under the sole effect of their mutual elastic interactions (Fig. 2.5). The procedure used for depicting the results is due to Püschl (1985). The initial orientations of the lines are reckoned by the angles ϕ_1 and ϕ_2 they make with the line direction of the incipient junction (see e.g., Fig. 2.6). Rotating the initial lines in their slip planes allows constructing orientation mappings for forest interactions in the parameter plane (ϕ_1, ϕ_2).

Figure 5.1*a* shows an orientation mapping for the interaction of two $\frac{1}{2}\langle 111\rangle\{110\}$ infinite lines that form $\langle 001\rangle$ junctions in bcc metals. Figure 5.1*b* shows a similar mapping for two $\frac{1}{2}\langle 110\rangle\{111\}$ slip systems that form Lomer locks in fcc crystals. The initial lines are segments of length 1 μm pinned at their ends, which cross each other at their midpoints.

The repeat distances of these mappings are given by the translations $(\phi_1, \phi_2) \rightarrow (\phi_1 \pm \pi, \phi_2 \pm \pi)$. Periodicities along the diagonals correspond to angle increments of π that change the signs of the interacting lines. The simulations results are almost generic, as they weakly depend on the Poisson ratio.

In Fig. 5.1, the domain of stable junctions is made up of a central lobe and its duplicates. The shape of the lobes is a consequence of the orientation dependence of the line energy. Their centres correspond to junctions of maximum strength formed on parent lines parallel to the junction direction. There are also repulsive states which, however, do not fill up the entire space between the lobes. At the border between junction lobes and repulsive interactions, one finds marginally stable attractive states that form very short junctions. There are also attractive

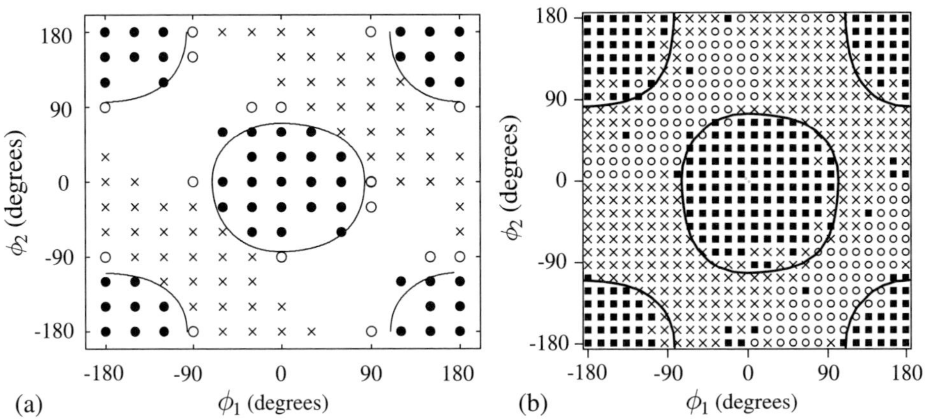

FIG. 5.1. Orientation mappings for dislocation reactions under zero external stress. The interacting lines make angles ϕ_1 and ϕ_2 with the junction direction. The thin lines show the boundaries for stable junction formation, as calculated from the change in total line energy (eqns. 1.4 and 1.5). (a) ⟨001⟩ junctions in bcc metals. The symbols indicate the nature of the final configurations, stable junction (•), marginal junction with very small length (∘) and bound crossed state (×). Repulsive interactions are found in empty spaces. From Wickham *et al.* 1999, courtesy K.W. Schwarz. (b) Lomer lock mapping for copper showing stable junctions (■), crossed states and marginal junctions (×) and repulsive states (∘). From Kubin *et al.* 2003.

states such that the junction is reduced to a point-like pinning point. These configurations, which were not previously mentioned in the literature, were called *bound crossed states* by Wickham *et al.* (1999). They occur when the quadruple node at the intersection of the parent segments is in equilibrium as soon as it is formed, which prevents the zipping of a junction. These crossed states are found in attractive areas of the mappings such that one parent segment at least has to be substantially distorted in its slip plane to line up along the direction of the incipient junction. The distortions involves curvatures and line tension forces that are strong enough to counteract the resolved attractive forces between the parent segments. In simulations carried out with segments of finite length, the domains of junctions and crossed states slightly expand when the segments length decreases. This is due to the logarithmic decrease of the outer cut-off radius in the line energy. Under an external stress, the junction lobes shrink because the weakest junctions become unstable (Wickham *et al.*, 1999).

5.2.1.2 Colinear interactions The colinear interaction is the interaction between dislocations of same Burgers vector gliding in a slip system and its cross-slip system (Fig. 2.3*d*). This interaction is found occasionally in the literature under the name of exchange interaction (Jackson, 1985) or reactive blocking

(Stach *et al.*, 2000). Attempts to measure its strength by latent hardening tests yielded contradictory results (Section 2.5.4). Orientation mappings clearly show that this interaction is by far the one leading to the most stable reaction (Madec *et al.*, 2003).

The two lines drawn in Fig. 5.2*a* glide in colinear slip systems and are about two cross each other. They have Burgers vectors of opposite sign, which are parallel to the intersection direction of their two planes. Figures 5.2*b* and 5.2*c* show the partial annihilation of the two parent segments, which is equivalent to the formation of a junction of null Burgers vector. Two new segments are formed, which are composed of two short portions of the parent lines, each in its own slip plane and with same Burgers vector, connected by a double node. The equilibrium configuration is reached when the forces on the double nodes are perpendicular to the intersection line, that is, when the short segments meet this line at right angles. Due to the small lengths of these segments, the critical stress to remobilize them is expected to be much larger than that of junctions. The process described in Fig. 5.2 was observed by *in situ* TEM by M.J. Mills (unpublished result, see SOM of Madec *et al.* 2003).

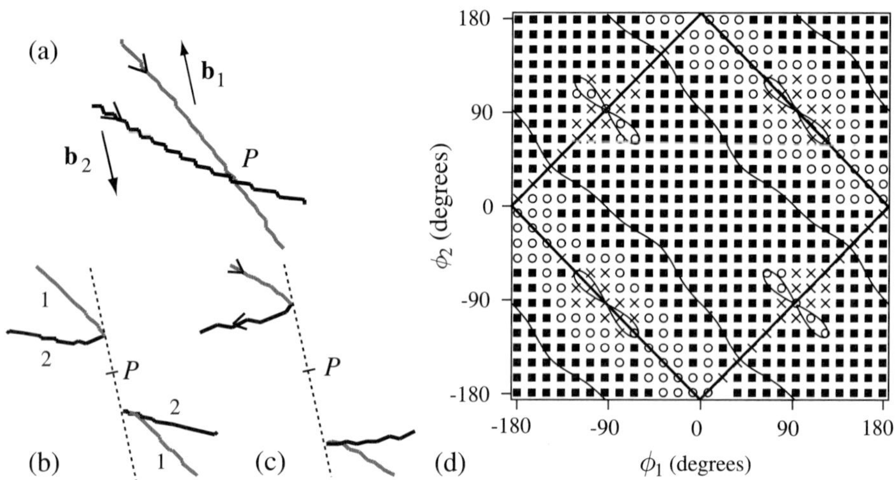

FIG. 5.2. Colinear annihilations in fcc crystals. (a) Two initially straight segments with Burgers vectors $\mathbf{b_1}$ and $\mathbf{b_2} = -\mathbf{b_1}$ are positioned close to their intersection point P. They have a length of 1 μm and are pinned at their ends. (b) During relaxation, annihilation starts at P and propagates along the intersection of the two slip planes. (c) The two nodes have reached their equilibrium position. (d) Orientation mapping for the colinear interaction. The symbols indicate the domains of annihilations (■), non-reacting attractive configurations (×) and repulsive states (○). The thick and thin lines correspond to boundaries predicted by isotropic elasticity calculations (see text). From Madec *et al.* 2003.

The strength of the colinear interaction is illustrated by Fig. 5.2*d*. The domain of colinear annihilations covers about 70% of the mapping, which is about twice more than the domain of Lomer lock formation (Fig. 5.1*b*). The lines drawn on this figure are calculated for straight segments using isotropic elasticity. The thick lines are the loci of configurations that are already in equilibrium when the two segments make contact; they indicate the presence of a narrow domain of crossed states. The condition for null interaction between the two initial lines is calculated using eqn. 5-35 of Hirth and Lothe (1992), which was derived by Kroupa (1961). The result is represented by the thin lines, which approximately separate attractive and repulsive domains. The domain of annihilations extends into the repulsive domain for the following reason. At short approach distances, the lines bend and twist in their slip planes in such a way that some parts experience attractive forces. In some cases, this is sufficient for triggering annihilation. By increasing the length of the segments from 1 μm to 30 μm, one obtains an almost perfect prediction for the domain of colinear annihilations. As with the junction lobes, the increase in the outer cut-off radius of the line energy makes the dislocations less flexible. Conversely, the fraction of repulsive segments that are annihilated becomes even more substantial when their lengths decrease well below 1 μm.

5.2.1.3 Other mappings A set of orientation mappings for dislocation reactions in fcc and bcc crystals was presented by Kubin *et al.* (2003). One can draw from these mappings a qualitative insight into the strength of the final configurations. For example, colinear annihilations constitute, in average, much stronger obstacles than Lomer locks. The Hirth locks, exhibit the smallest lobe and are, indeed, the weakest junctions. The reason is that they only draw their stability from the orientation dependence of the line energy.

The possible types of interactions and junctions configurations in bcc crystals were revisited by DD simulations (Madec and Kubin, 2004), taking into account slip on {110} and {112} planes. Each of the four Burgers vectors is at the intersection of three {110} or {112} planes, which leads to a wealth of possible configurations. From an energetic viewpoint and taking crystal symmetries into account, there are seventeen fundamental types of interactions, of which twelve are junction-forming reactions. The latter can be split into three categories, sessile junctions akin to Lomer locks, glissile junctions and junction configurations such that all segments are glissile in the same slip plane. An anomalous slip activity on such a plane is often observed in group-V bcc metals at low temperatures (Section 3.2.1).

Reactions that contribute to forest hardening were also examined in zirconium for the interaction between non-coplanar prismatic slip systems (Monnet *et al.*, 2004), in ice (Devincre, 2013), in magnesium with a threshold stress (Capolungo, 2011) and in two geological materials, MgO (Carrez *et al.*, 2005) and olivine (Durinck *et al.*, 2007).

5.2.1.4 Multi-junctions The density of junction lines formed during plastic flow between non-coplanar slip systems is quite significant. For instance, it amounts to almost 30% of the total density in fcc crystals (Devincre *et al.*, 2008). Therefore, mobile dislocations have a non-negligible probability of interacting with already existing junctions, which may have consequences on strain hardening (Section 5.2.2.2). In bcc molybdenum at high temperature, Bulatov *et al.* (2006) investigated by large-scale DD simulations an interesting configurations formed when three non-coplanar slip systems are simultaneously active. This configuration, which is depicted in Fig. 5.3*b*, was called a *multi-junction*.

In bcc metals, the screw direction exhibits a ternary symmetry. For example, the screw direction [111] is at the intersection of three slip planes, $(\bar{1}01)$, $(0\bar{1}1)$ and $(\bar{1}10)$. The attractive interaction between segments with Burgers vectors $\frac{1}{2}[111]$ and $\frac{1}{2}[1\bar{1}\bar{1}]$ gliding in the first two slip planes produces a binary junction with Burgers vector [100] and line direction [111]. By adding a segment with Burgers vector $\frac{1}{2}[\bar{1}\bar{1}1]$ in the third slip plane, $(\bar{1}10)$, one obtains the reaction

$$\frac{1}{2}[111] + \frac{1}{2}[1\bar{1}\bar{1}] + \frac{1}{2}[\bar{1}\bar{1}1] \rightarrow \frac{1}{2}[1\bar{1}1] \tag{5.1}$$

The reaction product is glissile; it can bow and expand in one of the parent slip planes, here $(\bar{1}01)$. Thus, when a mobile segment intersects a non-coplanar binary

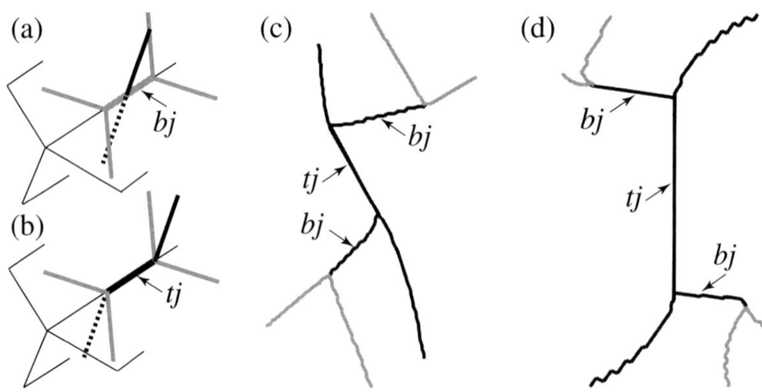

FIG. 5.3. Configurations of ternary junctions (see text for detail). (a) In a bcc crystal, two initial parent segments (grey lines) have formed a binary junction *bj*. A non-coplanar segment (black line) interacts with *bj*. (b) The reaction zips an axial ternary junction *tj*. (c, d) DD simulations of zigzag ternary configurations (from Madec and Kubin 2008). (c) In copper, the slip plane of the incoming segment does not contain the direction of the glissile junction *bj*. The two segments form a ternary junction *tj* (a Lomer lock), along a direction that is not colinear to *bj*. (d) Same mechanism in bcc Ta. The binary junction is glissile and the ternary junction, of mixed character, has a Burgers vector $\frac{1}{2}\langle 111\rangle$.

junction, an axial ternary junction is formed provided that the Burgers vectors of the three parent lines are different. This reaction leads to a new junction that zips completely (Fig. 5.3*b*), forming two quadruple nodes with the three parent segments. It efficiently minimizes the total line energy because the Burgers vector of the ternary junction, $\frac{1}{2}\langle 111\rangle$, is smaller than that of the binary junction, $\langle 001\rangle$. As a consequence, this type of reaction is more stable and should constitute a stronger obstacle to dislocation glide than usual binary junctions. This characteristic axial configuration was reproduced by MD simulations and identified by TEM in thin films extracted from a deformed $\langle 001\rangle$ molybdenum crystal (Bulatov *et al.*, 2006). Higher order configurations can also be considered, but their probability of occurrence decreases with increasing order.

The various types of ternary junction configurations formed in bcc and fcc crystals were examined in some detail by Madec and Kubin (2008). In fcc crystals, the $\langle 011\rangle$ screw directions contain only two non-coplanar $\{111\}$ slip systems, which prevents the occurrence of axial ternary configurations by the mechanism depicted in Fig. 5.3*b*. A simple axial configuration involving two slip planes and three Burgers vectors was nevertheless described by Schwarz (2003*b*). Another type of ternary configuration can be formed with a third dislocation moving in a non-coplanar slip system, provided that the binary junction is a glissile junction. The latter can bow towards the incoming segment and intersect it. The line direction of the binary junction is not colinear to that of the ternary junction, which is common to the slip planes of the incoming segment and of the glissile junction. Thus, the ternary junction cannot zip completely and a portion of the binary junction remains in the final configuration (Fig. 5.3*c*). We call this configuration a ternary zigzag configuration. Such configurations can occur in fcc crystals during stable deformation along $\langle 111\rangle$ and $\langle 001\rangle$ tensile axes. If cross-slip systems are activated, additional zigzag configurations are formed and axial configurations can also be obtained.

In bcc metals, taking into account $\{110\}$ and $\{112\}$ slip, 1,416 triplets of slip systems lead to the formation of axial configurations and 2,880 lead to the formation of zigzag configurations like that one shown in Fig. 5.3*d*. Considering only the slip systems that may be activated at high temperature during stable multislip deformation along [001] and [011] tensile axes, the ternary configurations can be classified into sixteen geometrically different groups (Madec and Kubin, 2008). Ten of them correspond to zigzag configurations and the other six to axial configurations.

5.2.2 The interaction coefficients

The strength of junctions was calculated analytically by several authors and DD simulations yielded the average strength α of a dislocation forest (Section 2.5.2). Upon moving up in scales, the first step consists in estimating the average strength of the six possible interactions between slip systems in fcc crystals (Fig. 2.3). In this section, all interaction strengths are determined for fcc metals, taking copper as a model material, and the state of the art regarding bcc metals is discussed. The next modelling step is discussed in Section 5.4.2.

5.2.2.1 Interaction strengths in fcc metals Reliable values for the interaction coefficients between non-coplanar slip systems can be obtained by devising model simulations from which the coefficients can be estimated one by one. In such simulations, a mobile density moves under stress through a random, constant three-dimensional density of forest dislocations ρ_f, with zero Schmid factor, with which it interacts by forming a given type of junction (Fig. 5.4). Under a constant imposed strain rate, the resolved critical stress on the primary system reaches a plateau value, τ_c, which is related to the strength of the investigated coefficient by $a^{1/2} = \tau_c/\mu b \rho_f^{1/2}$. Glissile junctions cannot bow under stress in these simulations; they have anyway zero Schmid factors. As 3D behaviour has to be reproduced, the implementation of such model simulations requires some care (Devincre *et al.*, 2006; Queyreau *et al.*, 2009).

Whereas the configuration shown in Fig. 5.4*a* looks familiar, the massive annihilations by colinear interactions in Fig. 5.4*b* lead to a strikingly different type of microstructure. As expected from Fig. 5.2*c*, the latter consists of small segments connected by double nodes. The interaction strengths between non-coplanar slip systems drawn from model simulations are shown in Fig. 5.5*a*.

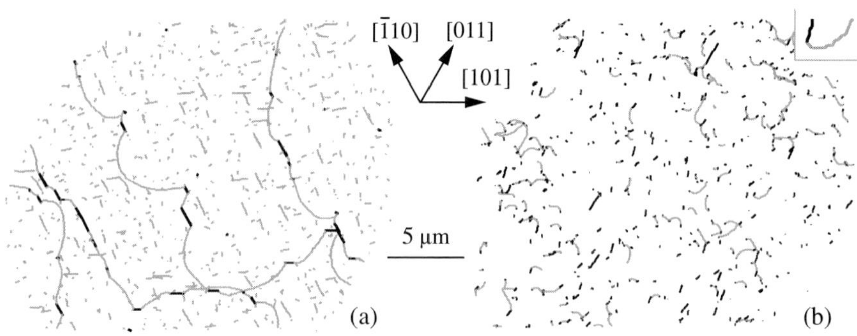

FIG. 5.4. Measurement of interaction strengths in fcc metals. Thin films of thickness 0.1 μm extracted from model 3D DD simulations. The forest density is 10^{12} m^{-2}. (a) Formation of Lomer locks (short straight black segments) by interaction of the $\frac{1}{2}[011](11\bar{1})$ primary slip system (long dark lines) with two forest slip systems, $\frac{1}{2}[110](\bar{1}11)$ and $\frac{1}{2}[\bar{1}01](111)$, seen as short grey lines. (b) Colinear interactions and annihilations. The same primary slip system interacts with a dislocation forest of its cross-slip system $\frac{1}{2}[011](1\bar{1}1)$. The inset (top right) shows a magnified view of one of the numerous double nodes left by colinear annihilations. After Madec *et al.* 2003.

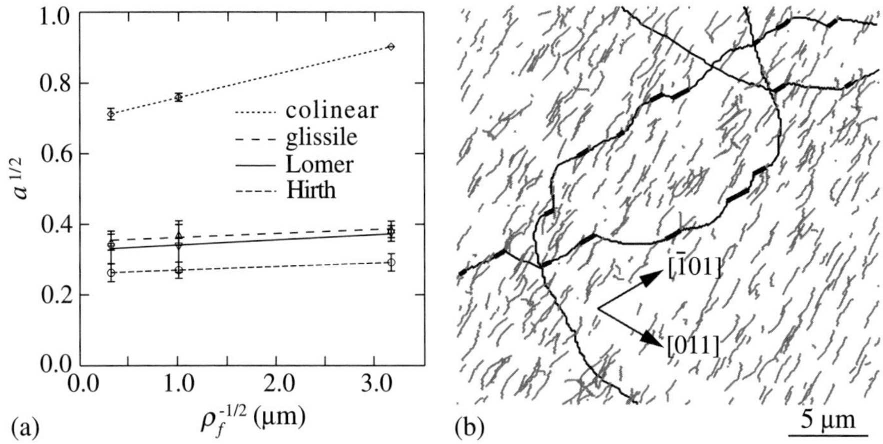

FIG. 5.5. (a) Measurements of interaction strengths ($a^{1/2}$) for forest dislocation densities ρ_f of $10^{11}, 10^{12}$ and $10^{13}\,\mathrm{m}^{-2}$ (updated from Devincre *et al.* 2006). (b) Model DD simulation of the strength of the ternary zigzag configuration shown in Fig. 5.3*c*. The primary system (long lines) is $\frac{1}{2}[011](1\bar{1}1)$ and the forest (short lines) is made up of glissile binary junctions. The ternary junctions (short black lines) are Lomer locks parallel to [$\bar{1}$01]. From Madec and Kubin 2008.

The interaction coefficients logarithmically decrease with increasing forest density (Fig. 5.5*a*), like the scalar interaction strength α (eqn. 2.21). This variation is particularly marked for the colinear interaction, which can be rationalized by noticing that the outer cut-off radius R_e of the critically bowed out configuration is inversely proportional to stress (eqn. 1.11), hence to $(a\rho)^{1/2}$. In contrast to previous calculations, the output of such simulations incorporate all the types of binary contact interactions mentioned in the previous sections and do not assume that junction configurations are symmetrical. Long-range interactions between all dislocation segments are also included. The most reliable values obtained for the interaction coefficients are given in Table 5.1. For comparison, this table also gives experimental values drawn from latent hardening tests on germanium crystals deformed in compression at 850 °C. The pitfalls of such tests (Section 2.5.4) were avoided by using the model by Alexander and Haasen (Section 3.4.5) to directly derive the coefficient values.

The colinear interaction is by far the strongest one and the interaction leading to the formation of Hirth locks is the weakest one. The interactions leading to the formation of Lomer locks and glissile junctions have similar strengths. Elastic calculations on symmetric (Dupuy and Fivel, 2002) and asymmetric (Picu and Soare, 2010) junction configurations yield opposite results regarding which type of junctions is the most stable.

TABLE 5.1. Interaction coefficients (a) and interaction strengths ($a^{1/2}$) for non-coplanar interacting slip systems in fcc crystals (after Kubin *et al.* 2008). The first two lines recall the conventional names of the secondary systems (cf. Section C.3) and the reaction products. The nomenclature of the coefficients derives from latent hardening tests. The values are given for a reference forest density $\rho_{fo} = 10^{12}\,\mathrm{m}^{-2}$; the uncertainty on the strengths is ±5%. The last line reproduces experimental results on Ge crystals by Alexander and Crawford (2000).

Slip system Reaction	critical Hirth lock	conjugate Lomer lock	no name glissile junction	cross-slip annihilation
Coefficient	a_1	a_3	a_2	a_{col}
a	0.07	0.122	0.137	0.625
$a^{1/2}$	0.26	0.35	0.37	0.79
$a_{Ge}^{1/2}$	0.24	0.37	0.31, 0.37	0.64

The self-interaction is responsible for strain hardening in stage I. It is not of the coplanar type, as it involves reactions between colinear jogs and primary dislocations (Section 2.6.2). DD simulations (Devincre *et al.*, 2007) and modelling (Kubin *et al.*, 2008) suggest that its strength is comparable to that of the Lomer lock. The microstructures formed during double coplanar slip are practically unknown; the similarities between the mechanical responses in stage I and during double coplanar slip suggest the occurrence of a similar strengthening mechanism. In tension, the $\langle 001 \rangle$ orientation is the only one for which colinear double slip is observed (Section 5.2.2.3) and for which a Hirth lock is formed (Section C.3). Hence, for the vast majority of single crystal orientations, the average strengths of coplanar and non-coplanar interactions exhibit reference values close to the scalar value $\alpha \approx 0.35$. In alloys, the interaction strengths can be modified by the coupling of forest strengthening with solute or particle strengthening (Section 5.3.1).

The model simulations discussed above can also serve to determine the strength of multi-junctions (Madec and Kubin, 2008). This is particularly easy in fcc crystals because the zigzag ternary configurations involve the usual types of junctions and are formed on glissile junctions. As a consequence, the obtained strengths are, at best, only slightly larger than that of a Lomer lock or a glissile junction. For the ternary configuration depicted in Fig. 5.5*b*, the interaction strength is 0.38, whereas the strength of the glissile junction is $a_2^{1/2} = 0.37$. Thus, ternary configurations induce practically no forest strengthening in fcc crystals. They nevertheless contribute to strain hardening, actually in a moderate manner, because they store three parent segments instead of two for binary junctions (Section 5.4.2).

5.2.2.2 Interaction strengths in bcc metals In bcc metals, the coefficients for forest interactions during {110} slip were estimated (Madec and Kubin, 2004; Queyreau *et al.*, 2009) and the full interaction matrix for {110} and {112} slip systems was recently determined for junctions and ternary configurations (R. Madec, to be published). The junction strengths are globally smaller than in fcc metals, as the Burgers vector of the $\langle 100\rangle$ junctions is larger than that of the parent segments, $\frac{1}{2}\langle 111\rangle$. The strengths of the colinear interactions do not differ much from the value found in fcc crystals. For ternary junctions, the situation is quite contrasted because of the multiplicity of possible configurations. For instance, the axial and zigzag configurations have maximum strengths of respectively 0.5 and 0.58 and minimum strengths around 0.3. Hence, taking into account the enhanced storage rate, strain hardening in multiple slip conditions may be larger or smaller than in fcc crystals depending upon the orientation considered and the type of loading.

5.2.2.3 Colinear double slip The high strength of the colinear interaction entails many consequences on microstructures and the mechanical response (see e.g., Section 2.6.2). We discuss here a major one, the selection of active slip systems for the high-symmetry loading axes $\langle 111\rangle$ and $\langle 001\rangle$ in fcc crystals. These orientations are stable in tension with, respectively, six and eight slip systems having same Schmid factor. Each of these slip systems is the cross-slip system of another one. These two orientations play an important role in the deformation of polycrystals because they serve as references for grains deforming in multislip conditions.

In contrast to what is frequently assumed, the number of active slip systems in $\langle 111\rangle$ crystals is only three. The inhibition of the missing slip systems can be understood as a consequence of the high strength of the colinear interaction. With a [111] loading axis, stable deformation is obtained by the symmetrical activation of three slip systems such that the sum of their Burgers vectors is [111] (see e.g., eqn. 5.1). There is, thus, no double colinear slip, which reduces the critical stress. Symmetrical triple slip producing glissile junctions occurs in two crystallographically equivalent variants (Fig. C.5). It was identified from slip trace analyses by Franciosi and Zaoui (1982). One or the other configuration is also obtained in large-scale DD simulations, usually with some unbalance between the most active slip systems and a small activity in the other ones (Fig. 5.6*a*).

Slip trace analyses on deformed [001] crystals are made difficult by the activation of cross-slip at low stresses. According to Honeycombe (1968), there is no less than four slip systems. The eight slip systems are in average active to the same extent. Locally, there seems to be traces of two slip planes and colinear double slip is activated (Franciosi and Zaoui 1982, Vorbrugg *et al.* 1971). Large-scale DD simulations carried out for this orientation yield various blends of the available experimental observations depending upon the initial configuration (Fig. 5.6*b*).

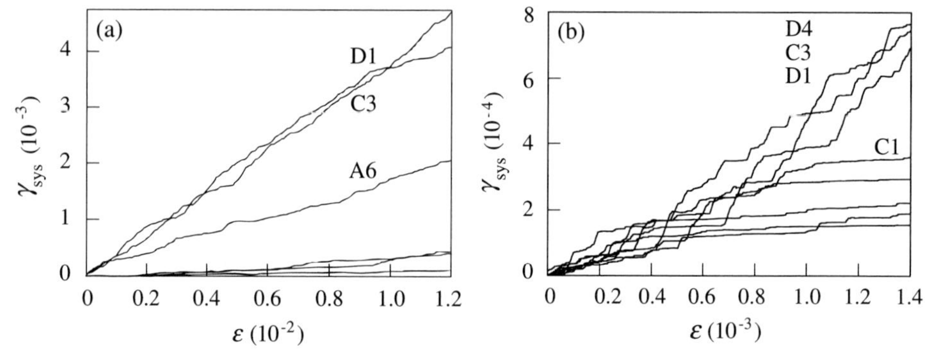

FIG. 5.6. Large-scale simulations of tensile deformation along high-symmetry orientations in fcc copper. The shear strain in each slip system is plotted as a function of the total axial strain. The notations for the slip systems are those of Schmid and Boas (Table C.1). (a) For ⟨111⟩, three non-colinear slip systems carry most of the deformation from the beginning of plastic flow. Their cross-slip systems are weakly active. (b) Typical slip geometry for [001] at small strains. The four more active slip systems correspond to two active slip planes and one set of double colinear slip. Four other slips systems are also activated to a lesser extent (Courtesy B. Devincre).

The activation of colinear slip in [001] crystals was observed in model simulations devised for the study of single slip under various loading axes (Section 2.6.2). Low-stress annihilations by cross-slip are replaced by a small initial density of edge colinear segments on the primary dislocation lines, with both signs in equal numbers (Devincre *et al.*, 2007). Fig. 5.7*a* shows a superjog that is extended by the motion of a primary segment, to which it is connected by a double node. When the lengths of the two segments become identical, the double node is pinned by two forces of same amplitude and opposite directions. Depending upon the local stresses, one or the other segment expands and multiply. This mechanism produces dislocation sources in the cross-slip plane.

The mechanism leading to colinear slip is rather complicated (see Devincre *et al.* 2007 for a detailed explanation). Whereas resistive superjogs like that shown in Fig. 5.7*a* can expand and multiply in the cross-slip plane, the superjogs with opposite sign are dragged along by primary lines. The successive intersections of primary attractive segments tends, however, to preferentially produce resistive superjogs and to reduce the density of non-resistive ones. As a result, the colinear density increases at the expense of the primary density until the two densities become equal to half the total density (Fig. 5.7*b*). This mechanism only operates within a few degrees from the [001] orientation. With increasing misorientation, the Schmid factor on the cross-slip plane decreases and the resistive superjogs lose their ability to multiply (Fig. 5.7*b*). Along the [$\bar{1}$11] direction, the magnitude of

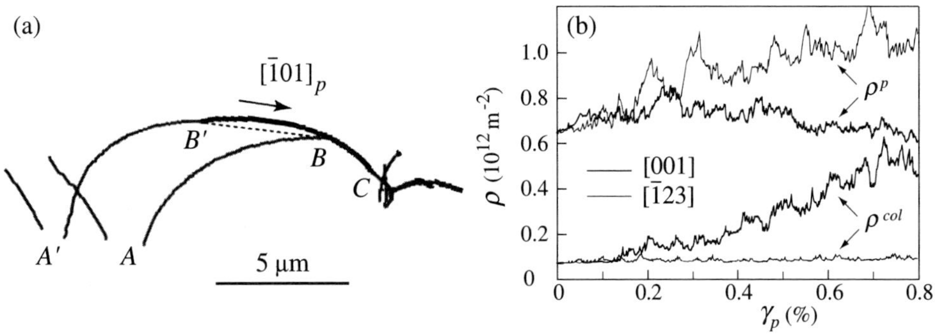

FIG. 5.7. Model simulations of primary $\frac{1}{2}[\bar{1}01](111)$ dislocations interacting with colinear superjogs. (a) Two superimposed configurations extracted from a simulation with [001] stress axis. The colinear superjog *BC* (black line) is connected to a primary line *AB*. Under stress, the double node *B* moves to *B′* along the projected direction of the primary Burgers vector and extends the colinear segment. At the next simulation step, the colinear segment *B′C* reaches a critical configuration and expands irreversibly in the cross-slip plane. (b) Evolution of the dislocation densities for two loading orientations. For $[\bar{1}23]$ the colinear density (ρ^{col}) remains stable. For the [001] orientation, colinear double slip sets in and the colinear density (thick curve) tends towards the primary density ρ^p (after Devincre *et al.* 2007).

the two Schmid factors is identical again, but the Schmid factor in the cross-slip plane has changed sign. The production of non-resistive superjogs is favoured and double colinear slip is not activated (Fig. 5.6*a*). Thus, this mechanism enforces the activation of coplanar double slip near the [001] loading axis. It also explains why a significant increase in strain hardening is experimentally recorded within a few degrees of the exact [001] orientation (Takeuchi, 1976).

More generally, the activation of colinear double slip is suppressed in all crystal structures, except for specific orientations where a density of superjogs with suitable sign can develop. In addition, symmetrical colinear double slip is necessarily activated under stress states such that the slip and cross-slip systems have the two highest Schmid factors. In fcc crystals, this case occurs under biaxial stress conditions (Figs. 5.14 and 5.17*a*).

5.3 Atomic-scale defects, precipitation strengthening

5.3.1 Dislocations and solute atoms

This section reviews DD simulations where moving dislocations interact with solute atoms. Such a problem is relevant of the atomic-scale and cannot be treated thoroughly by mesoscopic simulations. The KMC method provides an efficient means for connecting the two scales. For instance, a KMC scheme was used for estimating the velocity of a screw dislocation moving by the kink-pair mechanism in the presence of diffusing solutes (Deo *et al.* 2005*a*; Deo *et al.* 2005*b*, see Section 3.2.4.2). In addition, two studies on the influence of solute atoms on junction strength and the yield stress of fcc materials showed the occurrence of significant effects.

In the work by Chen *et al.* (2008*a*), solutes are described as point sources of expansion characterized by their size and diffusivity. A KMC simulation is coupled to a DD simulation in order to investigate how diffusing solutes modify the strength of a Lomer lock. For immobile solutes that have already segregated on the parent lines and the junction, the critical unzipping stress increases with solute size and concentration in the dislocation cores. However, for the same configuration, the critical stress significantly decreases when the junction is not yet formed in the initial condition. When the solute atoms diffuse along the lines during the unzipping process, their redistribution along moving parent arms with varying character induces a strengthening effect.

In a second study (Monnet and Devincre, 2006), immobile solute atoms are globally treated as inducing an alloy friction stress on dislocations, τ_F. It is shown that alloy friction modifies the equilibrium position of the triple nodes, thereby reducing the length of junctions and their strength. Model simulations confirm that the quantity μ/τ_F governs the decrease of the interaction coefficient of a single lock. Although this simulation and the previous one make use of different approaches to the interaction between dislocations and solute atoms, both of them show that these interactions are not simply additive.

So far, no attempt was made to revisit the classical elastic models for solute strengthening (Neuhäuser and Schwink, 1993; Argon, 2008), where junctions are treated as point obstacles. It could be interesting to make use of the wealth of information coming from atomistic simulations, to set local rules and test the connections with elastic string models and similar approaches (see, among many other publications, Rodary *et al.* 2004; Marian and Caro 2006; Proville 2009; Patinet and Proville 2011).

5.3.2 Dislocations and irradiation defects

The interactions between clusters of point defects and dislocations are of particular interest in the case in materials irradiated by high-energy particles. There is a strong pressure to develop a multiscale modelling framework for the post-irradiation deformation of nuclear materials (Samaras and Victoria, 2008). The challenges facing DD simulations are quite formidable (Stoller and Mansur, 2005), for the reasons mentioned in the previous section, and because of the multiplicity of small-scale interactions that have to be treated (Bacon *et al.*, 2009). Furthermore, post-irradiation deformation, which was studied until now, does not result in the same mechanical response as *in situ* deformation under irradiation (Singh *et al.*, 2007).

Irradiation in reactors by neutrons of energy larger than 0.1 MeV results in collision cascades, which produce a variety of point defect clusters like stacking fault tetrahedra (SFTs), vacancy loops, clusters of self-interstitial atoms (SIAs) and microvoids. With increasing doses, the post-irradiation mechanical response is characterized by a strong increase of the yield stress, an initial yield point and the occurrence of a Lüders-like behaviour (Section 2.7.2.1). According to the cascade-induced source hardening (CISH) model, the yield point reflects the unpinning of source segments from interstitial loops that are attracted by their stress field. These loops are located at a short distance of the lines but are not absorbed by them (Trinkaus *et al.* 1997*a*, 1997*b*, Singh *et al.* 1997). After the yield point, the removal of irradiation defects by their successive interactions with moving dislocations results in the formation of clear channels. This behaviour is typical of materials containing a microstructure that is destabilized by dislocation motion (Luft, 1991). It induces strain localizations and reduces the ductility (see Victoria *et al.* 2000 and Singh *et al.* 2002 for experimental examples).

The first set of DD simulations on these phenomena involved simplified local interaction rules inspired by current models, MD or KMC simulations. Glide or phenomenological climb rules were introduced in order to account for the small thickness (20–200 nm) of the channels. On these bases, the main features mentioned above could be reproduced in a generic manner (see for instance Diaz de la Rubia *et al.* 2000; Khraishi *et al.* 2002; Ghoniem *et al.* 2001; Huang and Ghoniem 2002; Singh *et al.* 2002).

Further DD simulations emphasize more accurate descriptions of the interactions between dislocations and point defect clusters. They complement comparisons between MD simulations and elastic models (Bacon *et al.*, 2009) and can treat dissociated dislocations (Marian *et al.*, 2009). The interaction of perfect dislocations with small interstitial loops is illustrated by Fig. 5.8.

The CISH model was examined by Rong *et al.* (2005) using the 2D simulation of Mohles (2001). Figure 5.8*a* shows the critical stress for moving an edge dislocation through the field of a row of polygonal, sessile interstitial loops of side *d*, located at a short distance below its glide plane. The geometric parameters are the effective size of equivalent circular loops, D_{eff}, their spacing *L* and their distance from the glide plane, *H*. These results, which follow well the predictions of elastic models, also confirm the simple treatment used in the CISH model, which is itself in agreement with experiment. Figure 5.8*b*, after Nogaret *et al.* (2008), shows the initiation of a clear channel by successive screws segments moving through a random 3D distribution of Frank loops. These loops are unfaulted by interaction with the moving segments and, as shown by MD simulations, take the Burgers vector of the primary dislocations. They can then be treated as interstitial prismatic loops. Upon contact, a moving screw dislocation absorbs the unfaulted

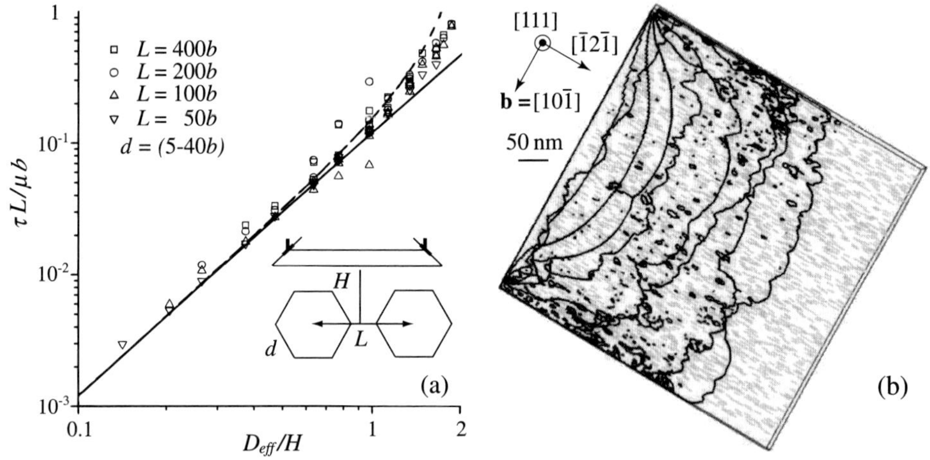

FIG. 5.8. Interactions of dislocations with interstitial loops in DD simulations. (a) Unpinning of a dislocation from a row of interstitial loops located below its slip plane. The geometric parameters are defined in the text; their values are given in the inset. Variation of the scaled critical unpinning stress τ with D_{eff}/H. The two lines show the predictions of two different elastic models. After Rong *et al.* 2005, courtesy D. J. Bacon. (b) Initiation of a clear band in a small copper platelet. Screw dislocations emitted by a source move across a distribution of Frank loops (grey contrasts) and unfault them (black loops). The last three emitted segments bow smoothly, as that they are gliding in an almost clear region. After Nogaret *et al.* 2008, courtesy T. Nogaret.

loops and forms three-dimensional helical turns. During further evolution, a screw segment is re-emitted into a new slip plane above the previous one, while the jogs are moved aside (cf. Fig. 5.8*b*). These processes lead to the progressive clearing of a band with increasing thickness, without any help from cross-slip or climb processes. Other discrete or collective mechanisms, like the effect of pile-up stresses, are also analysed in detail. The configurations obtained in this study realistically reproduce experimental observations. Hence, valuable information can be drawn from DD simulations provided that they incorporate a few rules derived from MD simulations.

5.3.3 Dislocation climb

In several 3D DD simulations climb is phenomenologically introduced through a mobility coefficient, which allows to move dislocations in a conservative manner in their climb plane (Section 4.3.2.1). As recalled by Bakó *et al.* (2011), more elaborated diffusion-based models were also proposed in the past for use in DD simulations. Two recent models and their first applications are outlined in this section.

Mordehai *et al.* (2008, 2009) implemented a physical approach to dislocation climb in a 3D lattice-based DD simulation. A continuous concentration of diffusing vacancies is considered. The climb rate arising from vacancy diffusion in the bulk is estimated by solving the second of Fick's equations, from which the concentration field in the presence of dislocations is derived. In this first step towards the future modelling of creep processes, several applications to elementary mechanisms are presented (Bardeen–Herring sources, shrinkage and coarsening of prismatic loops), in which dislocation lines climb but do not glide.

Dislocation glide was further introduced by Bakó *et al.* (2011) in a 3D nodal DD simulation, using a climb diffusion model similar to the one proposed by Mordehai *et al.* Glide is described by a mobility law for free-flight dislocation motion governed by phonon damping, for which the characteristic time is ten orders of magnitude smaller than the one associated with vacancy diffusion. Use is made of an adiabatic approximation that allows slip events to reach equilibrium between two slow climb steps. The coarsening kinetics of a small assembly of prismatic vacancy loops in aluminium was investigated in a quantitative manner, first by climb only and next by glide and climb. Glide allows the loops to move in their glide cylinders while they climb; when two attractive loops meet at the intersection of their glide cylinders, they can coalesce. As a result, dislocation glide is found to significantly increase the coarsening kinetics.

These two simulations are primarily intended to investigate the interactions of dislocations with irradiation defects. Although they require some fine tuning, in particular to treat the climb velocity of straight screw dislocation segments, the preliminary results are more than promising. When this becomes possible, it would also be interesting to revisit the rather ancient models for power-law creep in bulk materials. In every case, cross-slip of screw dislocations could also be taken into account.

5.3.4 Precipitation strengthening

Many models, which are mostly based on elasticity theory, were developed over the years for treating the various interactions between dislocations and precipitates. They are comprehensively discussed by Nembach (1996) and Argon (2008). In parallel with a similar work by Pretorius and Nembach (2001, 2004), Mohles and co-workers conducted a systematic investigation of precipitation strengthening in fcc alloys by DD simulations (see Section 4.5 for the method and its applications). In these simulations an initially straight dislocation, perfect or dissociated, moves by planar slip through a three-dimensional distribution of precipitates. The dislocation–precipitate interactions are modelled in 3D. The emphasis is on an accurate treatment of the dislocation self-stresses (Section 4.2.5.2) and the removal

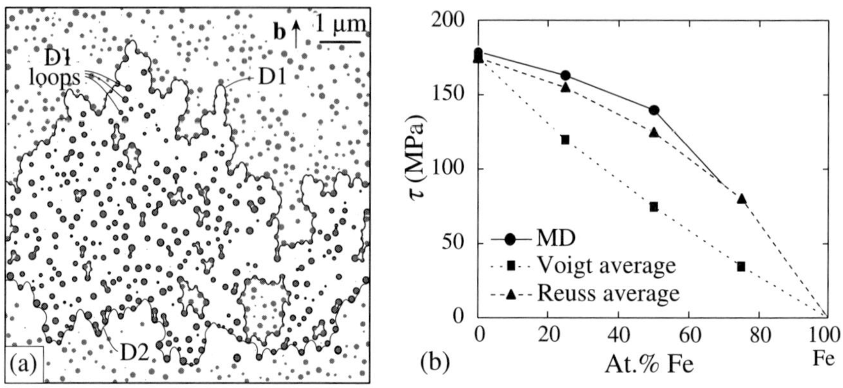

FIG. 5.9. (a) Equilibrium shape of a dissociated edge dislocation shearing ordered precipitates in a strongly overaged Nimonic superalloy. The Orowan loops left by the leading dislocation D_1 repel the trailing dislocation D2. From Mohles 2004*a*, courtesy V. Mohles. (b) Critical resolved stress for shearing nanometric Fe-Cu complexes (diameter 2.5 nm) as a function or iron content. MD simulation results (Kohler *et al.*, 2005) are compared to elastic calculations using the Voigt and Reuss isotropic averages. After Takahashi and Ghoniem 2008, courtesy A. Takahashi.

of adjustable parameters. In many (but not all) cases, the existing analytical models are verified or require small corrections.

Two studies deserve a particular mention, specifically the numerical derivation of a composition law for dispersion and size-mismatch strengthening (Mohles, 2003) and the investigation of a Ni-based superalloy with a small volume fraction of long-range ordered precipitates (Mohles 2004*a*, see also Pretorius and Nembach 2001). In this last investigation, which is illustrated by Fig. 5.9*a*, dislocation dissociation is taken into account and the yield stress is estimated as a function of the antiphase boundary energy (0.25 mJ m^{-2}), the mean particle radius and the volume fraction of particles. As was shown by Douin *et al.* (2007), dissociation may also influence the motion of dislocations in the matrix phase of $L1_2$ γ-γ' superalloys with high volume fractions of γ' precipitates.

In subsequent 3D DD simulations, use was made of the available methods for imposing rigorous boundary conditions and thermally activated cross-slip was sometimes implemented (Shin *et al.*, 2003; Vattré *et al.*, 2010). Khraishi *et al.* (2004) investigated the strengthening induced by a particle–matrix mismatch for a small concentration of spherical particles in a metal–matrix composite. Takahashi and Ghoniem (2008) examined the two mechanisms of Orowan looping and precipitate shearing (Fig. 5.9*b*). Boundary conditions based on the superposition principle were used in both studies. Numerical solutions were obtained with the help of a FE method in the former case and by a more flexible boundary element method in the latter one.

Several 3D DD simulations of elementary dislocation–precipitate interactions were further carried out in conditions quite similar to those used by Mohles. Rao *et al.* (2004) simulated in great detail the yield stress of model Ni_3Al-based superalloys with a medium-high volume fraction (40%) of the γ' phase. A pair of screw or mixed dislocations of same sign were sent from an edge of the simulation and sheared the precipitates or bypassed them by the Orowan bowing mechanism. The CRSS was examined as a function of a number of parameters like the APB energy, the precipitate size and shape, the coherency strain, the volume fraction (from 0.1 to 0.4) etc. A further study by Vattré *et al.* (2009), carried out at temperatures up to 1000 K extended the previous results to volume fractions between 40 and 70%. Particular attention was paid to the shearing of the γ' cuboids.

The nickel-based γ-γ' superalloys with high volume fraction of cuboidal γ' precipitates ($\approx 70\%$) draw their exceptional high-temperature properties from the confinement of dislocations in narrow (30–80 nm) channels of the γ phase, as well as from the anomalous strengthening of the long-range ordered γ' phase with increasing temperature (see Veyssière and Saada 1996 for a full review). To understand the orientation dependence of strengthening in superalloys, Vattré *et al.* (2010) carried out large-scale simulations at a temperature of 1123 K. The boundary conditions were solved using the Discrete–Continuous Model (Section 4.4.3). The objective was to explain why the [001] orientation exhibits a higher tensile yield stress and strain hardening rate than the [111] orientation in spite of a lower Schmid factor. The strengthening anisotropy was found to mainly result from differences in the dislocation densities stored at the γ-γ' interfaces. In addition, the mechanism leading to the so-called cube slip on {001} planes parallel to the interfaces was elucidated. Cube slip only occurs in [111] crystals and softens them; it produces coarse localized slip traces at specimens surfaces but cannot be detected by TEM observations (Bettge and Österle, 1999). It was identified as a composite slip process involving two interacting colinear {111} slip systems. These results were further incorporated into a micromechanical model for the strengthening anisotropy of superalloys (Vattré and Fedelich, 2011). In a second large-scale simulation, with periodic boundary conditions, Huang *et al.* (2012) revisited the behaviour of [001] superalloy crystals at high temperature.

5.4 Collective dislocation processes

5.4.1 Intermittency and avalanches

In this domain, the objectives of DD simulations consist in determining the dislocation mechanisms that govern pattern formation and dislocation avalanches in order to guide further dislocation-based modelling.

The detrimental effect of plastic instabilities on the ductility of metallic alloys was examined at the scale of dislocation avalanches by Csikor *et al.* (2007), to reveal the effect of large strain fluctuations on the shapes of deformed microspecimens. DD simulations were performed in various deformation conditions in a model aluminium material with linear dimensions in the micron range. These simulations make use of boundary conditions based on the superposition principle. The initial microstructure consists of a high density of Frank–Read source segments ($9\times10^{13}\,\mathrm{m}^{-2}$) pinned at their ends. The observed dislocation avalanches appear as crystallographic lamellar strain localizations with a fractal dimension close to two. The largest ones sweep the whole cross section of the simulated specimen.

The probability distributions for scaled avalanche strains reproduced in Fig. 5.10*a* include previous results by Dimiduk *et al.* (2006*b*). All distributions collapse into a single curve when taking into account that shear strains are proportional to b/L, where L is the specimen height (Section 2.7.2.3), and to a factor $(\Gamma + \Theta)/E$ that controls the amount of stress relaxation during avalanches. Γ is the effective stiffness of the simulated driving system, E the Young's modulus of the tested material and Θ is the continuous strain hardening rate. This scaling, which exhibits a cut-off value, was derived by Zaiser and Nikitas (2007) from a generic continuum model. The collapsed curves exhibit a stress exponent $\eta_s \approx 1.5$, even for bent specimens that contain geometrically necessary dislocations. These first DD results on dislocation avalanches illustrate the robustness of the critical size exponent.

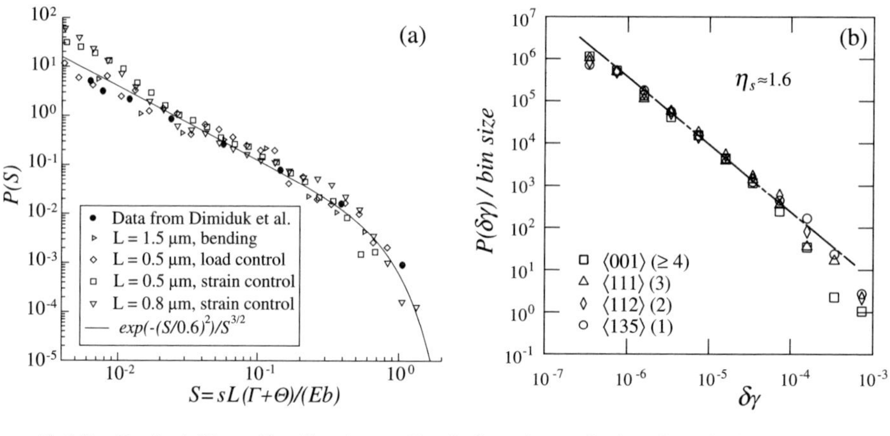

FIG. 5.10. Probability distributions P of the size of simulated dislocation avalanches. (a) Micron-sized crystals under various deformation conditions (inset). S is a scaled avalanche size (see text) such that all results collapse into the generic form given in the inset. From Csikor *et al.* 2007, courtesy D. Weygand. (b) Shear strain burst amplitudes in large-scale simulations of crystals strained in tension along various orientations. The number of active slip systems is given in parentheses. From Devincre *et al.* 2008.

In another study (Devincre *et al.*, 2008), large-scale DD simulations with periodic boundary conditions were used to investigate the specific mechanisms involved in dislocation avalanches. The linear dimensions of the primary cell were adjusted around a value of 5 μm in order to obtain dislocation glide paths of 50–100 μm (Section 4.4.1). The initial microstructure contained prismatic dislocation loops in density $2 \times 10^{12}\,m^{-2}$. Figure 5.10*b* shows typical probability distributions of shear strain bursts obtained in copper single crystals deformed in strain control along orientations favouring double and multiple slip, as well as single slip. The scaling exponent is $\eta_s \approx 1.6$.

Figure 5.11 shows an avalanche and a few aftershocks extracted from such simulations. The avalanche is initiated by a single primary segment (1), which is pinned by junctions at its ends. Once this segment reaches a critical shape (3), the junctions are unzipped, first at left and then at right, and the line expands until position (5). The area swept is about 85 μm² and an equivalent area is further swept until configuration (11) is reached. Junctions are particularly numerous along this last configuration, as it has reached regions where the forest density is higher than average. There are also a few gaps along the primary lines at places where cross-slip occurred. The strain deduced from the total swept area matches the total recorded strain. Hence, there is no overlap with other avalanches. The

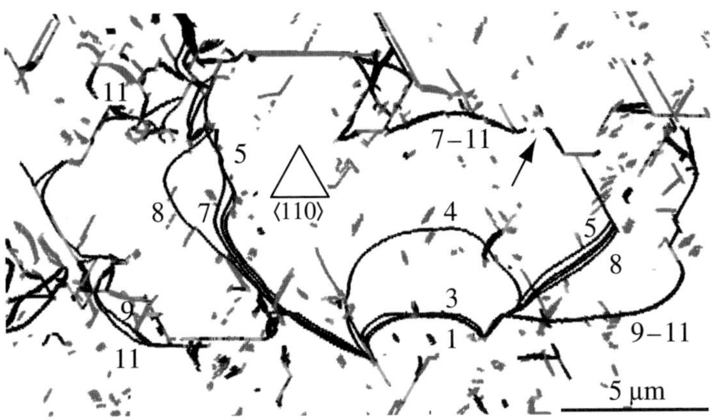

FIG. 5.11. A large avalanche on the [011](1$\bar{1}$1) slip system of a deformed [001] crystal, as seen in a thin (1$\bar{1}$1) film of thickness 0.25 μm extracted from a DD simulation. The expansion of the initially pinned segment (1) is materialized by eleven superimposed configurations taken at constant time intervals of 0.2 μs. The contour of the expanding avalanche is shown in black. The light grey lines are forest dislocations and junctions formed between primary and forest dislocations. The arrow points at a gap in the expanding line caused by a local cross-slip event. From Devincre *et al.* 2008.

moving segment expands under a stress of 5 MPa with an average velocity of about $4\,\mathrm{m\,s^{-1}}$, which peaks at about $25\,\mathrm{m\,s^{-1}}$ in almost dislocation-free areas and is about $80\,\mathrm{m\,s^{-1}}$ in dislocation-free areas.

These observations show that junction processes govern the initiation and termination of dislocation avalanches, which seems natural in a material where forest intersections govern plastic flow. This result applies as well to crystals deformed in the easy glide stage I, where colinear superjogs produced by screw dislocation annihilations play the role of forest segments (Section 2.6.2). Thus, the critical configurations consist of critically bowed segments and the spatio-temporal fluctuations at the origin of intermittency arise from the distributions of effective stresses on such segments and their evolution. Avalanches are arrested by interaction with large fluctuations of the stored density and the latter increases with the number of active slip systems. For instance, in deformation conditions with n symmetrically active slip systems ($n \geq 2$), the forest density associated with each slip system increases like $(n - 1)/n$ if one neglects the density of junctions (cf. eqn. 5.2). More generally, the number of active slip systems and their respective stored densities are responsible for the orientation dependence of the macroscopic strain hardening rate (Fig. 2.18).

Several quantities that are usually assumed to be continuous and uniform evolve by local and intermittent bursts at a fine scale. This is of course the case for the plastic strain, but also for the mobile density, which only exists during strain bursts, and for the stored density which only increases when there is a mobile density. This relation between mobile and forest densities justifies the assumption made in the storage–recovery model (Section 2.6.7), according to which there is always a 'reservoir' of mobile dislocations. It also justifies the assumption that the mobile density is negligible with respect to the stored density, since the latter is the sum of most of the previously stored mobile densities. There is also no strain hardening during a strain burst since forest dislocations are almost immobile. However, the average critical stress at which dislocation segments trigger successive avalanches steadily increases with increasing strain.

5.4.2 From intermittent to continuous flow

Since the first observations on ice crystals by Miguel *et al.* (2001), it was often argued that the traditional modelling of plastic flow as a continuous and uniform process should be reconsidered because what is needed is a proper mean-field theory of depinning transitions in dislocation systems. A multiscale approach was used for integrating intermittent behaviour into a continuum formulation at the scale of slip systems. The storage–recovery model was used for this purpose, as its two dislocation-based equations incorporate the adequate length scales.

The first equation is the generalized dislocation strengthening relation, eqn. 2.23. Equation 2.25 shows that this relation also yields the characteristic dimension of the microstructure when it is combined with the similitude relation. It contains six strengthening coefficients, of which the values were determined in Section 5.2.2.1.

When there is no dynamic recovery, the equation for the storage rate (eqn. 2.26) depends on six mean free paths, one for each type of interaction. The storage rate of mobile segments by forest interactions was modelled in the most general case by estimating the number of intersections and junction formation events per unit swept area (Kubin *et al.*, 2008). The treatment of junctions includes the extra storage induced by ternary junctions. This allowed deriving the mean free paths Λ_i for dislocations of slip system (i) interacting with non-coplanar slip systems. The results are analytical only for symmetrical deformation conditions (Devincre *et al.*, 2008)

$$\Lambda_i = K_{hkl}\frac{\mu b}{\tau_c^i} \qquad \text{with} \qquad K_{hkl} = \left[\frac{\bar{a}^{1/2}n(1+\kappa)^{3/2}}{p_o k_o(n-1-\kappa)}\right] \tag{5.2}$$

At left, τ_c^i is the critical stress for the onset of slip in slip system (i), which is given by eqn. 2.23, and K_{hkl} is an orientation-dependent mean free path coefficient. In the second equation, n is the number of active slip systems, which is orientation-dependent, and $\bar{a}^{1/2}$ is an average interaction coefficient.[10] The three quantities p_o, k_o and κ are dimensionless material constants for fcc materials. They are related respectively to the probability for forming a stable junction upon crossing a forest dislocation, to the average length of the stored segments and to the density of junctions in the microstructure.

The values of the mean free path coefficients K_{hkl} and of the three dimensionless coefficients p_o, k_o and κ were determined in several different manners and cross-checked by model and large-scale DD simulations. In the last case, the coefficient values were obtained by smoothening out the fluctuating simulation outputs. This procedure is illustrated by Fig. 5.12*a*, which shows that the coefficient p_o is actually a constant. To fix the ideas, the mean free paths constants accounting for forest interactions in high-symmetry orientations are $K_{001} \approx 4.6$, $K_{111} \approx 7.3$ and $K_{112} \approx 11$. Under a resolved stress of 10 MPa, these values are similar to those of the related mean free paths expressed in microns.

The recovery rate of dislocations contains an orientation-dependent recovery length y_{hkl} (eqn. 2.26), which incorporates the critical annihilation distance for cross-slip. The integration of this microscopic mechanism at the scale of a slip system involves, however, other features that are lumped into an unknown free parameter. For a given material and temperature, the orientation dependence of the recovery length is deduced from the experimentally measured stress τ_{III} for the $[\bar{1}23]$ orientation (Kubin *et al.*, 2009). Stage I and coplanar interactions between

[10] In this equation a misprint appearing in previous publications is corrected.

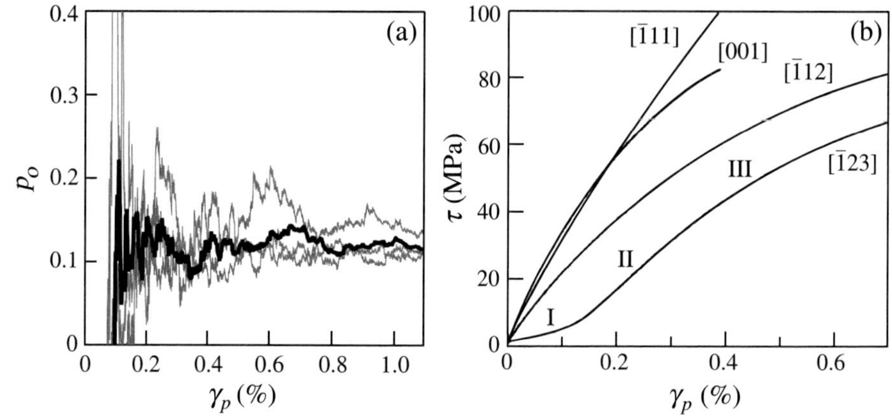

FIG. 5.12. (a) Measurement of the dimensionless constant p_o (see text) from the output of large-scale DD simulations. The value of p_o is drawn as a function of plastic shear strain for crystals deformed along [$\bar{1}$12], [$\bar{1}$11] and [001] loading axes (thin lines). The mean value is shown as a thick line fluctuating around an average constant value $p_o \approx 0.12$. (b) Solutions of the revisited storage–recovery model for copper crystals deformed at 300 K along a few selected orientations. From Devincre *et al.* 2008.

slip systems are not accounted for in the original storage–recovery model. They are included through two mean free path constants, which are estimated from the available data (Kubin *et al.* 2008).

Finally, the integration at the scale of a bulk crystal is performed as outlined in Section 2.6.7. The full set of equations of the storage–recovery model is solved on a meshed model specimen using a crystal plasticity FE code developed by Hoc *et al.* (2001). Figure 5.12*b* shows a few simulated stress–strain curves that can be compared to the experimental ones shown in Fig. 2.18. In addition to the first three hardening stages, which are reproduced without including any switch, some interesting features appear. For example, the crossing of [001] and [$\bar{1}$11] curves arises from the higher initial strain hardening rate of the [001] orientation and its lower critical stress for the onset of the dynamic recovery stage III. The evolution of shear strains and total dislocation densities in the active slip systems is also computed, as illustrated by Fig. 2.25. The output of this model is limited to the end of stage III in the monotonic deformation of fcc single crystals. Further extensions will allow removing more free parameters from current formulations of the storage–recovery model.

The connection between avalanche behaviour and mean-field approaches to the mechanical response can thus be performed in a physical manner via the dislocation mean free paths. If A is the area swept by an avalanche, its characteristic linear dimension is approximately $A^{1/2}$. As the probability distribution of avalanche sizes is bounded, it is always possible to define an average avalanche size $\mathcal{L}$ over a reasonably large number of bursts and under an approximately constant stress. The mean free path Λ can then be seen as a continuum length scale equivalent to $\mathcal{L}$. If it is so, one may expect that the two dimensions be proportional to each other. In consequence, the stress and orientation dependence of the dislocation mean free paths could be inherited from similar properties of the average avalanches sizes. Another possible consequence is discussed in the next section.

5.4.3 Dislocation patterns

Mesoscale simulations may provide guidelines for modelling the formation of dislocation patterns in fcc crystals, in practice of dislocation cells in multiple slip conditions (Fig. 1.9). To be considered as realistic, the simulated microstructures must follow the strengthening relation (eqn. 2.20) as well as the similitude relation (eqn. 2.24). Dislocation patterns are readily obtained in two-dimensional end-on DD simulations. The drastic simplifications made to implement dislocation properties other than long-range interactions result in a wide variety of 2D microstructures or slip patterns. The most elaborated simulations (Benzerga *et al.*, 2004; Bakó and Hoffelner, 2007) lead to interesting results but do not reproduce simultaneously the two relations mentioned above.

Owing to the small plastic strains that can be reached in large-scale 3D simulations, only embryonic cells are formed in fcc crystals deformed in multiple slip conditions. Whereas the value determined for the strengthening constant is fully consistent with experimental ones (Fig. 2.16), no verification of the similitude relation can be performed. Some interesting results were nevertheless obtained by investigating how simulated patterns are modified when specific dislocation processes are turned on or off (Madec *et al.*, 2002*b*). When dislocation interactions are truncated at distances larger than $\rho^{-1/2}$, pattern formation still occurs and no change in flow stress can be evidenced in the fluctuating stress–strain curves. In contrast, turning cross-slip off strongly reduces the tendency to pattern formation. A fully uniform microstructure is even obtained upon removing elastic interactions at distances smaller than $\rho^{-1/2}$. In this last case, cross-slip is also suppressed since strong local interactions are required for its activation (Section 2.4.4). Hence, short-range interactions and cross-slip appear as the main mechanisms governing dislocation cell formation.

Advantage was taken of 3D simulations to construct an end-on 2D simulation where the major dislocation processes are incorporated and calibrated in such a way as to represent an infinitely thin slice of a 3D crystal (for detail, see Gómez-García *et al.* 2006). These *2.5D* simulations were designed with the objective of obtaining dislocation cells that follow the similitude relation. The elementary simulation cell is a square of side L loaded in constant strain rate along the [0 1] axis

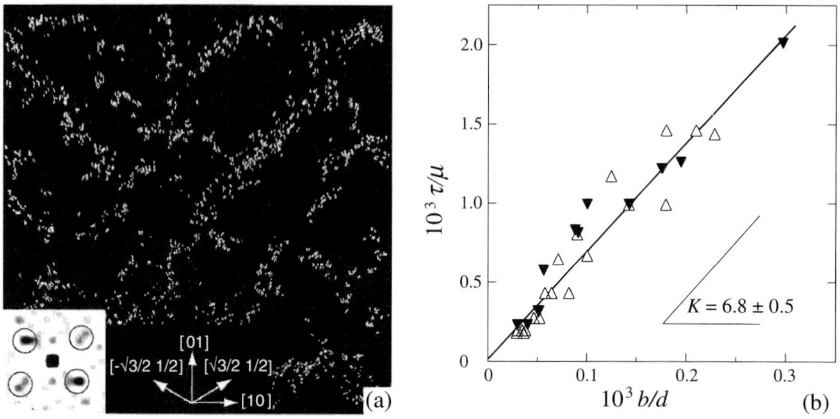

FIG. 5.13. 2.5D simulations of dislocation cells. (a) A pattern of dislocation cells with average wavelength $d = 2.5\ \mu m$ in an elementary cell of size $L = 14\ \mu m$. The glide directions of the two oblique slip systems are indicated and the Fourier transform of the microstructure is shown in the inset. (b) The dimensionless relation between stress and the inverse of cell size. All dislocation interactions are included in one set of simulations (△) and long-range interactions are removed in the other one (▼). From Gómez-García *et al.* 2006.

(Fig. 5.13*a*). Two symmetric slip systems are introduced and periodic boundary conditions are used. Specific properties mimicking 3D behaviour include in particular the formation of junctions and their destruction based on a critical stress criterion similar to those given by eqn. 1.17. In consequence, the strengthening relation is verified. It is assumed that the simulated area is crossed by a dislocation flux generated by dislocation sources in the adjoining volumes. Fresh dislocations with equilibrated signs are thus introduced in the two slip systems with a multiplication rate identical to the 3D one. These dislocations are inserted at random positions in order to mimic the spreading out of plastic activity induced by cross-slip.

Self-organized microstructures are formed when they satisfy periodic boundary conditions, that is, when their wavelength along the $\langle 01 \rangle$ axes is commensurate with the dimension L of the simulation cell (Fig. 5.13*a*). As illustrated by Fig. 5.13*b*, this occurs for stress and cell size values that follow the similitude relation with a coefficient value of $K \approx 6.8$. In another set of simulations, the long-range dislocation interactions are truncated beyond a radius $R_c = L/6$, smaller than the pattern wavelength. Dislocation patterns are still formed and their average wavelength remains unchanged.

In summary, 3D simulations confirm that cell formation is governed by short-range dislocation interactions and cross-slip (or climb). There is at present no model for pattern formation based on such processes, but the study of dislocation avalanches reported in Section 5.4.1 may suggest one. It was noted that moving dislocations are blocked by large positive fluctuations of the stored forest density. Thus, avalanches increase the amplitude of these density fluctuations as well as their ability to further block other avalanches. Dislocation patterns may emerge from such a positive feed-back loop. If it were so, they would inherit the similitude property from the average size of dislocation avalanches and the dislocation mean free paths defined in Section 5.4.2.

5.4.4 Patterning in cyclic deformation

In the cyclic deformation of fcc crystals, the most investigated microstructures are those of persistent slip bands (PSBs). Their main properties are recalled in Section 2.6.6. PSBs can develop over long distances and emerge at the free surfaces of crystals or external grains of polycrystals. As cyclic plastic flow is partly irreversible in PSBs, they produce an accumulation of elementary slip steps in the form of sharp persistent surface markings. The latter are preferential sites for the generation of micro-cracks. The modelling of this damage process is a major motivation for microstructural studies on the properties of PSBs (Mughrabi, 2009). The dislocation-based modelling of steady-state in these very regular patterns (Fig. 2.23*b*) and the reason why they follow the similitude relation also raise fundamental questions.

The saturation stress of PSBs is usually modelled in terms of stress contributions arising in the PSB-channels from the critical bowing of a screw dislocation, the passing stress of a screw dipole and the long-range internal stress field of the PSB-walls. There is, however, no agreement about the respective weights of these contributions, especially that of the long-range stresses. To solve these controversies, three different model DD simulations were performed on two interacting, attractive screw dislocations confined between simplified wall structures (Schwarz and Mughrabi, 2006; Křišťán and Kratochvíl, 2007; El-Awady *et al.*, 2007). Although the saturation stress was correctly reproduced, the three approaches did not lead to convergent conclusions about the magnitude of the long-range stresses.

Déprés *et al.* (2004) performed large-scale DD simulations of low-cycle fatigue with the objective of drawing numerical data and making comparisons with experimental results. Specifically, a surface grain in a polycrystalline AISI 316L stainless steel was submitted to cyclic equibiaxial stresses. In such loading conditions, which are the simplest ones that can be obtained in polycrystal grains, symmetrical slip is activated on a primary system and its cross-slip (i.e., colinear) system. The experimentally observed microstructure (Fig. 5.14*a*) is saturated and reflects the simultaneous formation of two types of PSBs sharing the same Burgers vector. In the DD simulations, the grain boundaries are assumed to be impenetrable to dislocations and image forces at the free surface are neglected. The simulated

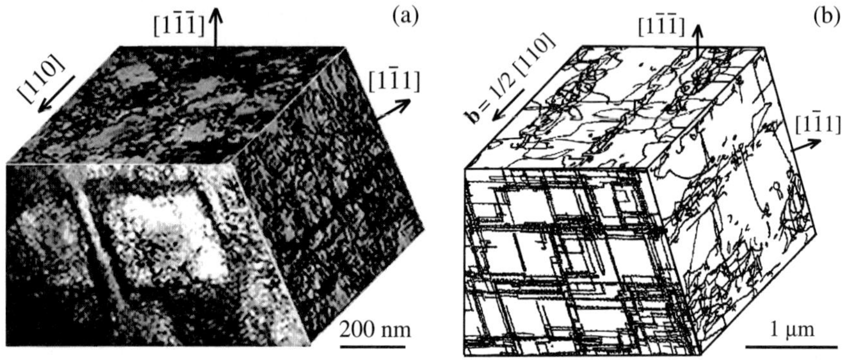

FIG. 5.14. Saturated microstructure in a surface grain of a cycled stainless steel. Symmetrical double colinear slip is activated on the $\frac{1}{2}[110](1\bar{1}\bar{1})$ and $\frac{1}{2}[110](1\bar{1}1)$ slip systems. (a) TEM observation after 100 cycles during thermo-mechanical fatigue (plastic strain amplitude $\delta\varepsilon_p \approx 0.5 \times 10^{-3}$). (b) Simulated microstructure in same loading conditions and after five cycles at 300 K. From Déprés *et al.* 2004, courtesy M. Fivel.

microstructure (Fig. 5.14*b*) is also saturated; it reproduces well the experimental cell-like pattern observed in the plane perpendicular to the Burgers vector, the dislocation density values at saturation and the characteristic widths of the walls and channels. Simulations of PSBs in single glide were also performed on the same grain for the sake of comparison.

The formation of the simulated patterns from a random initial density of sources was followed step by step. It involves cross-slip and related mechanisms that bear some analogy with those occurring during stage I (Section 2.6.2). In double colinear slip, the formation of the microstructure is accelerated by the large number of cross-slip events. In this work, further investigations were performed on the effect of grain size and imposed strain amplitude on the number and characteristic features of the localized bands. In a subsequent study (Déprés *et al.*, 2006), the persistent slip markings formed at the surface of the investigated grain were numerically recorded and analysed. The discussion of these results lead to a new approach to microcrack nucleation. A preliminary simulation was further performed on a rather challenging problem, strain localization in cyclically deformed precipitation-hardened alloys (Shin *et al.*, 2007).

5.4.5 Shock loading, high strain rates

According to Orowan's law (eqn. 1.21), the density of mobile dislocations and their velocity increase with increasing strain rate. Since the glide resistance of forest obstacles is mostly athermal, whereas the free-flight dislocation velocities are stress-dependent (eqn. 4.3), a transition is observed between quasi-static and dynamic deformation regimes. In large-scale DD simulations carried out with cell dimensions of 10–15 μm and imposed strain rates of about $20\,s^{-1}$, plastic flow is still obstacle-controlled. Under strain rates around $10^2\,s^{-1}$, the dynamic behaviour of the mobile dislocations largely prevails (Devincre and Kubin, 1997). In the conditions of the simulations performed by Wang *et al.* (2007*a*), inertial effects have to be accounted for when the strain rate reaches values of about $10^3\,s^{-1}$. When the average dislocation velocities attain a non-negligible fraction of the transverse shear wave velocity, c_t, relativistic effects have to be included. This relativistic regime seems to be rather unimportant in the simulations that are discussed in this section.

High-strain rates are usually imposed by shock loading, which induces the formation of a 1D compressive shock wave travelling with a supersonic velocity. Beyond a critical pressure, the Hugoniot elastic limit, plastic flow occurs at (or behind) the wave front by the homogeneous nucleation of dislocations and the multiplication of the initial density (cf. Fig. 5.15). The whole dislocation density further relaxes, as the pressure decays in the wake of the shear wave. Various microstructures are formed depending upon the imposed pressure and its duration. The comprehensive review by Meyers *et al.* (2009) gives a full account of experimental methods, microstructural and macroscopic observations as well as modelling in this domain.

Using the DMPP model, in which the DD simulation is embedded into a continuum mechanical framework, Shehadeh and co-workers (2005*a*, 2005*b*) examined the interaction of the high-pressure region of a shock wave with an initial dislocation density. The strain rates were in the range of 10^5–$10^7\,s^{-1}$ and the pulse durations were in the range of a few nanoseconds. The materials investigated

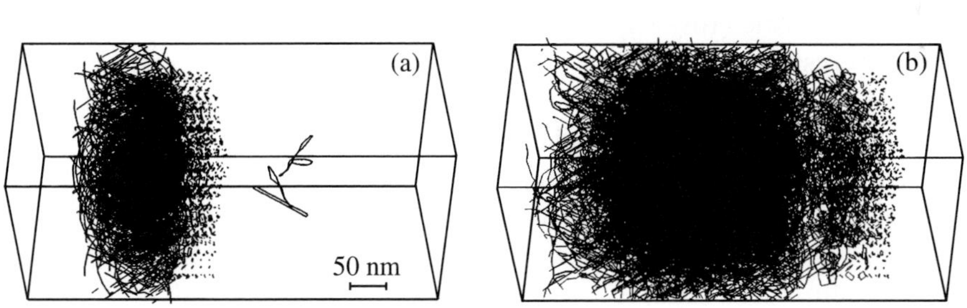

FIG. 5.15. Simulation of homogeneous dislocation nucleation at the front of a shock wave moving from left to right. The pressure is 35 GPa and the rise time of the shock wave is $\delta t = 50$ ps. (a) $t = 67$ ps. Dislocation loops are nucleated while the wave travels. A few pre-existing loops are seen ahead of the front. (b) $t = 122$ ps. The nucleated loops expand and interact at the rear of the nucleation region. Close to the wave front, there is a well-defined region of mixed homogeneous nucleation and heterogeneous multiplication. From Shehadeh *et al.* 2006.

were [001] copper and aluminium crystals and [001], [011] and [111] copper crystals. For a strain rate of $10^6\,\mathrm{s}^{-1}$ and an initial density of about $10^{11}\,\mathrm{m}^{-2}$, only 1.5% of the dislocations interacting with the shock front move at velocities larger than $0.8c_t$ and 92%, of which 25% are immobile, move at velocities smaller than $500\,\mathrm{m\,s}^{-1}$ (about $0.2c_t$ in copper). The average dislocation velocity, about $300\,\mathrm{m\,s}^{-1}$, decreases with increasing initial density since more mobile dislocations are carrying the imposed strain rate. The dislocation density increases with pressure and pulse duration and saturates within a time scale related to pulse duration. For strain rates above $10^6\,\mathrm{s}^{-1}$, crystallographic strain localizations develop. They form dislocation-dense microbands parallel to {111} slip planes, which were investigated more accurately at a strain rate of $10^8\,\mathrm{s}^{-1}$ (Shehadeh, 2012). Globally, all the described features qualitatively reproduce the experimental behaviour.

In another study (Shehadeh *et al.*, 2006), local criteria for homogeneous dislocation nucleation are introduced on the base of MD simulations, specifically a pressure threshold of 30 GPa and a critical shear stress of about 2 GPa. The nucleation mechanism is reproduced by introducing small dislocation loops at random positions in all slip planes, with a probability proportional to their Schmid factor. Huge dislocation densities, typically, $10^{16}\,\mathrm{m}^{-2}$, are nucleated in simulated events like that illustrated by Fig. 5.15. The comparison with MD simulation results looks promising, insofar as DD simulations may provide insights into an extended range of time and length scales.

In another set of simulations, Wang and co-workers investigated shockless deformation by imposing strain rates between 10^4 and $10^6\,\mathrm{s}^{-1}$. The objective was to examine the role of cross-slip, which seems to be ignored in current models for deformation at high strain rates. The simulations are performed with the PDD code. Dislocation inertia is taken into account (Wang *et al.*, 2007*a*) and use is made of a stochastic cross-slip rule (Wang *et al.*, 2007*b*). The simulated material is a copper crystal loaded along a [111] axis (Wang *et al.*, 2008) and along [100], [211] and [111] axes (Wang *et al.*, 2009). Periodic boundary conditions are applied to a cubic simulation cell with a side length of 5 μm.

A pattern of coarse, crystallographic slip bands containing finer slip bands appears at a strain rate of $10^5\,\mathrm{s}^{-1}$. This slip heterogeneity increases with increasing strain rate. There is evidence that double cross-slip is responsible for the growth of the fine slip bands and the expansion of the whole slip pattern. Thus, what is observed looks similar to the hierarchical slip structures and Lüders-like slip propagation described by Neuhäuser (1983). With increasing strain rate and for the three orientations considered, the total density increases owing to the formation of new slip bands. In parallel, the stress saturates at rather small strains that also increase with increasing strain rate. The microstructural observations reveal a complex interplay between several key quantities, which evolve with strain rate and depend on crystal orientation. These quantities are the mobile and transiently immobile dislocation densities, the dislocation velocities, the slip heterogeneity and the strain rate sensitivity. A framework based on the simulation results is proposed for developing dislocation-based constitutive formulations specific to high-strain rate deformation (Wang *et al.*, 2009).

5.5 Size effects in plasticity

5.5.1 Introduction

Size effects are induced by the interference of a length scale characterizing the dislocation microstructure with another dimension. The microstructural length scale can be the average length of dislocation segments, $\bar{\ell} = \rho^{-1/2}$, the characteristic dimension of a dislocation pattern and the mean free path of dislocations, which are all proportional to each other and are scaled by the Burgers vector. Gil Sevillano *et al.* (2001) distinguish two classes of size effects. Intrinsic size effects arise from decreasing microstructural lengths or sample dimensions; extrinsic size effects are associated with the development of mesoscopic strain gradients induced by structural heterogeneities or imposed by non-uniform deformation conditions. Composite size effects can also arise from the combination of several interfering length scales.

The search for strong materials and the development of nanotechnologies led to a wealth of studies on size effects and the realization that they are sometimes not so easy to understand and control in spite of their apparent simplicity. Several overviews provide a survey of techniques for testing and sensing plastic deformation at the micro- and nanoscales (see Dehm 2009, Lilleodden 2009, Legros 2010 and other articles in the same issues). Stresses in the range of a few GPas are readily obtained in nanostructured materials, albeit with a loss of ductility. Nevertheless, it is now possible to build ultra-strong materials that can durably sustain stresses amounting to a substantial fraction of the theoretical limit (Zhu and Li, 2010). In this context, the challenge for DD simulations is to control boundary conditions, devise realistic initial microstructures and find ways to take into account dislocation nucleation at sub-micron sizes. In what follows, a few examples are given of 3D DD simulations that contributed to the understanding of extrinsic size effects. The emphasis is on intrinsic size effects in thin films and in micropillars, which are the subject of many recent investigations.

5.5.2 A few examples

Within the classification given above, traditional strengthening mechanisms in structural materials, like the Hall–Petch effect in polycrystals or dislocation–precipitate interactions, are intrinsic size effects (see e.g., Arzt 1998). Of particular interest are the strong size effects obtained when dislocations are confined by precipitates in very narrow channels, as in γ-γ' superalloys (Section 5.3.4), or by very small grain sizes. In the last case, simulations of ultrafine-grained polycrystals deforming in multislip conditions involve a high computing cost. A preliminary 3D simulation was carried out by de Sansal *et al.* (2010). A 2.5D simulation (Section 5.4.3) was carried out by Lefebvre *et al.* (2007) in a model copper polycrystal with grain sizes in the range of 0.5–2 μm. Although significant microstructural changes occurred with decreasing grain size, the Hall–Petch relationship was not modified.

Strengthening by extrinsic size effects is often attributed to an additional density of GNDs that relaxes the stored elastic energy. In the continuum, it is described by phenomenological models for strain-gradient plasticity (Fleck *et al.*, 1994). These approaches and their physical relevance are discussed by Needleman and Gil Sevillano (2003) and in other articles of the same issue. Among typical examples, one finds the size effect in micro-bending, which was simulated by Motz *et al.* (2008) for single crystalline beams with thicknesses in the range of 0.5–1.5 μm. The bending moment, approximately scaled like the inverse of thickness (Fig. 5.16*a*), in consistence with previous experiments (Motz *et al.*, 2005). A model was suggested by the observation of arrays of parallel dislocations emitted by Frank–Read sources, which were blocked on their way to the central plane where the effective stress changes sign.

The indentation size effect (ISE) is another extrinsic size effect, which has been known for more than a century (Meyer, 1908). The overview by Gouldstone *et al.* (2007) can be consulted for a thorough discussion of experimental and modelling

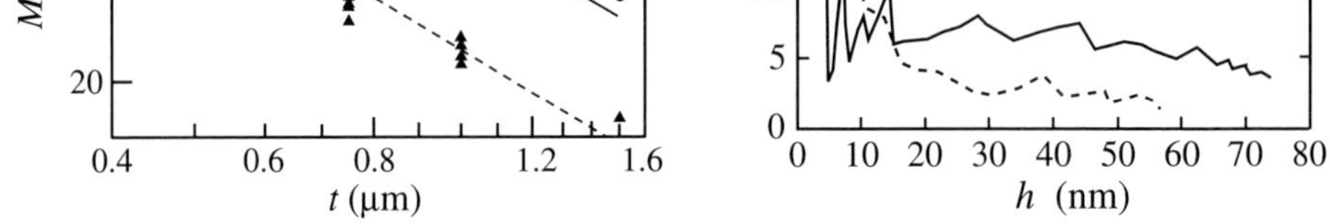

FIG. 5.16. (a) Size effect in microbending. The bending moment M normalized by the beam width B and the thickness t squared is drawn as a function of t. The two sets of simulated data correspond to different boundary conditions, (i) and (ii). They yield size exponents n close to –1. After Motz *et al.* 2008, courtesy C. Motz. (b) Indentation size effect. Variation of the hardness H as a function of the indentation depth h for a local rule implemented in DD simulations and for a GND-based criterion. Adapted from Fivel *et al.* 2011, courtesy M. Fivel.

issues on indentation across the scales. The relevant variables are the penetration depth h and the hardness $H = P/A$, where P is the load and A is the contact area. The ISE is characterized by an increase of hardness with decreasing sampled volume or, equivalently, with decreasing penetration depth. It starts manifesting itself for typically $h < 10\,\mu m$. For penetration depths smaller than a fraction of micron, the P–h curves exhibit a short elastic domain followed by plastic penetration bursts separated by intervals of elastic response. This so-called *pop-in* phenomenon is attributed to the nucleation of prismatic loops accommodating the elastic stresses below the indenter.

The multiscale approach to indentation presented by Chang *et al.* (2010) and Fivel *et al.* (2011) goes in separate steps from the atomic scale to the continuum. At the mesoscopic scale, DD simulations are carried out with boundary conditions based on the superposition principle. Figure 5.16*b* shows the ISE as reproduced by two different local rules. One is drawn from experiment and is such that the number of prismatic loops that are introduced per loading step matches the applied load. The second one derives from a GND-type criterion and assumes that the penetration of the indenter is accommodated by the plastic displacement of the generated dislocation loops. Realistic load versus penetration curves are obtained with the help of a storage–recovery model inserted into a crystal plasticity FE code. In addition, using the interaction strength values given in Table 5.1 allows reproducing the shape of the surface relief as observed by atomic force microscopy.

5.5.3 The silicon world

Dislocations are almost inevitably generated during the epitaxial growth and processing of single-crystal semiconducting films.[11] As dislocations are detrimental to the service properties of nano-devices, a considerable effort was devoted to investigations of the plastic response of thin epitaxial layers by experiment, modelling and simulation. A fully documented review by Fertig and Baker (2009) emphasizes the contribution of 3D DD simulations to the understanding of dislocation processes in thin semiconducting layers.

During epitaxial growth, the layers are submitted to equibiaxial stresses induced by a misfit strain ε_o. The strain energy linearly increases with the layer thickness h and is proportional to $\mu h \varepsilon_o$ per unit interfacial area. For a certain critical thickness, a further increase of elastic strain becomes energetically less favourable than the motion of a dislocation that threads through the layer and relaxes the elastic misfit strain. Analytical models were developed for predicting the critical stress at which a pre-existing or freshly nucleated threading dislocation starts moving (Matthews *et al.*, 1970; Freund, 1987; Nix, 1989). In these models

[11] See 'Modeling for the silicon world' by Schwarz *et al.* (2000).

and in what follows, it is assumed that the processing temperature is above the athermal transition temperature at which the lattice resistance vanishes. Hence, dislocation motion is obstacle-controlled.

Two types of layers are usually considered, an epitaxial film on substrate with a free surface, or a capped layer, that is, a film where the two surfaces are in epitaxy with identical substrates. DD simulations showed that the critical stress of an epitaxial layer with a free surface is close to one-half that of a capped layer with twice its height (Schwarz *et al.* 2004 and references therein). A simple reasoning yields an approximate general form for the critical stress, which also applies to multilayers. The elastic misfit strains induce a resolved stress τ on a threading dislocation of length h. Assuming a small rigid displacement of the line by δx, the mechanical work of this stress is proportional to $\tau b h \delta x$. A moving threading dislocation deposits on the substrate one or two misfit dislocations, which have an elastic energy proportional to $\mu b^2 \ell n(h/b)\delta x$ (eqn. 1.4). When the two energies are equal, the critical condition (index c) is written as

$$\tau_c \propto \frac{b}{h_c} \ell n \left(\frac{h_c}{b} \right) \tag{5.3}$$

Thus, the confinement size effect is roughly proportional to $1/h_c$ and can reach one or a few GPa when $h_c \approx 10\,\text{nm}$. As the critical (or channelling) stress τ_c is proportional to the critical misfit stresses and strains, the critical condition can be expressed in different manners. Ideally, the threading dislocation should move over long distances in such a way as to maintain a balance between driving and resistive forces, while the dislocations deposited on the substrate should accommodate the misfit. All simulations verify the predictions of analytical models for the critical stress. However, the channelling stresses are much higher in real films because of the strain hardening produced by simultaneously moving threading dislocations, which interact with each other and with misfit dislocations. This results in rather complex configurations of misfit dislocations, as illustrated by Fig. 5.17. In that case, the threading dislocations were produced in a controlled manner by helium implantation and annealing, in order to enhance the relaxation of the misfit strain by plastic deformation. The rounded contrasts in Fig. 5.17*a* are helium bubbles or platelets from which the threading dislocations were nucleated.

The strength of the interactions experienced by two threading dislocations are defined as the critical misfit stresses or strains needed to remobilize them in an otherwise perfect layer. Since analytical models cannot capture the exact shapes of interacting lines under stress, they predict significantly overestimated interaction strengths. DD simulations were carried out to determine realistic values (see Fertig and Baker 2009), mostly with the PARANOID code which accurately treats small curvature radii. Image forces were implemented in a rather simple manner, since Liu and Schwarz (2005) showed that they can be neglected beyond

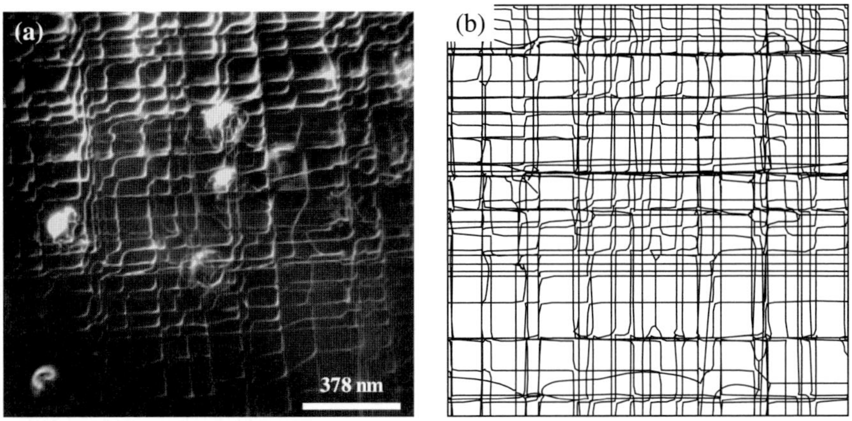

FIG. 5.17. Relaxed microstructure of misfit dislocations in a (001) strained epitaxial layer showing that a number of interactions took place during its formation, in particular colinear interactions. (a) TEM micrograph of a $Si_{0.8}Ge_{0.2}$ interface layer of thickness 104 nm. (b) Large-scale simulation of a relaxation process leading to the same final density for the same misfit strain. The scale is same as in (a). One-third of the initial density was annihilated by colinear reactions. From Schwarz *et al.* 2004, courtesy K. W. Schwarz.

a distance to the surface in the 10 nm range. This was further verified by full computations of the image forces.

The comprehensive study by Pant *et al.* (2003) can be summarized as follows. There are four possible interactions between intersecting threading and misfit dislocations. Attractive interactions result in the formation of junctions, crossed-states and colinear annihilations; they are, however, too weak to account for the observed strengthening. The interactions strengths of coplanar and intersecting threading dislocations are quite substantial, as some of them exceeded the average strength of a film with same thickness. The main conclusion is that the configurations of interacting lines strongly depend on thickness and so does their strength. Further, a full understanding of the mechanical response can be reached only by going beyond such model simulations.

In a next step, large-scale simulations incorporating a large initial density of threading dislocations on all slip systems were carried out to examine strain hardening, microstructure relaxation under stress and the Bauschinger effect observed during unloading. A study of microstructural relaxation effects in a strained layer (Schwarz, 2003*a*) yielded a hierarchy of interactions that block threads, specifically attractive interactions between threading dislocations, colinear annihilations with misfit dislocations and, to a smaller extent, other interactions. It was found that a sizeable fraction of the threading lines were not blocked by direct interactions with other dislocations but by stress fluctuations. The latter were caused by local excesses of misfit dislocations arising from the spatial distribution of the threading dislocations that created them. These fields over-relax the misfit strain, limit the residual stress and determine its evolution as a function of the total density. A comparison with experiments on Si-Ge layers (Schwarz *et al.* 2004, see Fig. 5.17) yielded an excellent quantitative agreement for the percentage of strain relaxation and the residual misfit strain as a function of the layer thickness.

Such simulations were pursued and expanded by Fertig and Baker (2009, 2010), leading to a general model for the plasticity of thin films, metallic or semiconducting. The model is based on dislocation interactions and emphasizes the effect of the fluctuating stress field associated with the misfit strain. In parallel, Schwarz and Tu (2009) proposed a continuum analytical model with length scales for the relaxation of a strained layer containing an initial density of threading dislocations. Two coupled rate equations express, respectively, the growth of the misfit array and the decay of the propagating density of threading lines caused by interactions with other lines and stress field fluctuations. All missing numerical information is drawn from DD simulations. A scaling analysis of the coupled equations shows that the residual strain decreases with increasing initial dislocation density. The kinetics of this decay is characterized by two regimes, a fast initial one followed by a slower one. The comparison with experiment, which is illustrated by Fig. 5.18*a*, was performed on $Si_{0.8}Ge_{0.2}$ specimens, for which the initial density is approximately known and the microstructure is well reproduced by DD simulations (Fig. 5.17). This type of model is potentially able to treat all other aspects of plasticity in thin films.

It is important to control not only the initial density of threading dislocations, but also their spatial distribution. This requires, however, establishing criteria for their heterogeneous nucleation. A typical procedure used for this purpose is as follows. A silicon substrate that contains stress concentrators serving as nucleation sites is processed and examined by scanning or transmission electron microscopy. In a DD simulation of this substrate, small dislocation loops are introduced and their critical size and location are fine-tuned until experimental observations are reproduced. In this way, nucleation criteria were determined for shallow trenches (Izumi *et al.*, 2005), pad corners (Kammler *et al.*, 2005) and nanoindentation pits (Fig. 5.18*b*).

In summary, DD simulations performed in close connection with experiment and observation turned out to be well suited for investigating classical dislocation problems in epitaxial semi-conducting films. The evolution of these simulations from a single threading dislocation towards a microstructure of interacting misfit and threading dislocations helped establishing predictive continuum models for the mechanical response. Further developments may allow better controlling stress concentrations and dislocation nucleation in semiconducting devices.

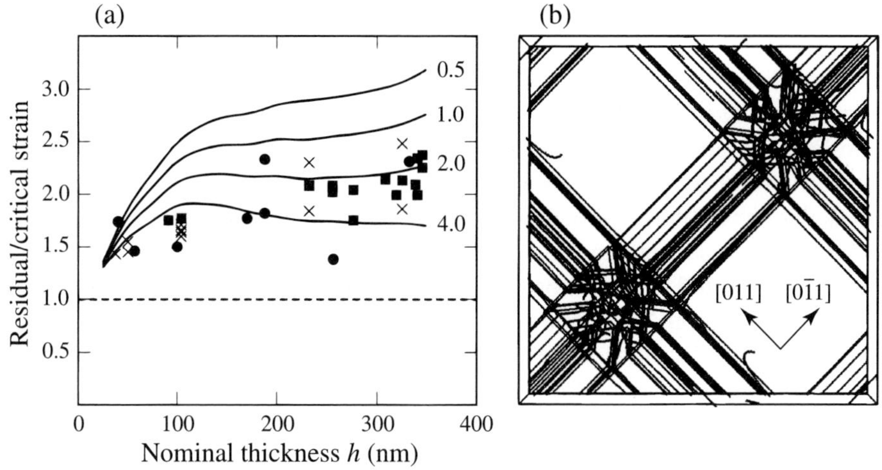

FIG. 5.18. (a) Ratio of residual to initial critical strain as a function of thickness in $Si_{0.8}Ge_{0.2}$ free-surface layers. The two filled symbols refer to different growth conditions. The open symbols refer to DD simulations (Schwarz *et al.*, 2004). The lines show the predicted ratios for different initial dislocation densities in units of 10^{12} m^{-2}. From Schwarz and Tu 2009. (b) DD simulation of the spatial distribution of dislocations nucleated from a square array of $\langle 110 \rangle$ pits in $Si_{0.85}Ge_{0.15}$ (top plan view). The size, depth and spacing of the pits are respectively 0.54, 0.25 and 1.5 μm. From Grydlik *et al.* 2012, courtesy F. Boioli.

5.5.4 Thin metallic films

Metallic multilayered composites are attractive materials, especially for their high strength and ductility. They were seldom examined by DD simulations, although the dislocation processes are akin to those examined in epitaxial layers. Akasheh *et al.* (2007) investigated the various interactions of threading dislocations with intersecting interfacial dislocations in the semi-coherent CuNi system and emphasized the need for flexible boundary conditions at the interfaces in order to obtain realistic strength values. These results were extended to the CuNiNb system that contains semi-coherent and incoherent interfaces (Zbib *et al.*, 2011).

Metallic thin films are in general polycrystalline with columnar grains of size comparable to their height. They can be free-standing, deposited on a surface or epitaxially grown. The free surfaces can be passivated (i.e., protected by a thin layer against chemical reactions with their environment) or unpassivated. These films undergo a size effect that arises from both confinement and grain boundary strengthening. At low stresses, dislocation multiplication occurs in the volume and dislocations are blocked by grain boundaries. Under high stresses, typically in the GPa range, two poorly known dislocation mechanisms emerge, which are difficult to handle in DD simulations. Grain boundaries may absorb dislocations and, in favourable cases, transmit them to a neighbouring grain. In addition, heterogeneous dislocation nucleation occurs at grain boundaries. DD simulations of the mechanical response of thin metallic films were reviewed and discussed by Fertig and Baker (2009).

Preliminary simulations of the mechanical response of single columnar grains were carried out using as initial configuration a constant density of Frank–Read sources with fixed ends. Grain boundaries were assumed to be impenetrable to dislocations (von Blanckenhagen *et al.* 2004, and references therein). The hypothesis of source-controlled deformation was found to yield partial agreement with experimental results. Simulations of a free standing columnar grain with top and bottom free surfaces were performed by Espinosa *et al.* (2005, 2006) on the base of literature data and their own deformation tests and TEM observations. Grain boundaries were assumed to be impenetrable to dislocations, but dislocation generation was assumed to occur only through grain boundary sources in constant density per unit grain boundary area. This last process was implemented by positioning a realistic density of conventional sources at grain boundaries, with statistically distributed lengths and locations. Such sources only emitted a single half-loop inside the grains. Ultra-fine grains with sizes d from 0.2 to 1 μm were investigated, with grain heights ranging from d to $5d$.

The simulated stress strain curves reproduce well the experimentally observed increase of yield stress and strain hardening rates for thicknesses smaller than 1 μm, with a size effect that scales like $1/h$. These trends are not followed by the simulated sample with aspect ratio $h/d = 5$, $h = 1$ μm, which exhibits a saturation of the yield stress and an enhanced strain hardening rate. The whole simulated behaviour is interpreted in terms of several factors. Given the stress, there is less grain boundary area in thinner films and a smaller number of active sources. In parallel, the dislocations have a higher probability to reach the free surfaces, thus producing less strain. For large aspect ratios, the dislocations are stored at grain boundaries. A microstructure of entangled dislocations develops inside the volume, inducing a back stress on the sources. As the simulation results critically depend on dislocation density, grain boundary sources are governing plastic flow. Although the proposed model is not fully quantitative, the main lines of the proposed interpretation should apply to polycrystalline films, as well as to other types of size effects in confined plasticity.

In the simulations by Zhou and LeSar (2012), a polycrystalline film is represented by a crystal split by barriers to dislocation glide into aggregates of nine square grains of size d and height h. Stress-free boundary conditions are implemented at the top and bottom surfaces and dislocation sources are pinned at their ends. The transmission of plastic flow through grain boundaries is modelled as the stress-assisted nucleation of Frank–Read sources caused by impinging dislocations on the other side of a grain boundary. The stress criterion is adjusted to reproduce the experimental stress–strain behaviour of copper films (Xiang and Vlassak, 2006).

The grain sizes and thicknesses range from 0.25 to 1.5 μm, which allows investigating aspect ratios in the range of $h/d = 0.17$–6. The total dislocation density and the density stored at grain boundaries were investigated as a function of grain size and height. Their evolution can be rationalized in terms of the free-surface area and the ratio of grain boundary area to volume. As expected, the yield stress tends to exhibit a $1/h$ scaling typical of thin films for the smallest aspect ratios and a Hall–Petch behaviour for the largest ones. The simulated stress–strain curves reproduce well the available experimental ones for all aspects ratios (Xiang and Vlassak, 2006; Gruber *et al.*, 2008).

These simulations show that a Frank–Read source with randomly positioned pinning points cannot operate inside a grain since its arms are blocked by the grain boundaries or escape through the free surfaces. Thus, the longest arm may form a single-ended source that accommodates the imposed strain rate during its lifetime. This suggests modelling the yield stress of free-standing films by a spiral source model analogous to the one derived by Zhou *et al.* (2011) for micropillars (Section 5.5.5.4). This model captures well the measured yield stresses and their large scatter in a range of aspect ratios limited by dislocation nucleation at high stresses and dislocation-strengthening at low stresses.

The objectives of DD simulations on thin metallic films look quite ambitious. Indeed, the investigated mechanical response ranges between that of free-standing films and that of ultrafine-grained polycrystals. More input is clearly needed for effectively treating the interactions of dislocations with grain boundaries and their stress dependence. The relative weights of dislocation multiplication at grain boundaries and in the volume is another point of concern. In particular, model DD simulations suggest that sources pinned by single jogs or junctions become less and less stable with decreasing film thickness (Lee *et al.*, 2011).

5.5.5 Small-scale pillars

5.5.5.1 Introduction Plastic flow in micropillars is discussed in Sections 2.7.2.3 and 5.4.1 in the context of dislocation avalanches. The discovery of a size effect in the compression of micropillars (Uchic *et al.*, 2004) somehow came as a surprise. In pure single-crystalline nickel micropillars with diameters ranging between 1 and 30 μm, the CRSS was found to vary as a power law of the diameter, with a size exponent between −0.6 and −0.7 (Dimiduk *et al.*, 2005). This showed that geometric confinement by free surfaces induces a strong intrinsic size-effect. A considerable amount of information on the deformation properties of micro- and nanopillars (that is, pillars with diameters in the sub-micron range) was further obtained through a synergy between mechanical testing, TEM observations and DD simulations. The experimental behaviour of all types of crystalline or amorphous micropillars is compiled and discussed by Greer and De Hosson (2011). The review articles by Uchic *et al.* (2009) and Kraft *et al.* (2010) discuss the deformation of fcc and bcc micropillars, with emphasis on studies carried out by two different groups.

Doubts were expressed about the reliability of compression tests performed on small-scale pillars and the intrinsic nature of the size effect. In consequence, the problems of sample fabrication by focused ion beam (FIB) milling and the drawbacks of compression testing lead to numerous investigations and discussions (Uchic *et al.*, 2009).

Radiation damage by Ga^{+} ions during FIB milling may strengthen the samples in various manners (Kiener *et al.*, 2007). Direct comparisons between FIB-prepared and electroplated or nanoimprinted micropillars led to the conclusion that the amount of irradiation strengthening is not substantial (Greer *et al.*, 2005; Greer and Nix, 2006; Jennings *et al.*, 2010; Dietiker *et al.*, 2011). In fact, tapered specimen geometries, which may be obtained irrespective of the preparation method, are more troublesome. They result in strain gradient formation and extrinsic strengthening.

The well-known difficulties associated with the compression of bulk specimens are recalled in Section C.3. Micro-compression tests using a flat punch indenter are affected by additional problems, notably an uncontrolled friction at the interface between the top surface and the indenter and a rigid connection of the bottom face to the softer substrate.

The first micro-tensile deformation tests were performed *in situ* in a scanning electron microscope by Kiener *et al.* (2008). The FIB-prepared copper microcrystals were deformed in single slip. Those having aspect ratios (gauge length to side length) larger than 2 exhibited almost no strain hardening and a moderate size effect (see Fig. 5.19). These results contributed to expand an already animated debate about possible extrinsic contributions to the size effect in compression (Kiener *et al.*, 2009*a*; Kiener *et al.*, 2009*b*; Uchic *et al.*, 2009; Shade *et al.*, 2009; Maass *et al.*, 2009; Senger *et al.*, 2011). The outcome of these discussions turned out to be that compression tests in single slip are associated with too signifiant lattice rotations (Section C.3) to be reliable. The possibility of relaxing the lateral constraint at the bottom face was debated (Kiener *et al.*, 2009*b*; Shade *et al.*, 2009), but experimentalists preferred to move to micro-compression tests on specimens with high-symmetry orientations, mainly $\langle 001 \rangle$, and large aspect ratios. Such specimens deform in a stable manner after an initial excursion in easy glide. For instance, the flow stresses of gold nanopillars in tension and compression become similar at strains of about 10% (Kim and Greer, 2009).

The concept of starvation was introduced by Greer *et al.* (2005) to explain the high stresses reached in gold nanopillars (a correction to these results is found in Greer *et al.* 2006). An *in situ* TEM nanoindentation test on nickel nanopillars performed by Shan *et al.* (2008) illustrates well the notion of starvation (Hemker and Nix, 2008). For a diameter of about 160 nm, the radiation damage layer and all dislocations are seen to escape at the surface upon loading.

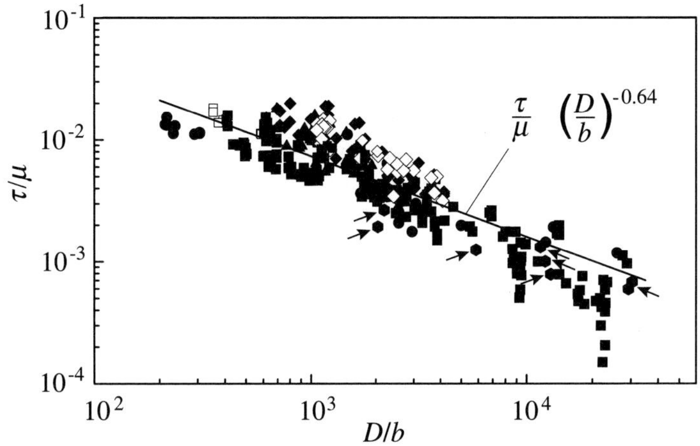

FIG. 5.19. Scaled data on resolved proof stresses versus diameter for Au, Cu, Ni and Al pillars fabricated by different methods and deformed in tension and compression. The arrows point at tensile data on copper in easy glide (Kiener *et al.*, 2008), which are clearly lying below the regression line. After Greer and De Hosson 2011, courtesy J. R. Greer.

This phenomenon of 'mechanical annealing' leaves the sample in a dislocation-starved state. Heterogeneous dislocation nucleation, which is observed at the flat contact surface with the nanoindenter, is held responsible for the very high subsequent yield stresses (well above 1 GPa). For larger diameters (390 nm), some defects remain in the samples (Shan *et al.*, 2008) and the remaining dislocations are principally lying on slip systems with zero Schmid factor (Greer and Nix, 2006). Thus, with decreasing diameters and increasing stresses, dislocation nucleation at the surfaces progressively replaces dislocation multiplication in the volume.

The next sections discuss the experimental aspects of the size effect, the contribution of DD simulations to its understanding and the available modelling approaches. The emphasis is on fcc small-scale pillars, where dislocation mechanisms are better understood (for other crystallographic structures, see Greer and De Hosson 2011).

5.5.5.2 Size effect and dislocation densities Compilations of the size effect in nano- and micropillars are found in the review articles quoted above and in an article by Dou and Derby (2009), which includes data on gold nanopillars with diameters down to 30 nm (gold presents the advantage of having non-chemically reactive surfaces). Figure 5.19 reproduces a compilation of the size effect; the reader is referred to the original publication (Greer and De Hosson, 2011) for references corresponding to the various symbols.

The best linear fit to the data yields a size exponent of about −0.64, similar to those obtained in other compilations. However, individual data sets exhibit exponent values ranging from −0.4 to about −0.9. As emphasized by Kraft *et al.* (2010), the definition of an average exponent value does not necessarily mean that it is universal or that there is a unique combination of active mechanisms all through the range of investigated diameters.

The large scatter of exponent values in Fig. 5.19 is partly due to experimental conditions. Given the density, statistical scatter is also expected to increase with decreasing sample size. In a pillar of diameter $D \approx 200\,\text{nm}$ and height $3D$, a dislocation segment of length D typically corresponds to a density of $\rho \approx 10^{13}$ m^{-2}. It follows that in such conditions the mechanical response may not be fully reproducible and can hardly be controlled. Actually, most of the scatter of the size exponent seems to arise from variations of the initial dislocation density. These densities are substantial and do not depend much on the sample preparation method. In FIB-prepared nickel micropillars, they amount to about $10^{13}\,\text{m}^{-2}$ (Norfleet *et al.*, 2008). In electroplated and FIB-prepared gold nanopillars, a rough estimate yields $10^{14}\,\text{m}^{-2}$ (Jennings *et al.*, 2010). Figure 5.20*a* shows a graph drawn from DD simulations performed in nickel (Rao *et al.*, 2008) with different initial densities. The scatter of the size exponent values looks similar to the one shown in Fig. 5.19. Hence, these results suggest that the initial density is the main factor governing the value of the size exponent.

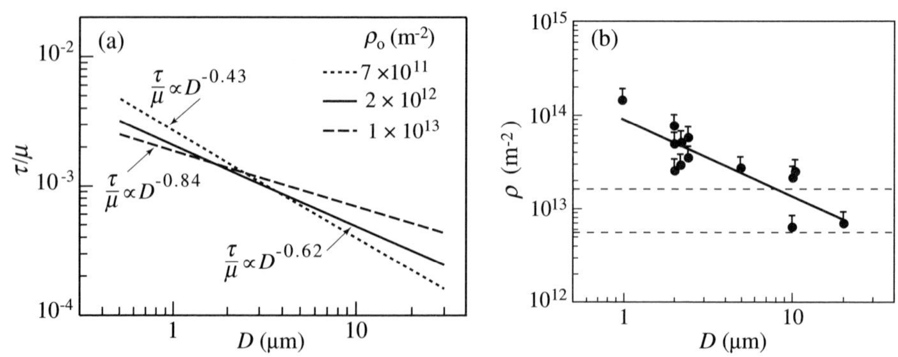

FIG. 5.20. (a) Size exponent in DD simulations of Ni micropillars containing different initial densities (ρ_o). The regression lines represent the dependence of the scaled resolved stress taken at 1% strain on simulation cell diameter (D). Adapted from Rao *et al.* 2008. (b) TEM measurement of total, almost saturated, dislocation densities in deformed Ni micropillars as a function of diameter. The error bars account for the unresolved density. The dashed lines define a typical range of initial densities. After Norfleet *et al.* 2008.

There is no strain hardening after yield for micropillar diameters down to about one-half micron, which suggests that a steady-state flow stress is reached. The careful TEM measurements by Norfleet *et al.* (2008) show that the total density is not much strain dependent but increases with decreasing diameter (Fig. 5.20*b*). The localization of plastic strain observed in these micropillars indicates that dislocations multiply by a source mechanism.

In micropillars, mechanical annealing induces the loss of part of the initial dislocation density. In particular, the escape of a Frank–Read source segment pinned at its ends leaves two single-arm segments pinned at one end and intersecting the surface at the other end. During its rotation, a single-arm (or spiral) source periodically reaches a position of minimum length that defines its critical stress. As in thin metallic films with columnar grains (von Blanckenhagen *et al.*, 2003; Zhou *et al.*, 2011), this source-truncation effect suggests a possible size-strengthening mechanism (Rao *et al.*, 2007). The stochastic source model proposed by Parthasarathy *et al.* (2007) examines a population of single-arm sources with randomly distributed pinning points. The critical stress of the average single-arm source with maximum length, the so-called weakest link, depends on the pillar diameter; it determines the contribution of dislocation sources to the CRSS. This model is further discussed in the next two sections.

5.5.5.3 DD simulations Ideally, DD simulations of small-scale pillars deformation require initial configurations that do not artificially constrain microstructural evolution under stress. Thermally activated cross-slip has to be implemented in order to obtain realistic pinning configurations for dislocation sources. Image forces on dislocations have to be taken into account for sub-micron pillar diameters because they affect the critical stress of single-arm sources pinned near the surface (Rao *et al.*, 2007). It is, therefore, necessary to implement boundary conditions upon investigating the mechanical response, especially when deformation is strongly non-uniform (see e.g., Gao *et al.* 2010, where the boundary conditions based on the superposition principle is implemented using a finite strain framework[12]).

The stochastic source model was investigated in two studies where the initial configurations consisted of random distributions of Frank–Read sources with fixed ends (Rao *et al.*, 2008; El-Awady *et al.*, 2009). The model material was nickel. In the first study, simplified rules are applied to dislocations interacting with the external surfaces. Deformation is carried out in compression on samples with cross dimensions in the range of 0.5–20 μm. Two size-dependent mechanisms are identified. At the yield stress, the Frank–Read sources transform into a population of single-arm sources with a probability that increases very fast with

[12] Finite strain calculations do not make use of the small strain approximation employed in Section C.2.2.

decreasing size. As illustrated above by Fig. 5.20*a*, the critical stresses at 1% strain reproduce well the experimental values of the size exponent. During plastic flow, the rotating mobile dislocations react with other dislocation lines, which induces the shutting-off of active dislocation sources and the activation of new sources with higher critical stresses. The contribution of dislocation strengthening to the yield and flow stresses decreases with decreasing sample size, while the contribution of single-arm sources increases. A continuous transition is thus predicted between obstacle-controlled and source-controlled behaviour.

A stochastic rule for cross-slip is implemented in the second study. The boundary conditions are based on the superposition principle and solved by a boundary element method (El-Awady *et al.*, 2009). The weakest link model is further revisited for pillar diameters from 0.25 to 5 μm. The initial distribution of the lengths of Frank–Read and single-arm sources is taken to follow a two-parameter Weibull distribution. For an initial source density of about $2\times10^{12}\,\mathrm{m}^{-2}$, the best fit of the Weibull distributions yields a size exponent of -0.69. Several scaling laws are examined at a fixed strain and strong correlations are evidenced, in particular between the average length of the active source segments and the pillar diameters.

In these two simulations, the large initial number of fixed pinning points reduces dislocation losses at the free surfaces and induces built-in correlations. It follows that the information drawn from the microstructures and their evolution is essentially qualitative. Nevertheless, the notions of source shut-off and weakest link activation make sense since they somehow describe a critical behaviour that should lead to dislocation avalanches.

Tang *et al.* (2008) investigated the deformation of $\langle 011\rangle$ submicron-sized pillars and mainly describe the behaviour obtained for diameter of 400 nm. Image forces are treated in the same manner as in epitaxial layers (Liu and Schwarz, 2005). This study demonstrates that it is possible to construct initial configurations containing no fixed pinning points. A microstructure containing a random distribution of dislocation lines, either straight of containing randomly distributed colinear superjogs, is fabricated first. As there is no cross-slip, the initial superjogs may represent the outcome of virtual cross-slip events. The initial microstructure is obtained by allowing this configuration to relax. It then forms a stable three-dimensional network with a rather high dislocation density ($1.9\times10^{14}\,\mathrm{m}^{-2}$). Fig. 5.21 schematically depicts a typical single-arm dislocation source formed under stress.

The rotating arm of the source, *OB*, is held by a node *O*, which is connected to an axial superjog *OA* with zero Schmid factor (grey lines). This superjog is stabilized by a junction *EF* formed with a neighbouring segment (black lines). The node *O* moves back and forth along the intersection *CD* of its $\{1\bar{1}1\}$ slip plane with the $\{111\}$ plane containing the superjog. When it reaches the surface, the source shuts down. The stress–strain curves exhibit large strain bursts corresponding to the generation of dislocations by similar sources and their shutdown. Between

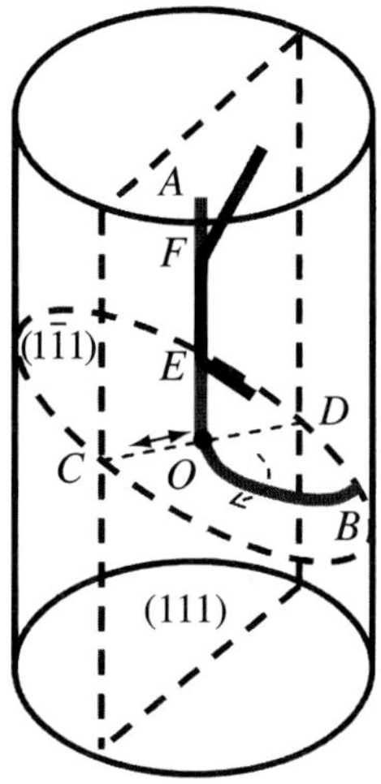

FIG. 5.21. Typical spiral source configuration under stress. It consists of a mobile arm *OB* pinned by an axial superjog *OA* (see text). From Tang *et al.* 2008, courtesy K. W. Schwarz.

these events, the specimen is in a state of starvation and almost elastic reloading sequences are observed.

In another set of simulations, which was carried out with the Weygand code, the boundary conditions are based on the superposition principle. The initial configurations first consisted of Frank–Read sources pinned at their ends (Senger *et al.*, 2008; Weygand *et al.*, 2008). In the simulations by Motz *et al.* (2009), a stochastic rule for cross-slip is implemented and the model material is aluminium. Random densities of dislocation loops with box-shaped distributions of diameters are constructed and further relaxed (Fig. 5.22*a*). DD simulations are carried out in displacement control on $\langle 100 \rangle$ micropillars with square cross sections of side $0.5-1$ μm and initial relaxed densities of 2×10^{13} m^{-2}. As illustrated by Fig. 5.22*b*, cross-slip induces a saturation of the dislocation density and considerably reduces the strain hardening rate.

Cross-slip also results in the natural formation of dislocation sources, which exhibit an amazing variety of pinning configurations. The latter are based on glissile and colinear reactions and some of them are quite stable. With decreasing diameters, Frank–Read sources are no longer found and plastic deformation is entirely carried out by spiral sources. Another simulation shows that the latter induce a strain rate dependence of the simulated flow stress (Senger *et al.*, 2011). This possibly arises because the rotation kinetics of single arm sources is governed by phonon damping.

In the work by Zhou *et al.* (2010, 2011), a large-scale simulation that mimics the deformation of a bulk crystal is first carried out. The initial configuration

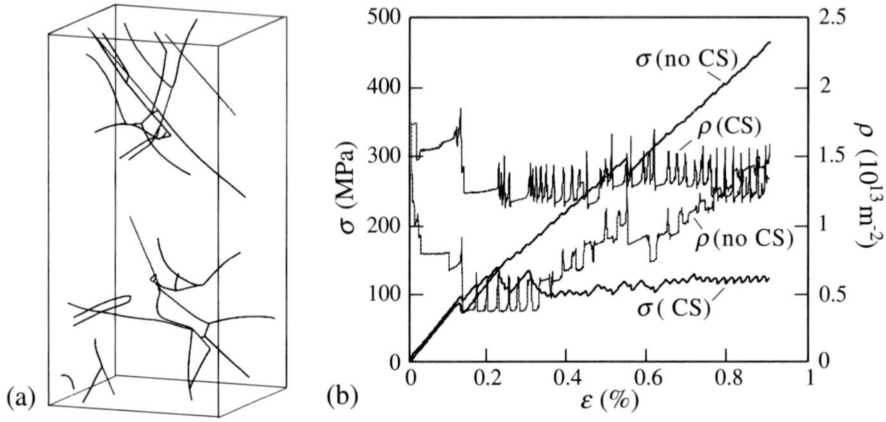

FIG. 5.22. (a) Relaxed dislocation configuration containing no fixed pinning point (density 2×10^{13} m^{-2}; cross-sectional side 0.75 μm). (b) Effect of cross-slip (CS) suppression on the axial stress–strain curve and dislocation density in a pillar with cross-sectional side 0.5 μm. After Motz *et al.* 2009, courtesy F. Motz.

consists of Frank–Read sources with fixed pinning points and a stochastic cross-slip rule is implemented. Microsamples with aspect ratios of 2 are extracted along various orientations from the unloaded (i.e., relaxed) final configuration. Their diameters are between 0.5 and 1 μm and the initial density is about 10^{13} m^{-2}. The boundary conditions are based on the superposition principle and solved by a boundary element method.

The mechanical response is found to be identical for single- and multiple-slip orientations, as the specimens oriented for single slip deform in multiple glide. This is due to the small aspect ratio and the small number of available sources (see Ng and Ngan 2008*a*). The size exponent value measured at an engineering strain of 1% is –0.64. As in Fig. 5.22*b*, the strain dependence of the dislocation density exhibits an initial decrease due to mechanical annealing, but instead of saturating the dislocation density further increases. Whenever a source stops operating or looses its pinning point at the surface, a new source with a higher critical stress is activated. The resulting strain hardening is interpreted in terms of the stochastic source exhaustion model developed by Parthasarathy *et al.* (2007).

In summary, the removal of fixed pinning points and the implementation of cross-slip allowed DD simulations to produce realistic source configurations and to give more substance to the mechanisms that control the yield and flow stresses. The picture that emerges is that of progressive transitions with decreasing diameters from bulk-like behaviour to a behaviour governed by the stochastic properties of Frank–Read sources and, further, of a decreasing number of single-arm sources. The effect of initial density on sample size is reasonably well reproduced.

Fig. 5.20*a*, which shows the simulated dependence of the size exponent on initial density at small strains (1%), exhibits an interesting feature. For large pillar diameters an increase of the initial density induces strengthening, as in bulk materials. Conversely, for diameters smaller than about 1 μm, softening is observed when the initial density increases. Fig. 3 of Motz *et al.* (2009) also shows such a softening for a pillar of size 0.5 μm and initial densities of 2 and 3×10^{13} m^{-2}. Thus, DD simulations suggest that a regime governed by the progressive exhaustion of semi-stable spiral sources may start manifesting itself for pillar diameters in the micron range.

Doubts were expressed about the occurrence of such a regime for pillar diameters in the sub-micron range (Lee and Nix 2011, see also Weinberger and Cai 2011 and Lee *et al.* 2011). The abundance of cross-slip events occurring in relaxed initial densities depends on the accuracy of the cross-slip rule. In addition, if dislocation dissociation could be explicitly taken into account, spiral sources might be found to be less stable and less numerous. As shown in the next section, these questions are related to another open problem, the experimental characterization of dislocation starvation and nucleation-controlled deformation.

5.5.5.4 Further experiment and modelling This section reports on quantitative results on source- and nucleation-controlled deformation in nanopillars and the modelling of these two regimes. Experimentation is reaching the domain of nanowires, which will allow testing predictions on dislocation nucleation processes drawn from atomistic simulations and theory.

We consider first micropillar diameters such that dislocation multiplication occurs by internal sources. A model for the mechanical response should include, as in bulk materials, evolutionary laws for the mobile and stored dislocation densities taking into account all relevant mechanisms. Such kinetic equations are usually complemented by an equation that expresses the splitting of the total strain rate into elastic and plastic strain rates (eqn. 3.7). Such a model was proposed by Malygin (2010); it also incorporates dislocation nucleation at free surfaces. In spite of the large number of parameters involved, reasonable trends are obtained and the size exponent values range between –0.62 and –1.

The stochastic source model (Parthasarathy *et al.*, 2007; Norfleet *et al.*, 2008) can also be taken as a starting point for further developments. This model was revisited by Zhou *et al.* (2011) and expressed in the following form

$$\frac{\tau}{\mu} \approx k\frac{b}{\lambda} + \alpha b\sqrt{\rho} \tag{5.4}$$

An irrelevant lattice resistance that appears in the original model is omitted. The first term at the right-hand side describes the average strength of a single-arm source; k is a known coefficient and λ is an effective source length that depends on total dislocation density and sample size. The second term, where $\alpha \approx 0.35$, describes dislocation strengthening (Section 2.5.2). In this simplified form, it is assumed that one or a few spiral sources govern the yield stress. The effective source length and its standard deviation are estimated from the statistical distribution of a single random spiral source in its slip plane. The size exponent value at yield and its dependence on dislocation density can then be determined.

Figure 5.23 shows that this model describes rather well the experimental value of the average size effect and its statistical scatter over a wide range of pillar diameters for a realistic initial density of about $10^{13}\,\mathrm{m}^{-2}$ (the reader is referred to the original publication for experimental references).

Although the agreement persists down to diameters of about 100 nm, the probability for finding a stable spiral source after mechanical annealing may be vanishingly small for such a small value. Following a model by Greer (2006), Zhou *et al.* (2011) estimate that for initial densities of 2×10^{12}, 2×10^{13} and $10^{14}\,\mathrm{m}^{-2}$, dislocation starvation should approximately occur for critical pillar diameters of, respectively, 1250, 400 and 180 nm.

The effect of initial density on the size effect of gold nanopillars with a high aspect ratio was investigated by Lee *et al.* (2009); it is illustrated by Fig. 5.24. A very ingenious methodology was devised to measure the effect of deformation and annealing on FIB-prepared ⟨001⟩ nanopillars with same initial microstructure.

All pristine pillars were initially prestrained by 35%, which introduced a dislocation density of $10^{15}\,\mathrm{m}^{-2}$. After remilling to a diameter of 300 nm, this density

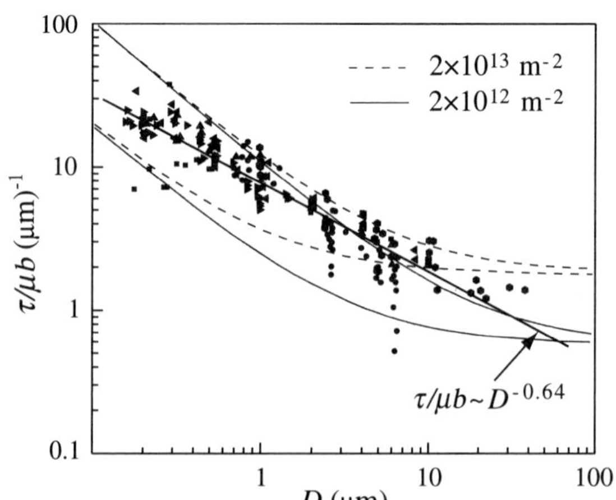

FIG. 5.23. Comparison of the size effect predicted by the stochastic single-arm source model with experimental results on compressed Ni, Al, Cu and Au small-scale pillars. The full and dashed curves show the standard deviations predicted by the model for two initial dislocation densities. After Zhou *et al.* 2011, courtesy C. Zhou.

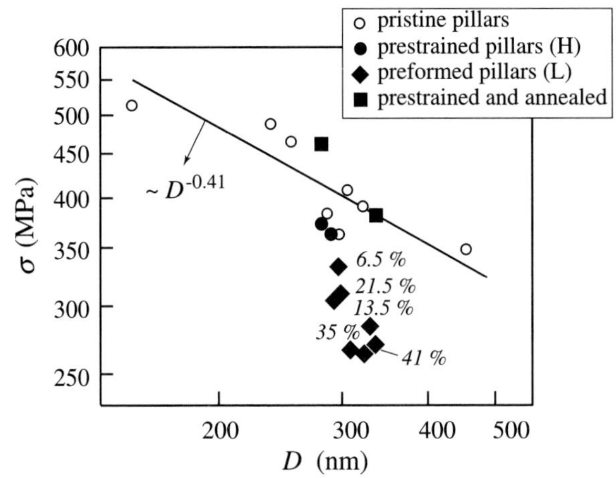

FIG. 5.24. Dependence of the true compressive stress at 5% strain on diameter in gold nanopillars. The pristine (as-prepared) nanopillars were submitted to various treatments (see text). The amount of deformation and the size exponent value are indicated. After Lee *et al.* 2009, courtesy S. W. Lee.

is drastically reduced to a value of about $10^{13}\,\mathrm{m}^{-2}$. Two such samples, with a high aspect ratio, are denoted H in Fig. 5.24; they exhibit the same proof stress as pristine pillars. Further deformation of samples with low aspect ratio (L) significantly reduces the 5% proof stress, which eventually tends towards a saturated value for a strain of about 35%.

TEM measurements yield a saturated dislocation density of about $10^{15}\,\mathrm{m}^{-2}$, which induces a strengthening in good agreement with the measured axial stress. Hence the nanopillars have transited from a dislocation exhaustion regime to a dislocation strengthening regime. Annealing removes the density introduced in the pristine pillars by the initial 35% prestrain and restores their high proof stress. The effect of prestrain is thus reversible and brings the specimens back to the dislocation exhaustion regime. These experiments confirm the duality of strengthening mechanisms depicted in Fig. 5.20*a*. The discussion provides arguments for dislocation starvation and heterogeneous surface nucleation in all pristine nanopillars.

Compression tests performed on bcc MoAl whiskers show that they are considerably softened by a prestrain (Section 2.7.2.3). *In situ* TEM deformation of such specimens was further carried out in tension, with a simultaneous recording of the stress–strain curves (Chisholm *et al.*, 2012). As-grown fibres with square cross sections of side 160 nm plastically collapse at high stresses. An example is also shown of an as-grown fibre which is not dislocation-free and breaks at a much lower stress after some plastic flow was produced by a spiral source. Specimens with same size prestrained by 9% provide perfect illustrations of starvation by mechanical annealing. In contrast, a more heavily prestrained sample (15%), with a size of about 320 nm, undergoes sustained plastic flow by profuse dislocation motion in a thick localized region (see videos 1 to 5).

The selective etching method (Bei *et al.*, 2007) was used to extract single-crystalline Al nanofibres from a lamellar Al_2Cu eutectic alloy (Mompiou *et al.*, 2012). The fibres exhibit semi-elliptical cross sections with effective diameters ranging from 120 nm to 1.2 μm. They were strained *in situ* in a scanning electron microscope and in a TEM. Their initial and final dislocation densities were measured by TEM.

The types of behaviour observed under stress range from abrupt fracture to more or less extended plastic flow (Fig. 5.25). In the ductile pillars, *post mortem* and *in situ* TEM observations show the occurrence of a highly localized deformation carried out by isolated spiral sources. The latter are formed during the initial motion of dislocations towards the surface. The *in situ* operation of a single source (see video S1) is examined in some detail. Cross-slip is seldom observed and the emitted dislocations transiently pile up before vanishing at the surface. The occurrence of a native oxide layer is responsible for this effect.

In situ stress measurements of local dislocation curvatures show that the critical stress of single-arm sources increases with decreasing distance of the pinning point from the surface. No evidence was found for dislocation starvation and the observations are consistent with a source exhaustion mechanism and the single-arm source model. The size exponent, as determined from both dislocation curvature and stress–strain data, is always close to –1.

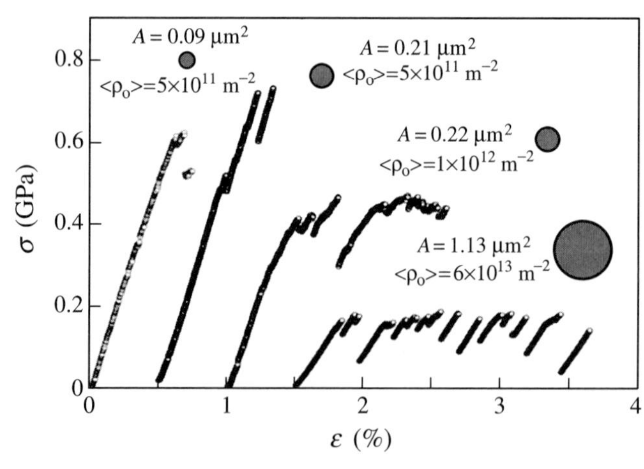

FIG. 5.25. Representative axial stress versus strain curves of aluminium single-crystalline nanofibres. Cross-sectional areas and initial dislocation densities are indicated. After Mompiou *et al.* 2012.

In the last two sets of experiments, the fibres with smallest diameters tend to be brittle since they are dislocation-free. They exhibit typical properties of nanowhiskers, which break at stresses that depend on the degree of perfection of their surfaces and may reach values close to the theoretical stress (Richter *et al.*, 2009). Nanopillars are prepared by different methods and contain an initial dislocation density. Since they exhibit ductility when they are starved, they are more suited for studies of heterogeneous dislocation nucleation at surfaces. The steady-state flow stress of electroplated copper nanopillars was found to exhibit a transition between a regime of relatively high activation volumes for large strain rates and a regime of significantly smaller activation volumes (Jennings *et al.*, 2011). For strain rates going from 10^{-3} to $10^{-1}\,\mathrm{s}^{-1}$, the critical diameter for this transition decreases approximately from 125 to 75 nm. The small activation volumes are found to be around $10b^3$, which is typical of surface dislocation nucleation (Zhu *et al.*, 2008). The higher values may arise from the sensitivity of single-arm sources kinetics to strain rate.

To examine the major factors governing starvation at the mesoscale, Nix and Lee (2011) proposed a very simple kinetic model expressing that the evolution of the mobile density ρ_m in nanopillars is governed by dislocation nucleation and annihilation at free surfaces. The evolution of the mobile density with time is written as a continuous rate equation

$$\dot{\rho}_m = \pi \frac{\omega_o}{b^2} \left(\frac{\tau}{\tau_{th}} \right)^m - \rho_m \frac{\bar{v}}{D} \tag{5.5}$$

The first term at the right-hand side is the rate of increase of the mobile density; it is obtained as follows. The thermally activated nucleation rate, which depends on the surface state, is taken by default as a power law of the stress scaled by the maximum shear strength of the surface, τ_{th}. Its prefactor is the nucleation frequency per unit atomic area, ω_o/b^2, times the surface πDh of a pillar of diameter D and height h. The increase in mobile density during the lifetime of the nucleated dislocation is the ratio of its average length, $\bar{\ell} = \pi D/4$, to the volume, $h\pi D^2/4$. The second term expresses the rate of density loss at the free surfaces (Tang *et al.*, 2007; Malygin, 2010). It is expressed as a negative storage rate determined by a mean free path $\Lambda = D$, that is, $d\rho/d\dot{\gamma}_p = -1/bD$, where $\dot{\gamma}_p$ is the plastic strain rate (Section 2.6.7). Its final form is obtained by developing $\dot{\gamma}_p$ with the help of Orowan's law (eqn. 1.21).

Eqn. 5.5 has a steady-state solution given by the conditions $\dot{\rho}_m = 0$, $\dot{\tau} = 0$. By inserting Orowan's law into the steady-state relation, one can verify that the steady-state flow stress depends on strain rate and exhibits a power law dependence on diameter with an exponent $-1/m$. The initial transient behaviour can also be solved by combining eqn. 5.5 with eqn. 3.7. Depending upon whether the initial density is smaller or larger than the steady-state density, a hardening or softening transient is obtained.

Experimental results are emerging on nanopillars and nanowhiskers with diameters below 100 nm (Richter *et al.*, 2009; Jennings *et al.*, 2011) and on nanowires with diameters in the 10 nm range (Zheng *et al.*, 2010). The processes of surface dislocation nucleation in such materials are, however, essentially known from theory and simulation. The current understanding and modelling in this domain is discussed in two review articles. Weinberger and Cai (2012) summarize the predictions of MD simulations on the plastic properties of nanowires. Weinberger *et al.* (2012) provide a broad overview of dislocation nucleation at surfaces based on static atomistic simulations and continuum modelling. In spite of the limited dimensions of the simulated diameters, a preliminary connection to experimental results and observations is attempted.

5.6 Concluding remarks

The applications of 3D DD simulations discussed in this chapter show that significant understanding can be achieved when a sustained effort is made with clearly defined objectives. Numerical results in agreement with experiment should lead, whenever possible, to modelling. Modelling is critical, as it allow establishing a connection between discrete and continuum approaches of plasticity. However, dislocation-based continuum models have a predictive value insofar as DD simulations can verify them and estimate unknown parameter values. The situation is less bright regarding the connection with the atomic scale; it follows that substantial progress will have to be made before fully predictive multiscale schemes can be constructed.

The existing DD simulations can nevertheless contribute to further developments, all the more as several of them are now easily accessible. A non-exhaustive list of challenging problems could include the formation of dislocation patterns in monotonic and cyclic deformation, the interaction of dislocations with atomic-scale defects and creep mechanisms, the composition of strengthening mechanisms, the expansion of the connection with continuum mechanics to damage and fracture, and further systematic studies on small-scale materials, engineering materials and geological materials.

APPENDIX A

THERMAL ACTIVATION OF DISLOCATION MOTION

In what follows we mainly consider materials exhibiting a negligible lattice resistance where dislocations overcome localized obstacles with the help of thermal activation. The case of dislocation motion by the kink-pair mechanism, which involves the competition of several activation sites along a dislocation line, is treated in Section 3.2.4.3.

Here and everywhere else, we adopt the following notations except when consecrated notations already exist. W is a mechanical work and E, or U, is an internal energy, for example an elastic strain energy or a core energy. In rate equations, activation enthalpies are denoted H and use is made of the Gibbs free energy $G = H - TS$ when the entropy term S matters or for more generality. Experimentalists measure activation enthalpies, as shown in Section A.1, whereas atomistic simulations yield activation enthalpies or internal energies. The entropy term is usually negligible below room temperature (Cottrell, 1953; Schoeck, 1980). The only contribution that matters arises from changes in the phonon spectrum at medium and high temperatures. It is described in terms of the temperature dependence of the shear modulus (Schoeck, 1965). The theory of thermally activated dislocation motion involves several complexities and subtleties; the references given in Section 1.2.2 may be consulted for a more complete approach.

A.1 Mesoscale framework

Force–distance diagrams, which are illustrated by Fig. A.1, were developed in parallel with early phenomenological models for the overcoming of localized obstacles. They allow visualizing the forces and energies involved during an activation event and their temperature dependencies (Kocks *et al.*, 1975; Argon, 2008). In what follows, the work ΔW of the effective force in Fig. A.1*b* is associated to an effective stress τ^* (eqn. 1.18), assuming that other stress components are not thermally activated. The activation energy then depends on τ^* and the thermodynamic system considered is localized around the activated event. Alternatively, the system considered is the whole sample and use is made of the resolved applied stress, which can be experimentally controlled. The two formulations lead to fundamentally identical analyses, but slightly different results. In both cases, a

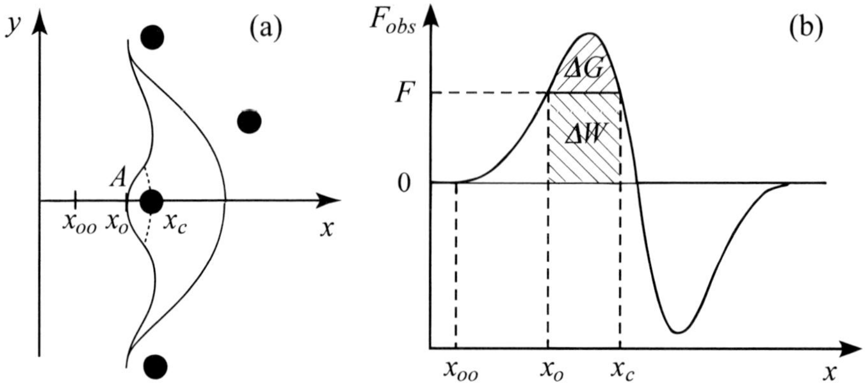

FIG. A.1. Thermally activated overcoming of localized obstacles (schematic). (a) A dislocation segment overcomes a distribution of obstacles. The equilibrium position of the dislocation far from the obstacle is x_{oo}. Under an effective force and the resistance of obstacle A, the equilibrium shape is represented by the full line cutting the x axis at position x_o. At the critical configuration (dashed curve) the position is x_c. After a successful activation event the dislocation moves until it is blocked by another obstacle. (b) A force–distance diagram. The resistance of the obstacle to dislocation motion F_{obs} is drawn as a function of its position x. At a small approach distance x_o, the resistive force is F and the free energy to be furnished by thermal activation is represented by the hatched area labelled ΔG. The area labelled ΔW represents the work of the effective force on the dislocation line.

better precision is obtained by making use of scaled stresses and energies to take into account the temperature dependence of the shear modulus.

The simple thermodynamic framework discussed in what follows is currently used at the mesoscale to describe the overcoming of localized obstacles by moving dislocations. It is consistent with Vineyard's harmonic transition state theory (see Rodney 2007). Care should be taken, however, upon extrapolating this framework to $T = 0\,\mathrm{K}$ (Alefeld, 1964).

Under a critical effective stress, a portion of mobile line of length ℓ_c reaches a critical configuration that usually depends on stress and temperature. The activation probability P per unit time at an absolute temperature T is written in the form

$$P = \nu^* \exp - \left(\frac{\Delta G(\tau^*)}{k_B T} \right), \tag{A.1}$$

where ν^* is an effective frequency that will be defined in what follows and ΔG is the difference of Gibbs free energies between the ground state and the saddle point.

ΔG depends on temperature and on the effective stress τ^*. This last dependency arises because the total height of the barrier, $\Delta G(0)$, is reduced by the mechanical work $\Delta W(\tau^*)$ done by the effective force $\tau^* b$ along the path leading to the critical configuration (Fig. A.1*b*). One thus has $\Delta G(\tau^*) = \Delta G(0) - \Delta W(\tau^*)$. At $T = 0\,\mathrm{K}$, there is no thermal activation and the effective stress reaches the maximum value $\tau^*(0)$ associated with the considered process. At a certain temperature T_a, which will be further discussed, the mechanical work vanishes because the total free energy of activation is entirely furnished by thermal fluctuations.

As the activation probability P represents the frequency of successful attempts, it is the inverse of the *waiting time* $t_w = 1/P$ for such an event. In the continuum, the dislocation line behaves as an elastic string. The attempt frequency, ν^*, is assimilated to the fundamental vibration mode associated with the critical activation length ℓ_c, $\nu_{\ell_c} \approx \nu_D b/\ell_c$, where ν_D is the Debye frequency (cf. eqn. 2.3). If the dislocation reaches the saddle point at its first attempt, it jumps by a distance d_{obs} towards the next obstacle. Its velocity is then $v_o \approx \nu_D b d_{obs}/\ell_c$. The dislocation velocity, $v = d_{obs}/t_w = P d_{obs}$, is then obtained from eqn. A.1 and

$$\frac{v}{v_o} = \frac{\dot{\gamma}}{\dot{\gamma}_o} = \frac{\dot{\varepsilon}}{\dot{\varepsilon}_o} = \exp - \left[\frac{\Delta G(\tau^*)}{k_B T}\right], \tag{A.2}$$

where dislocation velocities were transformed into shear and applied strain rates using Orowan's law. This expression can be rewritten in the form

$$\Delta G(\tau^*) = k_B T \ln\left(\frac{v_o}{v}\right) = k_B T \ln\left(\frac{\dot{\gamma}_o}{\dot{\gamma}}\right) = k_B T \ln\left(\frac{\dot{\varepsilon}_o}{\dot{\varepsilon}}\right) \tag{A.3}$$

Equation A.3 shows that the activation free energy is proportional to temperature but depends only in a logarithmic manner on dislocation velocity or strain rate.

The experimental analyses of thermally activated dislocation mechanisms are discussed in full detail by Caillard and Martin (2003); they are largely based on measurements obtained by performing differential tests. A small change in temperature under constant stress induces a change in applied strain rate that is approximated by the partial derivative $\partial \ln(\dot{\varepsilon})/\partial T|_{\tau^*}$ of eqn. A.2. The activation enthalpy $\Delta H(\tau^*)$ can be derived from this test, as can be verified by developing the previous expression and combining it with $\Delta H = \Delta G - T\Delta S$ and its partial derivative in same conditions. Similarly, a small change in applied strain rate at a constant temperature induces a stress change. The measurement of $\partial \ln(\dot{\gamma})/\partial \tau^*|_T$ yields access to a stress-dependent quantity, the *activation volume*

$$V(\tau^*) = -\left.\frac{\partial \Delta G}{\partial \tau^*}\right|_T \tag{A.4}$$

The *activation area*, $V/b = \mathcal{A}$ is related to the area swept by a dislocation during an activated event. It is usually taken as $\mathcal{A} \approx d_{obs}(x_c - x_{oo})$ (cf. Fig. A.1), where d_{obs} is assimilated to the average distance between obstacles along the lines. One may note that the actual activation length, ℓ_c, may sometimes be smaller than d_{obs}, as depicted in Fig. A.1*a*. The value of the activation volume yields an insight into the degree of localization of the obstacle governing plastic flow. It is also related to the *strain rate sensitivity* of the stress, $S = \partial\tau^*/\partial \ln \dot{\varepsilon}|_T$. This can be seen by taking the partial derivative of eqn. A.3 with respect to stress at constant temperature, from which one obtains $S = k_B T/V$.

A.2 Orders of magnitude

We now examine a few numerical values. Orders of magnitude for the waiting times can be obtained by taking a flight distance between obstacles $d_{obs} = Nb$, where N is an integer, and neglecting the free flight time of the dislocation between obstacles. The average dislocation velocity is then $\bar{v} = d_{obs}/t_w$. With $\bar{v} \approx 0.3\,\mu\text{m}\,\text{s}^{-1}$ in conventional deformation tests (Section 1.4.3) and $b \approx 0.3$ nm, $t_w \approx 10^{-3}N$. Since $N > 1$, the average waiting time cannot be smaller than 10^{-3} s and can reach values of the order of a few seconds for obstacle spacings in the micron range. They are smaller when there are competing nucleation sites along a dislocation line (Section 3.2.4.3).

It is usually assumed that $d_{obs} \approx \ell_c$, so that the prefactor v_o in eqn. A.2 reduces to $v_o \approx \nu_D b \approx c_t$, the velocity of transverse waves (see Section 2.2.2.3). One has typically $\nu_D \approx 10^{13}$ s and $b \approx 0.3$ nm, so that $v_o \approx 3000\,\text{m}\,\text{s}^{-1}$. As $\bar{v} \approx 0.3\,\mu\text{m}\,\text{s}^{-1}$, $\bar{v_o}/v \approx 10^{10}$ and eqn. A.3 can be rewritten in the approximate form

$$\Delta G(\tau^*) = \ln(v_o/v) = Ck_B T \approx 23k_B T \tag{A.5}$$

The first two equalities at the left hand side of eqn. A.5 generally apply to the overcoming of all types of localized obstacles within a broad temperature range. The logarithmic dependence of C makes it rather insensitive to dislocation velocities and strain rates in the quasi-static regime. This quantity can then be considered as quasi-constant. Nevertheless, the numerical value of C is modified when $d_{obs} \ll \ell_c$. As the relations given in eqn. A.5 are fully general, the discussion that follows applies to all types of materials. In particular, it applies to bcc metals, albeit with a slightly different value of C.

Assuming a negligible entropy contribution, eqn. A.5 allows performing quick estimates of the athermal transition temperature T_a at which the total activation enthalpy $\Delta H(0)$ is provided by thermal fluctuations. For instance a mechanism that is experimentally found to become athermal at $T_a = 300\,\text{K} \approx (1/40)\,\text{eV}$ is characterized by a total activation enthalpy $\Delta H(0) \approx 0.68$ eV. Close to $T = T_a$, the effective stress becomes very small and the activation volume becomes very large (eqn. A.4), so that the strain rate sensitivity k_BT/V is vanishingly small. At very low temperatures, the strain rate sensitivity also becomes vanishingly small, as there is no thermal activation at $T = 0\,\mathrm{K}$. In consequence, the strain rate sensitivity exhibits a maximum at an intermediate temperature.

We now consider deformation tests carried out under a constant strain rate at different temperatures. Still neglecting the entropy term, the derivative of eqn. A.5 with respect to temperature yields

$$-V\left(\frac{d\tau^*}{dT}\right)_{\dot{\varepsilon}} = Ck_B \approx const. \tag{A.6}$$

Thus, small activation volumes are always associated with strong dependencies of stress on temperature.

The effect of the high strain rates used in MD simulations can be approximately estimated as follows. The dislocation velocities typically achieve values of a few hundreds metres per second (Section 1.5), which are about 10^9 times larger than the average value $\bar{v} \approx 0.3\,\mu\mathrm{m\,s^{-1}}$ mentioned above for laboratory tests. It follows that the constant $C \approx 23$ is reduced to an effective value $C^* \approx 5$. Then, for a given total activation enthalpy, the athermal transition temperature increases up to $T_a^* = T_aC/C^*$. In consequence, in a MD simulation conducted at a nominal homologous temperature T/T_a, the effective behaviour corresponds to a much lower temperature, T/T_a^* and, hence, to a too high stress. This is why static simulations should be preferentially used for studies of thermally activated phenomena at the atomic scale. It is, however, sometimes possible to extract quasi-static data from the output of MD simulations (see, e.g. Monnet *et al.* 2010, and Section 3.2.4.2).

APPENDIX B

SELECTION OF MATERIALS CONSTANTS

Table B.1 presents a selection of material constants for dislocations dissociated into Shockley partials in compact planes. Two points that deserve discussion are detailed in Sections B1 and B2: one is the determination of stacking fault energies (SFEs) and plausible dissociation widths, and the other is the choice of adequate elastic constants for calculations and simulations on elastically anisotropic materials. Material constants for bcc metals are given in Table 3.1.

TABLE B.1. Material constants for some fcc, dc and hcp crystals. γ: intrinsic SFE; b: modulus of the Burgers vector; d_s/b: scaled dissociation width for screw dislocations. μ: shear modulus (Reuss average over compliances at $T = 300\,\text{K}$); K_s: energy coefficient for a screw dislocation in anisotropic elasticity; K_sb^3: pre-logarithmic energy factor of a screw dislocation in anisotropic elasticity ($T = 300\,\text{K}$). $\gamma/\mu b$: scaled SFE based on the Reuss average. See text for comments and references.

	γ (mJ m^{-2})	b (nm)	d_s/b	μ (GPa)	K_s (GPa)	K_sb^3 (eV)	$10^3\gamma/\mu b$
Al	122, 130	0.286	1.8–2.6	26.3	25.9	3.79	17.6
Ag	16.3, 18	0.289	7.13	25.6	26.6	4.01	2.20
Au	32, 33	0.288	(2.8)	24.2	24.7	3.69	4.57
Cu	45, 41	0.256	6–7	40.0	42.1	4.41	4.0
Ni	125, 110	0.249	≈ 5	77.8	78.6	7.58	6.0
Ge	78	0.40	5.99	53.1	52.1	21.0	3.66
Si	65	0.384	8.33	64.9	63.7	22.5	2.61
Be	390	0.228	–	150	148	11.1	11.6
Cd	15	0.298	12.42	24.1	26.9	4.43	2.09
Mg	30	0.321	(3.43)	16.6	16.6	3.43	5.63
Zn	35	0.266	9.77	47.3	51.5	6.06	2.78

B.1 Stacking fault energies, dissociation widths

In the first column the value of γ for Al is taken from an *ab initio* calculation by Woodward *et al.* (2008), which yields a range of dissociation widths for the screw dislocation. Values of γ for other fcc metals than Al, as well as for Ge and Si, are taken from a compilation of experimental TEM observations by Carter (1984). The second column of SFE values for fcc metals is taken from another *ab initio* calculation by Kibey *et al.* (2007). There is a quite good agreement between experimental and computed values.

For materials with large dissociation widths, such as Ag, Ge, Si, Cd and Zn, Püschl and Schoeck (1994) derived the values of d_s/b from SFEs using anisotropic elasticity. For fcc metals with narrower cores, such as Cu, Au and Ni, the screw dissociation widths are small and may elude TEM observation in weak-beam conditions (Au, Ni). Extrapolated values are obtained in anisotropic elasticity from the non-screw dissociation widths, which are larger, and the SFEs. It appears, however, that these estimates of d_s/b are systematically too small. In other terms, the dependence of dissociation widths on character is more isotropic than predicted. For instance in copper, the anisotropic prediction from the SFE is $d_s = 4.22b$, whereas Stobbs and Sworn (1971) experimentally found $d_s \approx 7b$. This last value is given in Table B.1 (right column). For Au, the uncertainty seems to be quite large. The experimental value $d_s = 2.8b$ given in parentheses in Table B.1 was extrapolated from the dissociation width of edge and mixed dislocations in weak-beam conditions (Jenkins, 1972) or drawn form high-resolution TEM observations (Balk and Hemker, 2001). As it practically matches the prediction of anisotropic elasticity (Püschl and Schoeck, 1994), this value may be considerably underestimated. Isotropic elasticity provides an upper bound of $d_s \approx 5b$.

These discrepancies were often attributed in the past to an overlap of the fields of the Shockley partials. They may also arise from the lack of an adequate definition of the dislocation core radii and the effect of the core fields on narrow dissociation widths (Henager Jr. and Hoagland, 2004). The reason why SFEs are measured with better accuracy than screw dissociation widths is twofold. First, edge dislocations have larger splitting widths, and second, various extended configurations of non-screw dislocations can be used to determine γ with an improved accuracy.

Embedded-atom potentials—in particular, those that are fitted to *ab initio* data (Mishin *et al.*, 2001)—are thought to yield realistic dissociation widths for fcc metals. The typical screw dissociation widths are then, indeed, substantially larger than those predicted by linear anisotropic elasticity. For copper, the value $d_s = 6b$ given in Table B.1 (left column) was yielded by a Mishin potential with $\gamma = 44.4\,\mathrm{mJ\,m^{-2}}$ (Rao *et al.*, 2010). The value given for Ni (right column) is taken from Rao *et al.* (1999). It was obtained using an EAM potential yielding $\gamma = 119\,\mathrm{mJ\,m^{-2}}$.

The values of γ for divalent hcp metals that preferentially glide in the basal plane are taken from a calculation by Legrand (1984) using pseudopotentials adapted to each material. The scaled dissociation widths are tentatively taken from an anisotropic estimate by Püschl and Schoeck (1994) based on Legrand's results. More recent *ab initio* calculations (Yasi *et al.*, 2009; Wu *et al.*, 2010) confirm that Be has a very high SFE, whereas Mg (and possibly other hcp metals) exhibits much lower values than is often thought. No cross-check of computed values with experimental ones is presently possible by lack of sufficiently accurate TEM observations.

In summary, the SFEs are well known in fcc and dc metals and only approximately in hcp metals. The dissociation widths are not precisely known in some materials, especially Au, where screw dislocations exhibit narrow cores.

B.2 Elastic constants, shear moduli

Isotropic and anisotropic elastic constants for selected elemental materials and a few compounds are tabulated in Appendix I of Hirth and Lothe (1992). Extended compilations of elastic stiffnesses and compliances for elements, alloys and compounds can be found in the metals reference book by Gale and Totemeier (2004). References for the temperature dependencies of these coefficients in fcc metals are given by Sauzay and Kubin (2011). The calculations of average and anisotropic elastic constants are detailed in Chapter 13 of Hirth and Lothe (1992).

The average shear modulus over compliances (Reuss average) is usually preferred to the Voigt average over stiffnesses for calculations of elastic interactions between dislocations. The Voigt average is preferred for calculations of local strains.

In anisotropic elasticity, the line energy takes on a form similar to that of the isotropic one (eqn. 1.4), with an energy coefficient K_s for screw dislocations replacing the isotropic shear modulus. For the materials listed in Table B.1, one has $K_s = [c_{44}(c_{11} - c_{12})/2]^{1/2}$. Scattergood and Bacon (1975, 1982, see also Bacon *et al.* 1979) showed that the anisotropic equilibrium shapes of curved segments or shear loops under stress are reasonably well reproduced if the energy coefficients are known for two orientations. Thus, they defined an effective shear modulus $K_A = K_s$ for screw dislocations and an effective Poisson's ratio $\nu_A = 1 - K_s/K_e$, where K_e is the anisotropic energy coefficient for edge dislocations. This explains why K_s is sometimes used in the literature as a substitute to average shear moduli (see Kocks *et al.* 1975, pp. 26–7). In general, this coefficient is an approximation of the true one because it is not calculated in the correct reference frame (Hirth and Lothe, 1992). However, it yields exact results for screw dislocations parallel to the dense directions of compact planes in fcc, dc and hcp structures. Values of K_s and ν_A for some fcc and bcc metals were tabulated by Bacon (1985).

Corrected coefficients for fcc metals, which take into account the effect of dislocation dissociation on the elastic energy of a perfect dislocation, were tabulated by Bacon (1978).

For large-scale DD simulations, two sets of isotropic constants can be used, the Reuss averages μ and ν and the Scattergood–Bacon constants, K_s and ν_A. In fcc crystals, use is also made of the isotropic shear modulus in $\{111\}$ planes, $\mu_{111} = (c_{11} - c_{12} + c_{44})/3$, which is also the shear modulus in the reciprocal $\langle 111 \rangle$ direction of bcc crystals. The approximation made upon replacing full anisotropic elasticity by isotropic elasticity can be visualized in DD simulations of elementary dislocation configurations performed by Han *et al.* (2003) on copper, Yin *et al.* (2010) in Mo and W, and in the particular case (Section 4.2.8) of α-Fe at high temperatures (Aubry *et al.*, 2011). A comparison between DD and MD results was also performed (Takahashi and Ghoniem, 2008). In all cases, the Voigt average yields very poor fits. The Reuss average and the Scattergood–Bacon approximation are acceptable at the dislocation scale, though the last one looks slightly better.

For almost isotropic materials such as Al, the three moduli μ, K_s and μ_{111} reduce to the Lamé coefficient c_{44}. For anisotropic materials such as Cu, where the Zener anisotropy ratio[13] is $A = 3.21$, their values are quite similar. For instance in copper at room temperature, $\mu = 40\,\text{GPa}$, $K_s = 42.1\,\text{GPa}$ (cf. Table B.1) and $\mu_{111} = 40.8\,\text{GPa}$. The differences between ν, the Reuss average of the Poisson ratio, and ν_A are somewhat larger. These isotropic moduli provide a better approximation of the dislocation fields than could be expected because the anisotropic fields depend on elastic constants in more than one direction (Bacon, 1985).

[13] In cubic crystals, the Zener anisotropy ratio is $A = 2c_{44}/(c_{11} - c_{12})$. For isotropic materials $(c_{11} - c_{12}) = 2c_{44}$ and $A = 1$.

APPENDIX C

SLIP IN SINGLE CRYSTALS

The plastic deformation of single crystals by dislocation glide exhibits characteristic features, in particular when dislocation–dislocation interactions govern plastic flow. This appendix presents basic definitions and elements related to the activation and evolution of slip systems, essentially in fcc crystals, during monotonic deformation under a constant imposed strain rate. The starting points are the Peach–Koehler force induced on a dislocation by an externally applied stress and the Schmid law. The lattice rotations are then examined, as well as their dependence on the initial orientation of the loading axis.

C.1 The Peach–Koehler force

To derive the Peach–Koehler force, one has first to define the direction of the Burgers vector in a consistent manner. The Final-Start, Right-Hand or *FSRH convention*, which is described in Fig. C.1, is the most commonly used one. An orientation is ascribed to the unit length of line direction $\boldsymbol{\ell}_u$ and a closed, right-handed path consisting of atomic steps, the *Burgers circuit*, is drawn around the dislocation core. The successive steps of this circuit are redrawn in a perfect crystal, starting from S and ending at a different position F. By convention, the Burgers vector is given by the vector $\mathbf{FS} = \mathbf{b}$ associated to the closure defect. Its sign depends on the orientation of $\boldsymbol{\ell}_u$. The positive normal to the slip plane is given by $\mathbf{n} = \boldsymbol{\ell}_u \times \mathbf{b}/|b|$.

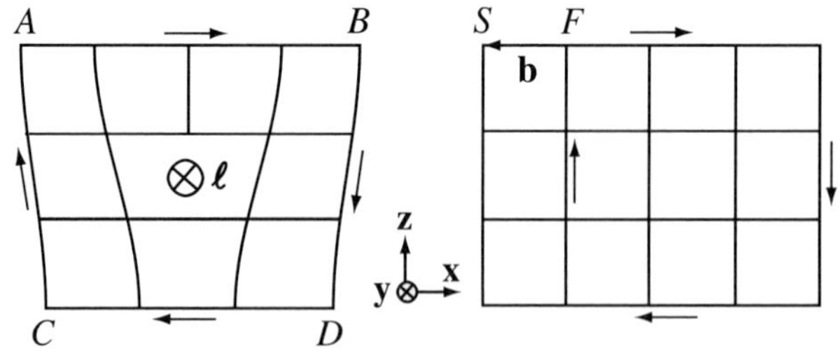

FIG. C.1. The FSRH convention.

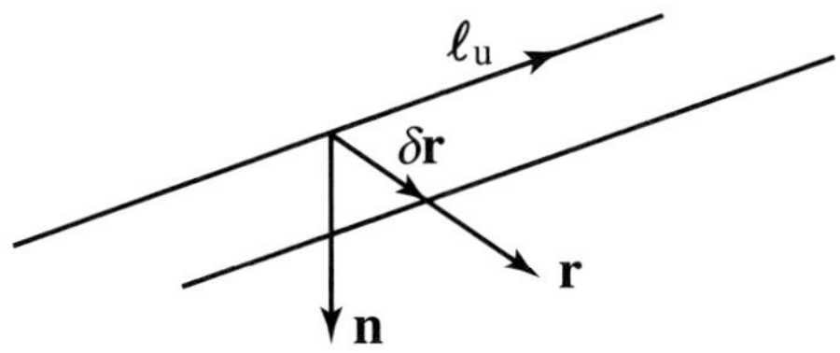

FIG. C.2. A small step of motion of a dislocation segment with unit line direction $\boldsymbol{\ell}_u$ shifts a Volterra cut by $\delta\boldsymbol{r}$.

For calculating the force on a dislocation, we consider a straight element of dislocation line with unit length $\boldsymbol{\ell}_u$, to which a stress tensor $\boldsymbol{\sigma}$ of any origin is applied. Through a Volterra process, the line performs a small step of motion $\delta\boldsymbol{r}$ along a unit direction $\mathbf{r}$ (Fig. C.2). The positive unit normal to the area swept, $\ell\delta r$, is taken as $\mathbf{n} = \boldsymbol{\ell}_u \times \mathbf{r}$. The sign of $\mathbf{n}$ ensures that the upper and lower surfaces of the Volterra cut are displaced with respect to each other by $\mathbf{b}$ in the proper direction. For example (Hirth and Lothe, 1992), if the cut δr is performed along the positive climb direction $\mathbf{z}$ in Fig. C.1, the volume change, $\delta V = -b\delta r(\mathbf{y}\times\mathbf{z}).\mathbf{x}$, must be negative. Indeed, $\delta V = -b\delta r$ per unit line, provided that the Burgers vector has the sign given by the FSRH convention. Changing the sign of the line direction does not affect the final result because the sign of the Burgers vector is also changed.

The stress tensor $\boldsymbol{\sigma}$ induces a force $\boldsymbol{F} = \boldsymbol{\sigma}.\mathbf{n} = \boldsymbol{\sigma}.(\boldsymbol{\ell}_u \times \mathbf{r})$ per unit surface of the Volterra cut. This force produces a relative displacement of the two lips by $\mathbf{b}$ and the resulting mechanical work per unit length of line is written

$$\delta W_1 = \boldsymbol{\sigma}.(\boldsymbol{\ell}_u \times \delta\mathbf{r}).\mathbf{b} = (\boldsymbol{\ell}_u \times \delta\mathbf{r}).(\boldsymbol{\sigma}.\mathbf{b}) = [(\boldsymbol{\sigma}.\mathbf{b}) \times \boldsymbol{\ell}_u)].\delta\mathbf{r} \qquad \text{(C.1)}$$

The work of the configurational force $\boldsymbol{F}$ applied to the core is written $\delta W_2 = \boldsymbol{F}.\delta\mathbf{r}$ per unit length. Since $\delta W_1 = \delta W_2$, the *Peach–Koehler force* is given by (Peach and Koehler, 1950):

$$\boldsymbol{F} = (\boldsymbol{\sigma}.\mathbf{b}) \times \boldsymbol{\ell}_u \qquad \text{(C.2)}$$

The stress per unit length of line is then $\boldsymbol{\tau} = \boldsymbol{F}/|\mathbf{b}|$. The cross-product in eqn. C.2 implies that the Peach–Koehler force is normal to the line, but not necessarily that it lies in the slip plane. It has no component parallel to the line because such a displacement produces no work. The Peach–Koehler force can also be derived within a continuum mechanical framework (see for instance Phillips 2001). The demonstration given above is adapted from the one given by Hirth and Lothe (1992) for a straight dislocation.

The glide and climb forces are obtained by projecting the Peach–Koehler force on the glide and climb directions, respectively. A simple manner to derive these forces consists in going back to the force per unit area of Volterra cut, $\boldsymbol{F} = \boldsymbol{\sigma}.\mathbf{n}$. The projection of this force on the direction of the Burgers vector, $\boldsymbol{F} = \boldsymbol{\sigma}.\mathbf{n}.\mathbf{b}$, yields the glide force acting on the dislocation line. In the same way, the projection on the unit normal, $\boldsymbol{F} = \boldsymbol{\sigma}.\mathbf{n}.\mathbf{n}$, yields the climb force. The glide and climb stresses, τ_g and τ_c are then conveniently expressed in component form by dividing the forces by the modulus of $\mathbf{b}$

$$\tau_g = (b_i/|\mathbf{b}|)n_j\sigma_{ij} \tag{C.3}$$

$$\tau_c = n_i n_j \sigma_{ij}, \tag{C.4}$$

where summation is carried out over repeated indices.

C.2 Schmid's law, lattice rotation

C.2.1 Schmid factor, Schmid's law

Assume now that a tensile or compressive stress $\sigma_{ii} = \sigma$ is applied to a crystal. The potential slip systems are defined by their normals $\mathbf{n}^s$ and their unit Burgers vectors (or slip directions) $\mathbf{b}^s/|\mathbf{b}|$. Equation C.3 yields a glide stress $\tau^s = (b_i^s/|\mathbf{b}|)\sigma n_i^s$. Denoting by φ^s and λ^s the angles between the stress axis and, respectively, the slip plane normal and the Burgers vector of slip system s, we have

$$\tau^s = \sigma \cos\varphi^s \cos\lambda^s = \mathcal{F}^s\sigma, \tag{C.5}$$

where $\mathcal{F}^s$ is the *Schmid factor* of slip system s. This quantity, which is derived here from the Peach–Koehler relation, relates the uniaxial stress to the resolved stress in a slip system. The term $\cos\varphi^s$ accounts for the ratio of the cross-sectional area normal to the specimen axis and along the slip plane, whereas $\cos\lambda^s$ represents the projection of the force in the unit slip direction (Fig. C.3*a*). A similar relation also exists between a shear strain in the slip plane γ^s and its contribution ε^s to the total uniaxial strain. During a small plastic strain increment, the mechanical work of the applied stress is $dW = \sigma\, d\varepsilon^s$. Conservation of energy implies that this work is identical to the work of the resolved stress, $\tau^s d\gamma^s = \mathcal{F}^s\sigma\, d\varepsilon^s$. Hence $d\varepsilon^s = \mathcal{F}^s d\gamma^s$.

The Schmid law was established by Schmid (1924) and extensively investigated by Schmid and Boas (1935). It states that slip starts in a given slip system when the resolved stress τ^s reaches a critical value τ_c^s. At yield, this stress is often called the *critical resolved shear stress* (CRSS). The Schmid law is well verified for slip on compact planes in crystals where dislocation–dislocation interactions govern plastic flow. It applies, in particular, to fcc crystals and the hcp crystals that preferentially glide in the basal plane. For {111} slip, the preferentially activated slip system at yield is the one with highest Schmid factor. For high-symmetry orientations, several slip systems can be simultaneously activated. The value of

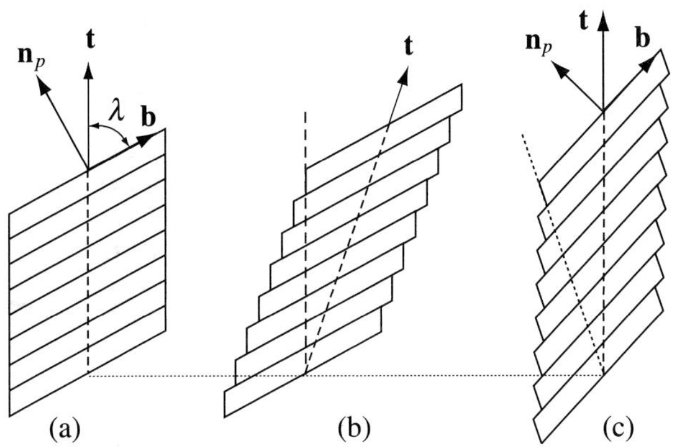

FIG. C.3. Single slip deformation of a single crystal in tension. The 2D view is a projection on the plane $(\mathbf{t}, \mathbf{b})$, where $\mathbf{t}$ is the tensile axis. $\mathbf{n}_p$ is the projected normal to the slip planes (see text). (a) Initial state. (b) The same crystal deformed in simple shear. (c) The same crystal deformed in tension along a fixed loading axis.

$\mathcal{F}$ is maximum when the two angles λ and φ are identical (eqn. C.5) and have a maximum value. This is achieved when the stress axis, the normal $\mathbf{n}$ and the Burgers vector $\mathbf{b}$ are coplanar with $\lambda = \varphi = 45°$. Then, $\mathcal{F} \approx 0.5$, which occurs to a very good approximation for $\langle 149 \rangle$ orientations. One may note that the Schmid law does not imply that the critical stresses are identical in all potentially active slip systems. Notable exceptions to this law, especially in bcc metals and alloys, were observed very early (see Section 3.2.3).

C.2.2 Lattice rotations

During plastic flow, the evolution of the Schmid factor is governed by the imposed boundary conditions and the activity of slip systems. This is schematically illustrated by Fig. C.3 in the case of single slip along a tensile direction $\mathbf{t}$.

In Fig. C.3, the inclined slabs represent atomic slip planes, which can slide with respect to each other by crystallographic shear. The crystallographic axes are defined by the Burgers vector $\mathbf{b}$ and the normal to the slip plane, $\mathbf{n}$, which is seen in projection as the vectors, $\mathbf{t}$, $\mathbf{b}$ and $\mathbf{n}$ are in general not coplanar.

In Fig. C.3*a*, the axis of the specimen is in its initial state. Figure C.3*b* shows that the same crystal is elongated by deformation in simple shear. The slip planes undergo a rotation with respect to the laboratory frame around an axis $\mathbf{r} = \mathbf{b} \times \mathbf{t}$, which tends to reduce the angle λ between $\mathbf{b}$ and $\mathbf{t}$. Although tensile elongation can be obtained by allowing the direction of the applied force to rotate, the direction of **t** is, however, fixed during conventional tensile tests (Fig. C.3*c*). This constraint implies that to obtain the same elongation of the axis as in Fig. C.3*b*, the same crystallographic shear should be accompanied by a stress-free rigid body rotation with a sign opposite to that of the previous rotation. Thus, in tension, the loading axis rotates relative to the crystallographic axes towards the direction of the Burgers vector, which modifies the angle λ as well as the angle φ between **t** and **n**. In compression, one can see easily that the slip planes rotate until they are perpendicular to the tensile axes after what no further deformation is possible. Thus, this rotation occurs around an axis $\mathbf{r} = \mathbf{n} \times \mathbf{t}$ that tends to reduce the angle φ between **n** and **t** (cf. Fig. C.4). These rotations of the loading axis with respect to the crystallographic axes result from the imposed boundary conditions (Section C.3) and modify the Schmid factors. Other types of boundary conditions would result in different rotations.

The example of Fig. C.3 shows that rigid body rotations have to be taken into account upon analysing tensile or compressive deformation tests in single crystals. The occurrence of these rotations is understood in simple terms by considering first familiar definitions associated with the deformation of continuous bodies (Rice, 1999; Rees, 2006). In the small strain approximation, a strain matrix ε_{ij} is defined from the matrix of displacement gradients $\partial u_i/\partial x_j$, where the indices i and j take the values 1, 2 or 3 in three dimensions and 1 or 2 in two dimensions. The strain matrix is decomposed into the sum of a diagonal matrix of elongations ε_{ii} and a symmetric, non-diagonal matrix of shears

$$\varepsilon_{ij} = \gamma_{ij}/2 = \frac{1}{2}(\partial u_i/\partial x_j + \partial u_j/\partial x_i), \tag{C.6}$$

where $i \neq j$ and γ_{ij} is called the engineering shear strain. These components describe pure shears, which transform a two-dimensional square into a parallelogram through two rotations $\theta_{ij} = \partial u_i/\partial x_j$. In the small strain approximation, the displacements are developed at first order and $\sin\theta_{ij} \approx \tan\theta_{ij} \approx \theta_{ij}$. Then, the sum of these angles is γ_{ij}. This definition of shears in eqn. C.6 omits other types of strains involving rotations. For example, rigid body rotations imply that $\partial u_i/\partial x_j = -\partial u_j/\partial x_i$, so that $\gamma_{ij} = 0$. Such rotations are, however, removed by an opposite rotation of the reference frame. The same holds for simple shears, where one direction of a square is invariant and there is only one rotation (Fig. C.3*b*). These shears can also be transformed into a pure shear by a suitable rotation. More generally, it is easily seen that shears such that the two rotations have different angles can always be decomposed into a simple shear and a rigid body rotation.

Actually, the full distortion of a small volume element includes translations, which involve no displacement gradient, strains and rotations. In matrix form and in the small strain approximation, this is written $e_{ij} = \varepsilon_{ij} + \omega_{ij}$. The rotation matrix

ω_{ij} is skew symmetric ($\omega_{ij} = -\omega_{ji}$, $\omega_{ii} = 0$), with components that vanish for pure shear

$$\omega_{ij} = \frac{1}{2}(\partial u_i/\partial x_j - \partial u_j/\partial x_i) \tag{C.7}$$

Rigid body rotations are now included, as ω_{ij} is non-zero when $\partial u_i/\partial x_j = -\partial u_j/\partial x_i$.

The kinematics of plastic flow by crystallographic shears is part of the continuum mechanical framework used to simulate dislocation-based plastic flow in crystals (Argon, 2008). In the theory of elasticity, the small strain approximation corresponds to linear elasticity. Many applications in solid mechanics imply dealing with finite (i.e., large) plastic strains, by opposition to the small strain approximation used above. In the second-order approximation, the treatment of elastic problems become more complex (Rice, 1999; Dunne and Petrinic, 2005).

C.3 Active slip systems in fcc crystals

We consider the standard stereographic triangle shown in Fig. C.4*a*. For orientations inside the triangle, the slip system with maximum Schmid factor that is activated at yield is $[\bar{1}01](111)$. This slip system is called the *primary* slip system. In most of the experimental literature on single crystal deformation, the lattice rotations are not determined. To bypass this difficulty, experimentalists are using simple geometric relations that yield an ideal evolution of the loading axis, from which the resolved stress–strain curves can be determined (Reid 1973; Hosford 1993, 2005). The predicted rotations of a tensile loading axis with respect to the crystallographic axes are shown in Fig. C.4*a*.

The predicted behaviour in tension (Fig. C.4*a*), complemented by some experimentally observed features (see for instance Takeuchi 1975), is as follows. Figure C.5, which is further discussed here, can help visualizing the activated slip systems for high-symmetry orientations.

- *For stress axes between* [001] *and* $[\bar{1}11]$, the primary slip system has, by reason of symmetry, same Schmid factor as a secondary system, $[011](\bar{1}\bar{1}1)$, which is called the *conjugate* system. Symmetric double slip occurs and the sum of the two rotations towards [011] and $[\bar{1}01]$ for the primary system is $[\bar{1}12]$. Thus, the stress axis moves towards this last direction and stops when reaching it. The $[\bar{1}12]$ orientation is a stable one, because all misorientations caused by local fluctuations are damped out and neighbouring orientations are attracted to it.
- *For low-symmetry orientations* inside the stereographic triangle, primary slip is activated and the stress axis rotates towards $[\bar{1}01]$. This leads to an easy glide stage, or stage I, in the resolved stress–strain curves (Section 2.6.2). The dislocation density in the primary slip system increases and self-hardening increases

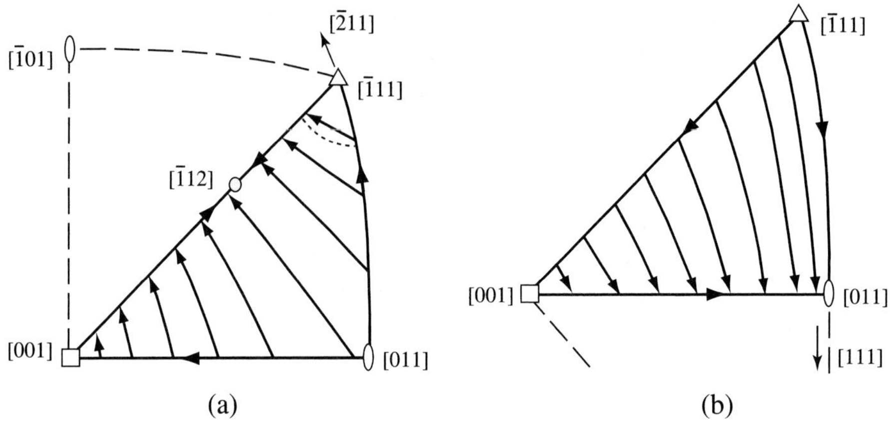

FIG. C.4. Rotations of the stress axis with respect to crystallographic axes in the $[001]-[011]-[\bar{1}11]$ stereographic triangle (after Reid 1973). The primary slip system is $[\bar{1}01](111)$. (a) The stress axes rotate towards the primary Burgers vector in tension. (b) they rotate towards the slip plane normal in compression. For an activated $[111](\bar{1}01)$ primary slip system in bcc metals, (a) represents the rotations in compression and (b) the rotations in tension (see text).

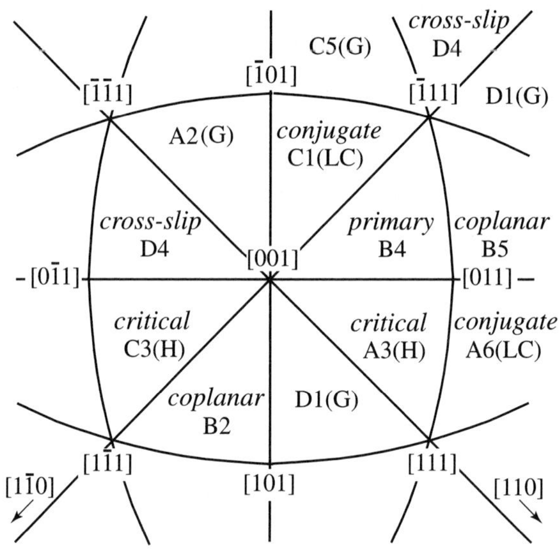

FIG. C.5. Interactions of the primary slip system with other slip systems having maximum Schmid factors in neighbouring stereographic triangles. The Schmid and Boas notations for these slip systems (Table C.1) are indicated. LC, H and G denote Lomer–Cottrell locks, Hirth locks and glissile junctions formed during these interactions.

its critical resolved flow stress, all the more as the initial orientation is far from the [001] – [$\bar{1}$11] zone axis. The conjugate system initially has a smaller Schmid factor than the primary one, which increases with increasing lattice rotation. As a result, the conjugate slip system can be activated at some point and double slip sets in.[14] The dislocations of the conjugate system have, however, a small mean free path because of the large primary density and they only produce a small fraction of the total shear. During this second stage, or stage II, the rotation of the loading axis is still governed by the primary shear to a good approximation (cf. Fig. 2.25). A small transient may occur when the [001] – [$\bar{1}$11] zone axis is eventually reached; it is followed by symmetric double slip along this zone axis.

- *For stress axes between* [011] *and* [$\bar{1}$11], the [$\bar{1}$10](111) slip system, which is *coplanar* to the primary system, is activated by reason of symmetry and double coplanar slip should occur in principle (Fig. C.4*a*). For orientations close to [011], the stress axis migrates towards the direction that is the sum of the two Burgers vector directions, [$\bar{1}$01] + [$\bar{1}$10] = [$\bar{2}$11], that is, towards the [$\bar{1}$11] corner. Such evolution is thought to be unstable because any positive strain fluctuation in the primary slip system increases its Schmid factor. A small amount of double coplanar slip is, however, observed to occur in crystals that are initially oriented with sufficient precision (Jackson, 1968; Takeuchi, 1976). Thus, deformation along this zone axis may be weakly metastable.
- *For stress axes between* [011] *and* [001], the secondary system having same Schmid factor as the primary one is the *critical* slip system, [101]($\bar{1}$11). However, symmetric double slip on this binary axis is unstable with respect to small orientation fluctuations and the primary slip system is always activated in tension, in contrast to what is usually depicted in textbooks (Fig. C.4*a*).
- *The* [011] *stress axis* is unstable with respect to orientation fluctuations because all arrows in Fig. C.4*a* are pointing away from it. Depending upon the initial misorientation of the crystal, the stress axis moves towards the primary Burgers vector, [$\bar{1}$01] or along the [011] – [$\bar{1}$11] zone axis.
- *For the* [$\bar{1}$11] *stress axis*, there are six potentially active slip systems with same Schmid factors, of which three are the cross-slip systems of the three others. For reasons that are discussed in Section 5.2.2.3, three slip systems are activated with different Burgers vectors such that their sum is [$\bar{1}$11]. Thus, there are no lattice rotations for this stress axis. The non-activated slip systems are the cross-slip systems of the active ones. This corner orientation is stable and attracts all initial orientations within a certain range that is depicted by the dotted curve in

[14] According to the convention adopted here, double slip is asymmetric unless stated otherwise. To avoid confusion, the term 'duplex slip' is not used.

Fig. C.4*a*. The active slip systems are different on both sides of that boundary. In consequence, stress axes that reach it from outside on the two neighbouring binary axes, cannot further move unless new slip systems are activated.

- *For the* [001] *stress axis*, the situation is similar to that of the $[\bar{1}11]$ corner orientation, but far more complicated. Of the eight slip systems that initially have same Schmid factor no less than four are activated and different sets of slip systems are active in different volume elements (Section 5.2.2.3). The basin of attraction of this corner orientation is restricted to a few degrees (Takeuchi, 1976).

In tension and for low-symmetry orientations, double slip on primary and conjugate slip systems leads to a deformation stage II, which is characterized by a strong work hardening rate arising from the formation of LC locks. This explains why a large number of studies were devoted to the study of this particular reaction. At and around the corner orientations, $\langle 001 \rangle$ and $\langle 111 \rangle$, the behaviour is rather complex. These high-symmetry orientations are particularly important. They represent two extreme cases of multiple slip conditions that set upper and lower bounds to the mechanical response of polycrystals (Kocks, 1960).

During plastic flow, the primary slip system interacts with other slip systems, which are the most highly stressed in neighbouring stereographic triangles. The portion of stereographic plot shown in Fig. C.5 summarizes the traditional names of these slip systems in fcc crystals, their shorthand notation according to Schmid and Boas (1935) and the corresponding dislocation reactions occurring with the primary slip system. Glissile junctions are only formed at, or near, the [001] or $[\bar{1}11]$ corners and the corresponding interacting planes bear no conventional name. As shown in Section 5.2.2.1, all types of interactions do not have exactly the same strength. Thus, the critical stresses for the activation of secondary slip systems may slightly differ from each other.

In compression, the lattice rotation associated to each active slip system occurs towards the normal to the slip plane and changes sign with respect to tension (Fig. C.4*b*). Binary zone axes and corner orientations that are stable or unstable in tension become respectively unstable and stable in compression. For low-symmetry orientations, primary slip is followed by double slip on the critical system and symmetric double slip along the stable [001] – [011] zone axis, towards the stable [011] corner orientation.

It was noted very early that it is difficult to obtain uniform uniaxial deformation in compression. In single glide, the rotation of the single crystal axis with respect to the loading axis (Fig. C.3*c*) is accommodated by deformation bands perpendicular to the primary slip plane, the kink bands (see Section 3 of Clarebrough and Hargreaves 1959 and Magid *et al.* 2009). In tension, uniform deformation can be obtained over a large part of the gauge length, which has an aspect ratio of typically 10. In compression, a friction between compression platens and the specimen may hinder shear displacements parallel to the contact surface, which increase with increasing sample height. In addition, there are always two zones where the length of primary slip lines is limited by the platens, which induces further rotation gradients and accommodation by the activation of secondary slip systems (cf. Fig. 1 of Weiss and Marsan 2003). Although these problems can be partially solved by lubricating the contact surfaces and increasing the gauge length, some friction is nevertheless necessary to prevent too long specimens from buckling. Thus, obtaining reliable stress-strain behaviour and realistic microstructures in stage I requires a careful optimization of sample geometry and friction at the top and bottom surfaces (Taylor and Farren, 1926). The situation is less critical for well-oriented crystals with stable high-symmetry orientations, provided that they have a large enough aspect ratio and are carefully aligned. For these reasons, ductile materials are tested in tension whenever possible.

The reciprocal lattice of the fcc lattice is body-centred cubic. Hence, the plane normals in one become Burgers vectors in the other and reciprocally. As a consequence, the lattice rotations occur in opposite directions in fcc crystals and in bcc crystals above the athermal transition temperature. It follows that Figs. C.4*a* and C.4*b* illustrate respectively the compressive and tensile rotations associated to the primary slip system $[111](\bar{1}01)$ in the bcc lattice, which is the reciprocal of the primary slip system in the fcc lattice.

The notations of Schmid and Boas for slip systems and the correspondence with crystallographic notations are given in Table C.1. In the shorthand notations, the capital letter denotes a slip plane normal and the number a Burgers vector. For instance B4 is here the primary slip system, D4 its cross-slip system and B5 one of its coplanar slip systems.

TABLE C.1. The notations of Schmid and Boas for slip systems in fcc crystals. **n** is a slip plane normal and **b** a Burgers vector.

Notation	A2	A3	A6	B2	B4	B5
n	$[\bar{1}11]$	$[\bar{1}11]$	$[\bar{1}11]$	$[111]$	$[111]$	$[111]$
b	$[0\bar{1}1]$	$[101]$	$[110]$	$[0\bar{1}1]$	$[\bar{1}01]$	$[\bar{1}10]$

Notation	C1	C3	C5	D1	D4	D6
n	$[\bar{1}\bar{1}1]$	$[\bar{1}\bar{1}1]$	$[\bar{1}\bar{1}1]$	$[1\bar{1}1]$	$[1\bar{1}1]$	$[1\bar{1}1]$
b	$[011]$	$[101]$	$[\bar{1}10]$	$[011]$	$[\bar{1}01]$	$[110]$

APPENDIX D

FROM γ-SURFACE TO PEIERLS STRESS

The concept of generalized stacking fault energy surface (or γ-*surface*), which was introduced by Vitek (1968), provides a simple bypass to full DFT-based dislocation core calculations (see Vitek and Paidar 2008 for references to various materials). A perfect crystal is cut along a crystallographic plane (Fig. D.1). The upper half is sheared rigidly with respect to the lower half by an arbitrary fault vector $\mathbf{u}$ of the cut plane. The fault that is generated is called a *generalized stacking fault* (GSF).

When $\mathbf{u}$ sweeps all possible directions in the unit cell of the cut plane, the surface giving the surplus energy per unit area, $\gamma(\mathbf{u})$, is the γ-surface for the plane considered. An interfacial restoring force per unit surface (i.e., a stress) $\boldsymbol{\tau}(\mathbf{u})$, is associated to each fault vector. It is given by

$$\boldsymbol{\tau}(\mathbf{u}) = -\nabla(\gamma(\mathbf{u})) \tag{D.1}$$

The GSF energy surface is minimum at the origin ($\boldsymbol{\tau} = 0$), where the stacking is perfect. It has the symmetries and periodicities of the cut plane.

The γ-surface is currently computed using atomistic simulations or *ab initio* methods based on the DFT framework. Relaxation is authorized along the direction normal to the cut plane, but not in directions parallel to it. The surface that is generated may have metastable minima for intermediate values of the shear vector, $\mathbf{u}_s$. At such points, the restoring stress $\boldsymbol{\tau}$ vanishes and the perfect dislocation can split into two partials separated by a metastable stacking fault of energy $\gamma(\mathbf{u}_s)$.

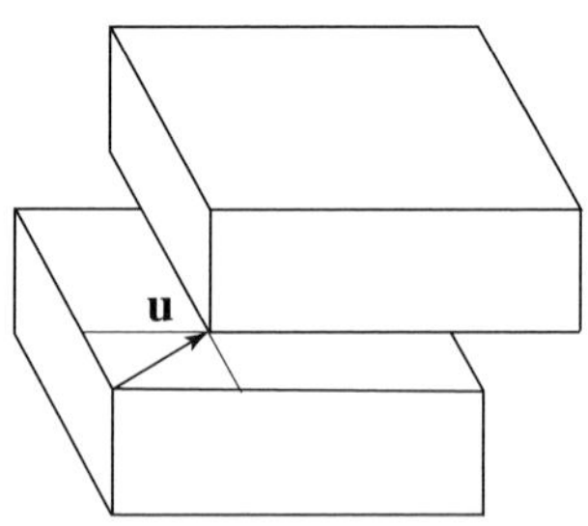

FIG. D.1. Shear displacement of two half-crystals relative to each other defining a generalized stacking fault.

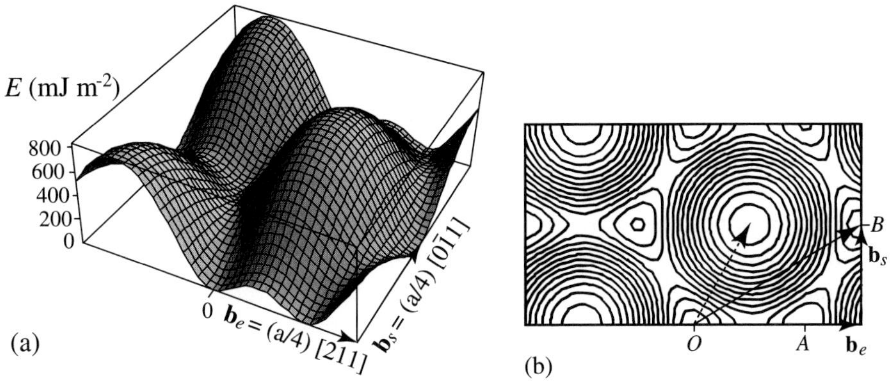

FIG. D.2. (a) Atomistic γ-surface for a $(\bar{1}11)$ plane in copper. The repeat unit of the surface is defined by the two translations $\frac{a}{2}[211]$ and $\frac{a}{2}[0\bar{1}1]$. The half-sum of these two vectors is the Burgers vector of the perfect dislocation, $\mathbf{b} = \frac{a}{2}[101]$; $\mathbf{b_e}$ and $\mathbf{b_s}$ are its edge and screw components. (b) Iso-energy contours. The dashed and full arrowed lines show shears leading respectively to a maximum energy configuration and to a metastable one. The low energy path leading to the intrinsic stacking fault involves successive shifts along **OA** and **AB** (courtesy S. Brochard).

This situation is illustrated in Fig. D.2*a* by a $(\bar{1}11)$ γ-surface in fcc copper computed using a phenomenological many-body potential (Aslanides and Pontikis, 1998). The high-energy peaks correspond to shears leading to stacking faults such that two identical atomic layers are on top of each other. The metastable positions corresponding to intrinsic stacking faults are reached via a minimum energy path consisting of a first shear towards a metastable position, $\mathbf{OA} = \frac{a}{6}[211]$ in Fig. D.2*b*, which forms a first Shockley partial. The second shear $\mathbf{AB} = \frac{a}{6}[1\bar{1}2]$ forms the second Shockley partial. Fig. D.2*a* can be compared to a first-principles calculation of the same γ-surface that was performed by Lu *et al.* (2000).

In bcc metals, the γ-surfaces exhibit no local minima in the planes where the core of screw dislocations spreads out (see Fig 3.5 and Vitek and Paidar 2008). In such cases, however, a reduction of the total core energy can be obtained by defining a generalized type of planar dissociation (Vitek, 1968; Duesbery and Vitek, 1998). The latter consists of particular types of partial dislocations called fractional dislocations. The Burgers vectors of these dislocations can be irrational and they are separated by ribbons of constant but unstable SFE. The main difference with respect to classical dissociations is that the restoring force being non-zero, the stacking fault ribbon is not stress-free. It exerts on the bordering partials a force that depends on the fault width.

Care must be taken when using γ-surfaces in covalent crystals, for instance in silicon. Indeed, dangling bonds and core reconstruction processes (Section 3.4.3) are not accounted for during the rigid shearing of the two halves of the crystal with respect to each other.

Two quantities are of general interest in sections of γ-surfaces along a given fault direction. One is the steepest positive slope, which corresponds to a maximum absolute value of the restoring force per unit area. This quantity is the maximum stress required for shearing the crystal in a block-like manner, that is, the *theoretical strength* along the considered direction of the slip plane. The second quantity is the maximum energy, or *unstable stacking fault energy*, γ_{us}, which is involved in the model for dislocation nucleation at crack tips developed by Rice and co-workers (see Rice and Beltz 1994). In Fig. D.2*a*, this maximum is met along the $[12\bar{1}]$ direction, which is perpendicular to the Burgers vector; its energy is about $0.8\,\mathrm{mJ\,m^{-2}}$.

GSF energy surfaces furnish simple means for comparing the responses of *ab initio* calculations to those of semi-empirical potentials, for instance in bcc metals (Frederiksen and Jacobsen, 2003; Ventelon and Willaime, 2010) and fcc metals (Boyer *et al.*, 2004). The concept of γ-surface can also be extended to generalized planar fault energies, for instance twin-like shears in consecutive or non-consecutive planes (Van Swygenhoven *et al.*, 2004; Ogata *et al.*, 2005; Li *et al.*, 2009).

A last application of γ-surfaces is concerned with the search for improved 2D solutions of the classical *Peierls–Nabarro model* (Peierls, 1940; Nabarro, 1947). This semi-continuum model treats dislocation cores confined in a single slip plane and extended in only one direction. It provides analytical values for their equilibrium width and the Peierls stress. The model was developed during a time span of fifty years, reviewed several times (Nabarro, 1997*a*; Joós and Duesbery, 1997; Hirth and Lothe, 1992; Argon, 2008) and further reviewed in the context of its extension to 2D models (Lu, 2005; Bulatov and Cai, 2006; Schoeck, 2005).

The basic idea of the Peierls–Nabarro (PN) models is illustrated by Figure D.3, which shows the planar core of an edge dislocation. The total energy of the defected crystal is split into two parts. The core energy is estimated from the definition of a continuous distribution of the relative displacements between the two half-crystals, $\mathbf{u}(x, y)$ in 2D. The energy of the two semi-infinite regions above and below the core region also depends on $\mathbf{u}$ and is treated by linear elasticity. Taken alone, the core contribution to the total energy favours narrow cores, which minimize large misfits between the two half-crystals, whereas the elastic energy favours extended cores that limit large elastic distortions. The solution $\mathbf{u}$ to the problem is obtained using minimization procedures for the total energy in 2D, taking into account whenever possible that the displacements in the elastic and core regions must match at their interface. In 1D, $u(x)$ is given by an integro-differential equation and the restoring stress profile $u(x) = -du/dx$ is assumed to be sinusoidal with period b. An analytical solution was derived by Peierls (1940) and further

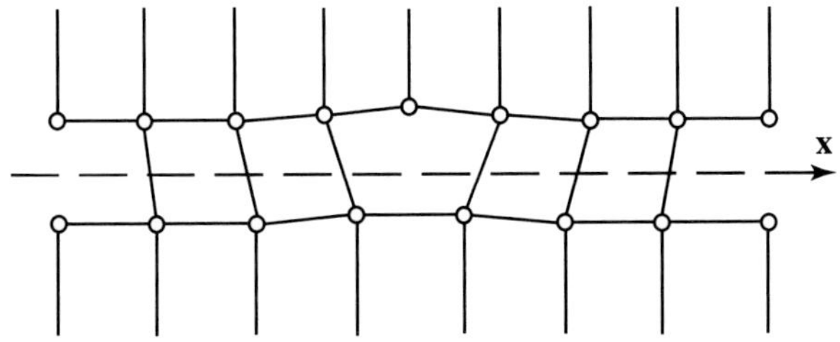

FIG. D.3. Schematic view of the planar extension of the core of an edge dislocation along its slip plane in a square lattice. The dashed line is the glide plane of the dislocation.

developed by Nabarro (1947), which yielded simple expressions for the core width and the Peierls stress.

In spite of several improvements, the 1D PN model is fundamentally handicapped by its fully continuum description of the core. Indeed, the effect of lattice periodicity only appears in the phenomenological restoring stress, which induces severe limitations (Lu 2005; Schoeck 2005). The model is not really predictive and becomes all the more inaccurate as dislocation cores are narrow. Nevertheless, it keeps a heuristic value. Calculated Peierls stresses for various materials were compiled by Nabarro (1997*b*), Wang (1996) and Takeuchi and Suzuki (1988). In some cases, they reasonably agree with expectation, which may seem surprising for such a simple model.

The development of a generalized 2D PN model became possible when *ab initio* γ-surfaces could be calculated and employed for estimating the misfit energy. This allowed the influence of the lattice periodicity on the misfit energy to be better taken into account (see Schoeck 2005 for a simple, efficient implementation and Bulatov and Cai 2006).

The core configurations obtained by full atomistic or *ab initio* simulations correspond well to those obtained by the 2D PN model provided that the corresponding γ-surface is properly implemented and use is made of anisotropic elasticity. However, the obtained configurations do not depend only on the SFE, they are also markedly influenced by the fine structure of the γ-surface (Schoeck, 2006). Thus, it is preferable to employ DFT γ-surfaces, especially for narrow cores. This is exemplified by the cases of aluminium (Schoeck, 2002) and palladium (Schoeck, 2001), where the estimated widths compare well with the result of full DFT calculations. Hence, the question of planar dissociations widths is basically solved.

The question of expanding the PN model to curved dissociated dislocations has been the subject of several attempts because of its relevance to the problem of homogeneous or heterogeneous nucleation of dislocation loops. A convincing implementation of *ab initio* γ-surfaces in the generalized PN model, using anisotropic elasticity, was performed by Yang *et al.* (2008) for dislocation loops in Cu and Al. It paves the way for further developments.

As far as Peierls stresses are concerned, the generalized PN model has limitations, which were recalled by Schoeck (2005). As already mentioned, the cores of strongly covalent materials cannot be adequately described by a γ-surface. For metallic materials, the core energy is treated as a local function of the displacement and this approximation is no longer valid for the strong distortions and large gradients involved in narrow cores. Nevertheless, the generalized PN model allows obtaining useful information in complex materials that are not suited for full DFT calculations. It was applied to minerals of the earth mantle, where the Peierls stresses in potentially active slip systems are poorly known (Carrez *et al.* 2007; Ferré *et al.* 2008; Metsue *et al.* 2010). In the last two studies, the method employed consists in coupling the PN model to an element-free FE Galerkin model (Denoual 2007). In this last reference, the flexibility of the method is illustrated by a calculation of the Peierls stress in tantalum, using an *ab initio* γ-surface to describe the faulted layers of the extended core. As is usual for calculations performed in bcc metals, the obtained Peierls stress is substantially larger than the experimental one (Section 3.2.3).

APPENDIX E

KINK-PAIR MODELS

E.1 Dislocations and Peierls potentials

This section outlines methods for estimating in elastic terms the energetics of the kink-pair mechanism for a dislocation moving over a periodic Peierls barrier in metallic materials. More detailed accounts can be found in Kocks *et al.* (1975) and Caillard and Martin (2003). These calculations make use of various potential energy profiles, sinusoidal, anti-parabolic, camel-hump or more complicated ones. For example, the sinusoidal potential is written

$$E(y) = E_o + \frac{\Delta E}{2}\left[1 - \cos\left(2\pi y/h_o\right)\right], \tag{E.1}$$

where E_o is the energy at the bottom of a Peierls valley, y is the direction normal to the Peierls relief in the slip plane (Fig. E.1) and h_o is the periodicity in the same direction.

A straight dislocation experiences a resistive stress $\tau_P = -dE(y)/bdy$, which is maximum when $y = h_o/4$. Hence, the Peierls stress is

$$\tau_P = \pi \frac{\Delta E}{bh_o} \tag{E.2}$$

Other potentials yield similar results. Under an applied stress, τ, a straight dislocation moves upwards on the Peierls profile and reaches an equilibrium position y_o given by $\tau b = -[dE(y)/dy]_{y_o}$. Its energy is then $E = E(y_o)$.

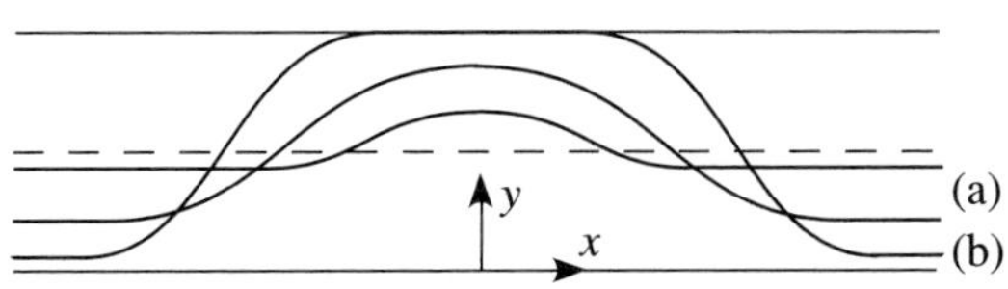

FIG. E.1. Top view of a periodic Peierls energy relief with energy minima (thin lines) and a maximum (thin dashed line). (a) Two critical bulge configurations with curvature radii that decrease with increasing temperature and decreasing stress. (b) Geometric low-stress, high-temperature critical configuration.

E.2 High-stress solutions

We now consider high-stress critical configurations, which form bulges like those shown in configurations (*a*) of Fig. E.1. The length of an element dx of line is $dx[1 + (dy/dx)^2]^{1/2}$. The energy fluctuation ΔU required for going from the initially straight, infinite configuration at $y = y_o$ to a bowed configuration is obtained by integration, taking into account the work of the applied stress τ

$$\Delta U = \int_{-\infty}^{+\infty} \left\{ E(y)\sqrt{1 + \left(\frac{dy}{dx}\right)^2} - E(y_o) - \tau b(y - y_o) \right\} dx \qquad \text{(E.3)}$$

This expression can only be solved numerically. It is often simplified by assuming that $dy/dx \ll 1$, as done in the vibrating string model (Section 2.2.2.3). This leads to

$$\Delta U = \int_{-\infty}^{+\infty} \left\{ E(y)\left[1 + \frac{1}{2}\left(\frac{dy}{dx}\right)^2\right] - E(y_o) - \tau b(y - y_o) \right\} dx \qquad \text{(E.4)}$$

Denoting by $\Phi(y) = E(y) - E(y_o) - \tau b(y - y_o)$ the effective potential under stress, eqn. E.4 is rewritten

$$\Delta U = \int_{-\infty}^{+\infty} \left\{ \Phi + \frac{1}{2}\left(\frac{dy}{dx}\right)^2 E(y) \right\} dx \qquad \text{(E.5)}$$

The saddle point configuration is among all possible solutions of eqns. E.3 or E.4 the one that minimizes the activation energy. It can be obtained from a variational argument as a solution of an Euler–Lagrange differential equation. In the present case, a simple integral solution exists, which is given by the Beltrami identity. If $f(y, y_x)$ is the functional form to be integrated, this solution is written $f - (dy/dx)[\partial f/\partial(dy/dx)] = const.$ (see Weisstein 2002 for a demonstration). Applying this result to eqn. E.5, one obtains

$$\Phi = \frac{1}{2}\left(\frac{dy}{dx}\right)^2 E(y), \qquad \text{(E.6)}$$

where the constant vanished upon integration.

An equivalent approximate result can be obtained by using the concept of line tension. Energy minimization is then equivalent to a condition of equilibrium between the local forces on the dislocation, specifically the line tension, the applied force and the resistive Peierls force

$$\Gamma\frac{d^2y}{dx^2}+\tau b-\frac{dE(y)}{dy}=\Gamma\frac{d^2y}{dx^2}-\frac{d\Phi}{dy}=0 \tag{E.7}$$

In this equation, the curvature radius appears in the approximated form of a second derivative, which is valid when $dy/dx \ll 1$. Integration with respect to y, assuming a constant line tension and with $2d^2y/dx^2 = d(dy/dx)^2/dy$, yields

$$\Phi=\frac{1}{2}\left(\frac{dy}{dx}\right)^2\Gamma\approx\frac{1}{2}\left(\frac{dy}{dx}\right)^2E_o \tag{E.8}$$

The second equality is valid in the line tension approximation if one assumes that the fluctuations of the Peierls potential are small compared to the energy at rest of the dislocation. Then, $E(y) \approx E_o$ and the constant line tension has same formal expression as the line energy per unit line. Simple relations are obtained by inserting the second equality of eqn. E.8 into eqn. E.5. The integration variable is taken as the height y of each elementary segment, which varies between the initial and critical values y_o and y_c. With $dx = (dx/dy)dy = (E_o/2\Phi)^{1/2}dy$ and $E(y) \approx E_o$, one obtains

$$\Delta U=2\sqrt{2E_o}\int_{y_o}^{y_c}\sqrt{\Phi}\,dy \tag{E.9}$$

The factor of two arises from the fact that the integration is carried out twice, on the left side and on the right side of the configuration. The energy of a single kink under zero stress, U_k, follows by setting $\tau = 0$ and integrating over one period h_o of the Peierls potential. Equation E.9 reduces to

$$U_k=E_o\int_0^{h_o}\sqrt{\frac{E(y)}{E_o}-1}\;dy \tag{E.10}$$

Upon integration with various simple potentials, one always finds

$$U_k\approx h_o(E_o\Delta E)^{1/2} \tag{E.11}$$

The stress dependence of the activation energy, $\Delta U(\tau)$, is obtained by integrating eqn. E.9 for successive values of y_c. For example, with the antiparabolic potential one obtains

$$\Delta U=2U_k\left(1-\frac{\tau}{\tau_P}\right)^2 \tag{E.12}$$

Other Peierls potentials yield similar forms with different exponents or more complex forms. At low stresses, eqn. E.12 yields $\Delta U \approx 2U_k$, which is the energy of two non-interacting kinks. This type of solution, in which the interaction between kinks is neglected as a result of the line tension approximation, is not valid for small kink separations (Section E.3).

To bypass the problem of fitting Peierls potentials to experimental results, Kocks, Argon and Ashby (1975) proposed a general parametric form for the activation enthalpy $\Delta H(\tau)$

$$\Delta H = \Delta H_o \left[1 - \left(\frac{\tau}{\tau_P} \right)^p \right]^q, \tag{E.13}$$

where ΔH_o is the total activation enthalpy. The two constants p and q are usually such that $0 < p \leq 1$ and $1 \leq q \leq 2$. Within these ranges of values, it is possible to reasonably fit both the high and low stress parts of the activation energy for the interaction of a dislocation with localized obstacles.

E.3 Kink-pairs at low stresses

When the critical configuration consists of a geometric kink-pair, as in configuration (*b*) of Fig. E.1, the interaction between the two kinks has to be taken into account for small kink-pair widths. Such small stress configurations can be treated easily (Seeger and Schiller 1962). We consider a kink-pair of width w larger than its height h_o and the width of a single kink. The total formation energy is the sum of the energies of the two kinks and of their interaction energy. It is written here as an activation enthalpy to keep same notations as in Section 3.2.4.2. As the two kinks are of atomic dimension, they cannot be treated like two interacting dislocation segments. The interaction energy is inversely proportional to the distance w, as for two electrical point charges of magnitude h_o and of opposite sign, and

$$\Delta H_{tot} = 2U_k - k\mu b^2 \frac{h_o^2}{w} \tag{E.14}$$

The factor μb^2 is introduced for the sake of dimensionality and $k \approx 1/8\pi$ (Hirth and Lothe, 1992). The activation enthalpy under stress, $\Delta H(\tau)$, is obtained by subtracting the work of the resolved stress, $\Delta W \approx \tau b h_o w$, during the formation of this configuration

$$\Delta H(\tau) = 2U_k - \mu b^2 \frac{h_o^2}{8\pi w} - \tau b h_o w \tag{E.15}$$

The critical configuration is again obtained by minimizing the activation energy with respect to the unknown kink-pair width w. A straightforward calculation yields

$$w_c = \left(\frac{\mu b h_o}{8\pi\tau}\right)^{1/2} \tag{E.16}$$

The activation energy is eventually written

$$\Delta H(\tau) = 2U_k - (h_o b)^{3/2}\left(\frac{\mu\tau}{2\pi}\right)^{1/2} \tag{E.17}$$

These relations are confronted to atomistic results in Section 3.2.4.2. As they are no longer valid for critical bulge configurations, the question was raised of a possible transition between low and high stress kink-pair regimes (Section 3.2.4.4). However, this question cannot be answered within a framework based on empirical Peierls potentials.

E.4 The kink-diffusion model

E.4.1 Steady-state conditions

This section discusses the steady-state motion of a perfect dislocation by the kink-pair mechanism, from which the existence of two distinct regimes is predicted. Although the demonstration is quite general, it is of particular interest in the presence of a strong secondary Peierls force as in covalent materials. In such a case, the velocity of the expanding kinks is much lower than in metals.

The lateral expansion of single kinks is limited to a maximum distance $X/2$, which is governed either by their annihilation with kinks of opposite sign or by fixed obstacles on the dislocation lines (Fig. E.2).

A single kink has a life-time t_k, during which it propagates along a distance $X/2$. Its velocity is given by

$$v_k = X/2t_k \tag{E.18}$$

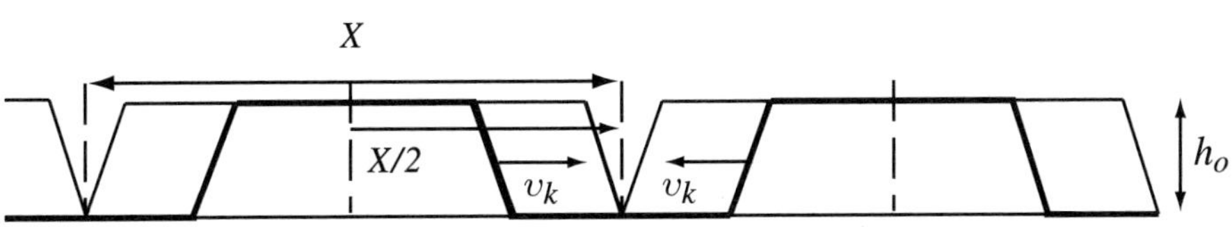

FIG. E.2. The expansion and mean free path, X, of kink-pairs of height h_o.

In steady-state, the motion of a dislocation line with a constant velocity implies that every time a pair of moving kinks is blocked or annihilated, a nucleation event occurs. If J is the nucleation rate of kink-pairs per unit time and length of dislocation line, the steady-state condition requires that one kink-pair is nucleated during the time interval t_k, taking into account that the length available for kink-pair nucleation is in average half the mean free path X. Thus,

$$Jt_kX/2 = 1 \tag{E.19}$$

From eqns. E.18 and E.19, we obtain

$$JX^2 = 4v_k \tag{E.20}$$

The dislocation velocity v_d is of the form $v_d = h_o/\tau_{nucl}$, where τ_{nucl} is the time interval between two successive kink-pair nucleations. It takes a time interval t_k to create a new dislocation segment of length X, but during this time interval a new kink-pair is generated on the expanding segment, in average at a time $t_k/2$ after it was generated. Thus, the periodicity of kink-pair nucleation events is $t_k/2$. According to eqn. E.19, $\tau_{nucl} = t_k/2 = 1/JX$ and the dislocation velocity is given by

$$v_d = h_oJX \tag{E.21}$$

When the mean free path of kinks is governed by strong obstacles with spacing L, one has $L = X$ and, according to eqn. E.21, the dislocation velocity is written

$$v_d = h_oJL \tag{E.22}$$

When the mean free path of kinks is governed by their annihilation with kinks of opposite sign, the combination of eqns. E.20 and E.21 yields an expression for the dislocation velocity that does not contain the free-flight distance of the kinks

$$v_d = 2h_o\sqrt{Jv_k} \tag{E.23}$$

Thus, two different regimes are defined, a *length-dependent* regime and a *kink-collision* regime. This last regime prevails for long segments or large kink densities.

E.4.2 Dislocation velocities

Owing to the high strength of the covalent bond, the critical kink-pairs can be assumed to have geometric shapes (Fig. E.2). Critical bulges may only form under very high stresses, because the lines are restricted to stay in the deep and narrow minima of the primary and secondary Peierls valleys. Thus, an insight into the formation energy of geometric kink-pairs can be obtained without assuming any shape for the Peierls potential. The kink-diffusion model by Hirth and Lothe (1992) is based on a diffusive treatment of the kink velocity over the secondary Peierls relief. Kinks of atomic height are treated like point defects undergoing thermally activated motion with a migration energy $W_m \gg k_B T$.[15] This allows defining the jump frequency of a kink over the secondary Peierls barrier by

$$f = \nu_D \exp - \left(\frac{W_m}{k_B T} \right) \tag{E.24}$$

The velocity of a kink submitted to a small external force $F = \tau b$ is calculated using Einstein's derivation for point defect drift in the linear regime. The mechanical work per atomic step of motion of a kink of height h_o is $\tau b h_o a/2$, assuming that the saddle point configuration is situated at one-half of the barrier width a. The latter is taken as the reconstructed period along the line (Section 3.4.3). One defines activation energies for jumps in the direction of the applied stress (+) or in the reverse direction (−) by $W_m^{\pm} = W_m \pm \tau b h_o a/2$. If f^+ and f^- are the corresponding jump frequencies, the forward and reverse kink velocities are given by $v_k^{\pm} = a f^{\pm}$ and the net kink velocity is $v_k = (v_k^+ - v_k^-)$. The exponential term in the rate equation (eqn. E.24) can be expanded with respect to stress when $\tau b h_o a/2 \ll k_B T$. As a result, one obtains a linear stress dependence of the net kink velocity

$$v_k = \frac{\tau b h_o}{k_B T} \nu_D a^2 \exp - \left(\frac{W_m}{k_B T} \right) = D_k \frac{\tau b h_o}{k_B T}, \tag{E.25}$$

where D_k is the kink diffusivity.

The nucleation rate of kink-pairs can be estimated in an approximate manner and within specific conditions for the two particular cases of low and high stresses. What is called high stress in this context is in fact a regime of interacting geometric kinks similar to the low stress regime in metals (Section E.3). In the present case, however, it is necessary to account for the fact that kink-pairs are nucleated in the presence of the secondary Peierls potential. At low stresses, the formation energy of kink pairs is derived from the equilibrium concentration of non-interacting single kinks under zero stress. Identical expressions for the dislocation velocities are obtained in the two cases and we restrict ourselves here to the low stress regime. A derivation of the velocities in the high stress regime is given by Hirth and Lothe (1992) and Caillard and Martin (2003).

[15] Strictly speaking, the formation and migration energies of kinks are free energies. Entropy terms are, however, not explicitly included in the kink-diffusion model (see Section 3.4.6).

In thermal equilibrium and under zero stress, the concentration of kinks of a given sign, c, is estimated by a traditional derivation for point defect concentrations. Denoting by F_k the (free) energy for kink formation, one has

$$c = c^+ = c^- = \frac{1}{a} \exp - \left(\frac{F_k}{k_B T} \right) \tag{E.26}$$

At small stresses and for a sufficiently large critical kink-pair width, the kink–kink interaction can be neglected. The concentration of kink-pairs, c_{kp}, is half the value given above for single kinks. At or near thermal equilibrium, the average distance between kink-pairs, X, is given by $X = 1/c_{kp} = 2/c$ and

$$X = 2a \exp \left(\frac{F_k}{k_B T} \right) \tag{E.27}$$

The kink-pair nucleation rate follows from eqn. E.20

$$J = \frac{v_k}{a^2} \exp - \left(\frac{2F_k}{k_B T} \right) \tag{E.28}$$

In the length-dependent regime, the dislocation velocity is obtained by inserting into eqn. E.28 the value of v_k given by the last equality in eqn. E.25 and importing the result into eqn. E.22

$$v_d = \frac{\tau b h_o^2 L}{a^2 k_B T} D_k \exp - \left(\frac{2F_k}{k_B T} \right) \tag{E.29}$$

Taking into account that the kink diffusivity involves the migration energy (eqn. E.25), the total activation energy in this regime is $2F_k + W_m$, that is, the sum of the formation energy of a kink-pair and the migration energy of a single kink.

In the kink-collision regime, the dislocation velocity is obtained by combining eqns. E.25, E.28 and E.23

$$v_d = \frac{2\tau b h_o^2}{a k_B T} D_k \exp - \left(\frac{F_k}{k_B T} \right) \tag{E.30}$$

In this regime, the total activation energy, $F_k + W_m$, is the sum of the formation and migration energies of a single kink. The full expressions for the two velocities are given by eqns. 3.11 and 3.12.

The transition between the two regimes is obtained by setting $X = L$ in eqn. E.27 for the average distance between kink-pairs in thermal equilibrium

$$L^* = 2a \exp\left(\frac{F_k}{k_B T^*}\right), \tag{E.31}$$

where T^* is the crossover temperature and L^* the corresponding segment length. Thus, with increasing segment length, the velocity should linearly increase in the length-dependent regime and saturate to a constant value when the kink-collision regime is reached. In parallel, the total activation energy should decrease by F_k at this transition. Actually, whereas the first prediction is confirmed by experimental results, the second one is not (Sections 3.4.7 and 3.4.8).

In strongly covalent materials, the domain of validity of the kink-diffusion model is assumed to extend up to very high stresses, typically a fraction of the Peierls stress, beyond which the kink-pairs have to be treated as small bulges. Although the Peierls stress is not accurately known in silicon, experimental results show that the total activation energy becomes stress dependent typically around $10^{-3}\mu$, well below the high stress limit. This is interpreted as arising from elastic interactions between kinks in narrow critical kink-pairs. As a consequence, the formation energy of kink-pairs should preferably be taken in the stress-dependent form proposed by Seeger and Schiller (eqn. E.17).

Alternative models to the kink-diffusion model make use of Peierls potentials. They also account for the secondary Peierls barrier and the kink migration energy (Indenbom *et al.* 1992; Iunin and Nikitenko 2001). The possible presence of point defects along the dislocations lines and their interaction with moving kinks was examined by several authors (Celli *et al.* 1963, Rybin and Orlov 1970, Iunin and Nikitenko 2001).

BIBLIOGRAPHY

Ackland, G.J., Mendelev, M.I., Srolovitz, D.J., Han, S., and Barashev, A.V. (2004). Development of new interatomic potentials for phosphorous impurities in α-iron. *J. Phys.: Condens. Matter*, **16**, 2629–2642.

Aifantis, E.C. (1986). On the dynamical origin of dislocation patterns. *Mater. Sci. Eng. A*, **81**, 563–574.

Akasheh, F., Zbib, H.M., Hirth, J.P., Hoagland, R.G., and Misra, A. (2007). Dislocation dynamics analysis of dislocation intersections in nanoscale metallic multilayered composites. *J. Appl. Phys.*, **101**, 084314(1–10).

Akhtar, A. and Teghtsoonian, A. (1971). Plastic deformation of zirconium crystals. *Acta Metall.*, **19**, 655–663.

Alberdi, J.M. (1984). *Grandes deformaciones plásticas en frio de cobre y aluminio (torsión)*. Ph.D. thesis, Faculty of Sciences, University of Navarra, San Sebastian, Spain.

Alefeld, G. (1964). Rate theory in solids at low temperatures. *Phys. Rev. Lett.*, **12**, 372–375.

Alexander, H. (1986). Dislocations in covalent crystals. In *Dislocations in Solids* (ed. F.R.N. Nabarro), Volume 7, pp. 115–234. North-Holland, Amsterdam.

Alexander, H. and Crawford, J.L. (2000). Latent hardening of germanium crystals. *Phys. Stat. Sol. (b)*, **222**, 41–49.

Alexander, H. and Haasen, P. (1968). Dislocations and plastic flow in the diamond structure. *Solid State Phys.*, **22**, 27–158.

Alexander, H., Kolar, H.R., and Spence, J.C.H. (1999). Kinks on partials of 60° dislocations in silicon as revealed by a novel TEM technique. *Phys. Stat. Sol. (a)*, **171**, 5–16.

Alexander, H. and Teichler, H. (1991). Dislocations. In *Electronic Structure and Properties of Semiconductors* (ed. W. Schröter), Volume 4, Materials Science and Technology, pp. 249–319. VCH, Weinheim (FRG).

Alshits, V.I. (1992). The phonon-dislocation interaction and its role in dislocation dragging and thermal resistivity. In *Elastic strain fields and dislocation mobility* (ed. V.I. Indenbom and J. Lothe), Volume 31, Modern Problems in Condensed Matter Science, pp. 625–697. Elsevier Science B.V., Amsterdam.

Alshits, V.I. and Indenbom, V.L. (1986). Mechanisms of dislocation drag. In *Dislocations in Solids* (ed. F.R.N. Nabarro), Volume 7, pp. 43–111. North-Holland, Amsterdam.

Ambrosi, P., Göttler, E., and Schwink, Ch. (1974). On the dislocation arrangement and flow stress of [111]—copper single crystals deformed in tension. *Scripta Metall.*, **8**, 1093–1098.

Amodeo, J., Carrez, P., Devincre, B., and Cordier, P. (2011). Multiscale modelling of MgO plasticity. *Acta Mater.*, **59**, 2291–2301.

Ananthakrishna, G. (2007). Statistical and dynamical approaches to collective behaviour of dislocations. In *Dislocations in Solids* (ed. F.R.N. Nabarro and J.P. Hirth), Volume 13, pp. 81–223. Elsevier B.V., Amsterdam.

Anongba, P.N.B., Bonneville, J., and Martin, J.L. (1993*a*). Hardening stages of [112] copper single crystals at intermediate and high temperatures—I. Mechanical behaviour. *Acta Metall. Mater.*, **41**, 2897–2906.

Anongba, P.N.B., Bonneville, J., and Martin, J.L. (1993*b*). Hardening stages of [112] copper single crystals at intermediate and high temperatures—II. Slip systems and microstructures. *Acta Metall. Mater.*, **41**, 2907–2922.

Aono, Y., Kitajima, K., and Kuramoto, E. (1980). Plastic deformation of Fe-N single crystals in the temperature range between 4.2 and 300 K. *Scripta Metall.*, **14**, 321–324.

Aono, Y., Kuramoto, E., and Kitajima, K. (1982). Fundamental plastic behaviors in high-purity bcc metals (Nb, Mo, Fe). In *Strength of Metals and Alloys, Proc. ICSMA 6* (ed. R.C. Gifkins), Volume 1, Oxford, pp. 9–14. Pergamon Press.

Argon, A.S. (2008). *Strengthening Mechanisms in Plasticity*. Oxford Series in Materials Modelling (ed. A.P. Sutton and R.E. Rudd). Oxford University Press.

Argon, A.S. (2009). A new model of dynamic recovery for Stage III of pure fcc metals without cross-slip. *Int. J. Mater. Res.*, **100**, 489–493.

Argon, A.S. and Haasen, P. (1993). A new mechanism of work hardening in the late stages of large strain plastic flow in f.c.c. and diamond cubic crystals. *Acta Metall. Mater.*, **41**, 3289–3306.

Arsenlis, A., Cai, W., Tang, M., Rhee, M., Oppelstrup, T., and Hommes, G. *et al.* (2007). Enabling strain hardening simulations with dislocation dynamics. *Model. Simul. Mater. Sci. Eng.*, **15**, 553–595.

Arsenlis, A., Parks, D.M., Becker, R., and Bulatov, V.V. (2004). On the evolution of crystallographic dislocation density in non-homogeneously deforming crystals. *J. Mech. Phys. Solids*, **52**, 1213–1246.

Arsenlis, T., Bulatov, V., Cai, W., Hommes, G., Rhee, M., and Tang, M. (2012). ParaDiS. http://paradis.stanford.edu.

Arzt, E. (1998). Size effects in materials due to microstructural and dimensional constraints: A comparative review. *Acta Mater.*, **46**, 5611–5626.

Aslanides, A. and Pontikis, V. (1998). Atomistic studies of dislocation cores in aluminium and copper. *Comput. Mater. Sci.*, **10**, 401–405.

Aubry, S., Fitzgerald, S.P., Dudarev, S.L., and Cai, W. (2011). Equilibrium shape of dislocation shear loops in anisotropic α-Fe. *Model. Simul. Mater. Sci. Eng.*, **19**, 065006(1–14).

Bacon, D.J. (1967). A method for describing a flexible dislocation. *Phys. Stat. Sol. (b)*, **23**, 527–538.

Bacon, D.J. (1978). The effect of dissociation on dislocation energy and line tension. *Phil. Mag. A*, **38**, 333–339.

Bacon, D.J. (1985). Dislocations in anisotropic elastic media. In *Fundamentals of Deformation and Fracture* (ed. B.A. Bilby, K.J. Miller, and J.R. Willis), pp. 401–405. Cambrige University Press.

Bacon, D.J., Barnett, D.M., and Scattergood, R.O. (1979). Anisotropic continuum theory of lattice defects. *Progr. Mater. Sci.*, **23**, 51–262.

Bacon, D.J., Kocks, U.F., and Scattergood, R.O. (1973). The effect of dislocation self-interaction on the Orowan stress. *Phil. Mag.*, **28**, 1241–1263.

Bacon, D.J., Osetsky, Y.N., and Rodney, D. (2009). Dislocation-obstacle interaction at the atomic level. In *Dislocations in Solids* (ed. J.P. Hirth and L. Kubin), Volume 15, pp. 1–90. Elsevier B.V., Amsterdam.

Bacon, D.J. and Vitek, V. (2002). Atomic-scale modelling of dislocations and related properties in the hexagonal close-packed metals. *Mater. Trans. A*, **33**, 721–733.

Baird, J.D. and Gale, B. (1965). Attractive dislocation intersections and work hardening in metals. *Phil. Trans. Roy. Soc. A*, **257**, 553–590.

Bakó, B., Clouet, E., Dupuy, L.M., and Blétry, M. (2011). Dislocation dynamics simulations with climb: Kinetics of dislocation loop coarsening controlled by bulk diffusion. *Phil. Mag.*, **91**, 3173–3191.

Bakó, B. and Hoffelner, W. (2007). Cellular dislocation patterning during plastic deformation. *Phys. Rev. B*, **76**, 214108(1–6).

Balk, T.J. and Hemker, K.J. (2001). High resolution transmission microscopy of dislocation cores in gold and iridium. *Phil. Mag. A*, **81**, 1507–1351.

Basinski, S.J. and Basinski, Z.S. (1979). Plastic deformation and work hardening. In *Dislocations in Solids* (ed. F.R.N. Nabarro), Volume 4, pp. 261–362. North-Holland, Amsterdam.

Basinski, Z.S. (1959). Thermally activated glide in face-centred cubic metals and its application to the theory of strain hardening. *Phil. Mag.*, **4**, 393–432.

Basinski, Z.S. (1974). Forest hardening in face centred cubic metals. *Scripta Metall.*, **8**, 1301–1308.

Basinski, Z.S. and Basinski, S.J. (1964). Dislocation distribution in deformed copper crystals. *Phil. Mag.*, **97**, 51–80.

Basinski, Z.S. and Basinski, S.J. (1992). Fundamental aspects of low amplitude cyclic plastic deformation in face-centred cubic crystals. *Progr. Mater. Sci.*, **36**, 89–148.

Bassani, J.L. and Wu, T.-Y. (1991). Latent hardening in single crystals II. Analytical characterization and predictions. *Proc. R. Soc. Lond. A*, **435**, 21–41.

Bay, B., Hansen, N., Hughes, D.A., and Kuhlmann-Wilsdorf, D. (1992). Evolution of f.c.c. deformation structures in polyslip. *Acta Metall. Mater.*, **40**, 205–219.

Becker, R. and Orowan, E. (1932). Über sprunghafte Dehnung von Zinkkristallen. *Z. Phys.*, **79**, 566–572.

Beer, G., Smith, I.M., and Duenser, C. (2008). *The boundary element method with programming*. Springer-Verlag, Wien.

Bei, H., Shim, S., George, E.P., Miller, M.K., Herbert, E.G., and Pharr, G.M. (2007). Compressive strengths of molybdenum alloy micro-pillars prepared using a new technique. *Scripta Mater.*, **57**, 397–400.

Bei, H., Shim, S., Pharr, G.M., and George, E.P. (2008). Effects of pre-strain on the compressive stress-strain response of Mo-alloy single-crystal micropillars. *Acta Mater.*, **56**, 4762–4770.

Bennetto, J., Nunes, R.W., and Vanderbilt, D. (1997). Period-doubled structure for the 90° partial dislocation in silicon. *Phys. Rev. Lett.*, **79**, 245–248.

Benoit, W., Gremaud, G., and Quenet, B. (1993). Dislocations and kink dynamics in f.c.c. metals studied by mechanical spectroscopy. *Mater. Sci. Eng. A*, **164**, 42–57.

Benzerga, A.A., Bréchet, Y., Needleman, A., and Van der Giessen, E. (2004). Incorporating 3D mechanisms into 2D dislocation dynamics. *Model. Simul. Mater. Sci. Eng.*, **12**, 159–196.

Bettge, D. and Österle, W. (1999). "Cube slip" in near-[111] oriented specimens of a single-crystal nickel-based superalloy. *Scripta Mater.*, **40**, 389–395.

Bharathi, M.S., Lebyodkin, M., Ananthakrishna, G., Fressengeas, C., and Kubin, L.P. (2002). The hidden order behind jerky flow. *Acta Mater.*, **50**, 2813–2824.

Biget, M. and Saada, G. (1989). Low-temperature plasticity of high-purity α-titanium single crystals. *Phil. Mag. A*, **59**, 747–757.

Bitzek, E., Brandl, C., Derlet, P.M., and Van Swygenhoven, H. (2008). Dislocation cross-slip in nanocrystalline fcc metals. *Phys. Rev. Lett.*, **100**, 235501(1–4).

Bonneville, J. (1985). *Étude des paramètres d'activation du glissement dévié dans le cuivre*. Ph.D. thesis, École Polytechnique Fédérale de Lausanne.

Bonneville, J. and Escaig, B. (1979). Cross-slipping process and the stress-orientation dependence in pure copper. *Acta Metall.*, **27**, 1477–1486.

Bonneville, J., Escaig, B., and Martin, J.L. (1988). A study of cross-slip activation parameters in copper. *Acta Metall.*, **36**, 1989–2002.

Bonneville, J. and Vanderschaeve, G. (1986). A study of cross-slip in the fcc structure. In *Strength of Metals and Alloys, Proc. ICSMA 7* (ed. H.J. McQueen, J.P. Bailon, J.I. Dickson, J.J. Jonas, and M.G. Akben), Volume 1, pp. 9–14.

Boyer, R.D., Li, J., Ogata, S., and Yip, S. (2004). Analysis of shear deformations in Al and Cu: Empirical potentials versus density functional theory. *Model. Simul. Mater. Sci. Eng.*, **12**, 1017–1029.

Brenner, S.S. (1956). Tensile strength of whiskers. *J. Appl. Phys.*, **27**, 1484–1491.

Brinckmann, S., Kim, J.-Y., and Greer, J.R. (2008). Fundamental differences in mechanical behavior between two types of crystals at the nanoscale. *Phys. Rev. Lett.*, **100**, 155502(1–4).

Brown, L.M. (1964). The self-stress of dislocations and the shape of extended nodes. *Phil. Mag.*, **10**, 441–466.

Brown, L.M. (1967). A proof of Lothe's theorem. *Phil. Mag. A*, **15**, 363–370.

Brown, L.M. and Ham, R.K. (1971). Dislocation-particle interactions. In *Strengthening Methods in Solids* (ed. A. Kelly and R.B. Nicholson), pp. 9–135. Applied Science Publishers, Barking (UK).

Brunner, D. (2010). Temperature dependence of the plastic flow of high-purity tunsgten single crystals. *Int. J. Mater. Res.*, **101**, 1003–1013.

Brunner, D. and Glebowsky, A. (2000). Analysis of flow-stress measurements of high-purity tungsten single crystals. *Mater. Lett.*, **44**, 144–152.

Bulatov, V.V., Abraham, F.F., Kubin, L.P., Devincre, B., and Yip, S. (1998). Connecting atomistic and mesoscale simulations of crystal plasticity. *Nature*, **391**, 669–672.

Bulatov, V.V. and Cai, W. (2002). Nodal effect in dislocation mobility. *Phys. Rev. Lett.*, **89**, 111501(1–4).

Bulatov, V.V. and Cai, W. (2006). *Computer Simulations of Dislocations*. Oxford Series in Materials Modelling (ed. A.P. Sutton and R.E. Rudd). Oxford University Press.

Bulatov, V.V., Cai, W., Fier, J., Hiratani, M., Pierce, T., Tang, M., Rhee, M., Yates, K., and Arsenlis, T. (2004). Scalable line dynamics in ParaDis. http://www.stanford.edu/~caiwei/papers.html.

Bulatov, V.V., Hsiung, L.L., Tang, M., Arsenlis, A., Bartelt, M.C., and Cai, W. *et al.* (2006). Dislocation multi-junctions and strain hardening. *Nature*, **440**, 1174–1178.

Bulatov, V.V., Justo, J.F., Cai, W., and Yip, S. (1997). Kink asymmetry and multiplicity in dislocation cores. *Phys. Rev. Lett.*, **79**, 5042–5045.

Bulatov, V.V., Justo, J.F., Cai, W., Yip, S., Argon, A.S., Lenosky, T. *et al.* (2001*a*). Parameter-free modelling of dislocation motion: The case of silicon. *Phil. Mag. A*, **81**, 1257–1281.

Bulatov, V.V., Rhee, M., and Cai, W. (2001*b*). Periodic boundary conditions for dislocation dynamics simulations in three dimensions. In *Multiscale Modeling of Materials—2000* (ed. L. Kubin, R. Selinger, J. Bassani, and K. Cho), Volume 653, MRS Symposium Proceedings, pp. Z1.3(1–6). Materials Research Society, Warrendale, PA (USA).

Cai, W. (2012). DDLab. http://micro.stanford.edu/ caiwei/CSD-Book.

Cai, W., Arsenlis, A., Weinberger, C.R., and Bulatov, V.V. (2006). A non-singular theory of dislocations. *J. Mech. Phys. Solids*, **54**, 561–687.

Cai, W., Bulatov, V.V., Chang, J., Li, J., and Yip, S. (2003). Periodic image effects in dislocation modeling. *Phil. Mag.*, **83**, 539–567.

Cai, W., Bulatov, V.V., Chang, J., Li, J., and Yip, S. (2004). Dislocation core effects on mobility. In *Dislocations in Solids* (ed. F.R.N. Nabarro and J.P. Hirth), Volume 12, pp. 1–80. Elsevier B.V., Amsterdam.

Cai, W., Bulatov, V.V., Justo, J.F., Argon, A.S., and Yip, S. (2000). Intrinsic mobility of a dissociated dislocation in silicon. *Phys. Rev. Lett.*, **84**, 3346–3349.

Cai, W. and Weinberger, C.R. (2008). Surface-controlled dislocation mutliplication in metal micropillars. *PNAS*, **105**, 14304–14307.

Caillard, D. (2010*a*). Kinetics of dislocations in pure Fe. Part I. In situ straining experiments at room temperature. *Acta Mater.*, **58**, 3493–3503.

Caillard, D. (2010*b*). Kinetics of dislocations in pure Fe. Part II. In situ straining experiments at low temperature. *Acta Mater.*, **58**, 3504–3515.

Caillard, D. (2011). An in situ study of hardening and softening or iron by carbon interstitials. *Acta Mater.*, **59**, 4974–4989.

Caillard, D. and Couret, A. (1996). Dislocation cores and yield stress anomalies. In *Dislocations in Solids* (ed. F.R.N. Nabarro and M.S. Duesbery), Volume 10, pp. 69–134. Elsevier Science B.V., Amsterdam.

Caillard, D. and Martin, J.-L. (2003). *Thermally Activated Mechanisms in Crystal Plasticity*. Pergamon Materials Series (ed. R.W. Cahn). Pergamon (Elsevier), Amsterdam.

Canadinc, D., Sehitoglu, H., Maier, H.J., and Kurath, P. (2008). On the incorporation of length scales associated with pearlitic and bainitic microstructures in a visco-plastic self-consistent model. *Mater. Sci. Eng. A*, **485**, 258–271.

Capolungo, L. (2011). Dislocation junction formation and strength in magnesium. *Acta Mater.*, **59**, 2909–2917.

Carrez, P., Cordier, P., Devincre, B., and Kubin, L.P. (2005). Dislocations reactions and junctions in MgO. *Mater. Sci. Eng. A*, **400–401**, 324–328.

Carrez, P., Ferré, D., and Cordier, P. (2007). Implications for plastic flow in the deep mantle from modelling dislocations in $MgSiO_3$ minerals. *Nature*, **446**, 68–70.

Carter, C.B. (1984). What's new in dislocation dissociation. In *Dislocations 1984* (ed. P. Veyssière, L. Kubin, and J. Castaing), pp. 227–251. Editions du CNRS, Paris.

Castaing, J., Veyssière, P., Kubin, L.P., and Rabier, J. (1981). Plastic deformation of silicon between 300 °C and 600 °C. *Phil. Mag. A*, **44**, 1407–1413.

Celli, V., Kabler, M., Ninomiya, T., and Thomson, R. (1963). Theory of dislocation mobility in semiconductors. *Phys. Rev.*, **131**, 58–72.

Chang, H.-J., Fivel, M., Rodney, D., and Verdier, M. (2010). Multiscale modelling of indentation in fcc metals: From atomic to continuum. *C.R. Physique*, **11**, 285–292.

Chaussidon, J., Fivel, M., and Rodney, D. (2006). The glide of screw dislocations in bcc Fe: Atomistic static and dynamic simulations. *Acta Mater.*, **54**, 3407–3416.

Chaussidon, J., Robertson, C., Rodney, D., and Fivel, M. (2008). Dislocation dynamics simulations of plasticity in Fe laths at low temperature. *Acta Mater.*, **56**, 5466–5476.

Chen, Q., Liu, X.-Y., and Biner, S.B. (2008*a*). Solute and dislocation junction interactions. *Acta Mater.*, **56**, 2937–2947.

Chen, Z., Chu, K.T., Srolovitz, D.J., Rickman, J.M., and Haataja, P. (2010). Dislocation climb strengthening in systems with immobile obstacles: Three-dimensional level-set simulation study. *Phys. Rev. B*, **81**, 054104(1–12).

Chen, Z., Lu, G., Kioussis, N., and Ghoniem, N. (2008*b*). Effect of the local environment on the mobility of dislocations in refractory bcc metals: Concurrent multiscale approach. *Phys. Rev. B*, **78**, 134102(1–5).

Chevy, J., Fressengeas, C., Lebyodkin, M., Taupin, V., Bastie, P., and Duval, P. (2010). Characterizing short-range vs. long-range spatial correlations in dislocation distributions. *Acta Mater.*, **58**, 1837–1849.

Chisholm, C., Bei, H., Lowri, M.B., Oh, J., Syed Asif, S.A., and Warren, O.L. *et al.* (2012). Dislocation starvation and exhaustion hardening in Mo alloy nanofibers. *Acta Mater.*, **60**, 2258–2264.

Chiu, Y.-L. and Veyssière, P. (2008). Dislocation self-organization under single slip straining and dipole properties. *Mater. Sci. Eng. A*, **483–484**, 191–194.

Chomel, P. and Cottu, J.P. (1982). Influence of substitutional elements on flow stress in iron at low temperature. *Acta Metall.*, **30**, 1481–1491.

Chopard, B. and Droz, M. (1998). *Cellular automata modeling of physical systems*. Cambridge University Press, Cambridge.

Christian, J.W. (1982). Some surprising features of the plastic deformation of body-centered cubic metals and alloys. *Metall. Trans. A*, **14**, 1237–1256.

Cizek, P. (2010). Dislocation boundaries and disclinations formed within the cube-oriented grains during tensile deformation of aluminium. *Acta Mater.*, **58**, 5280–5833.

Clarebrough, L.M. and Hargreaves, M.E. (1959). Work hardening of metals. *Progr. Mater. Sci.*, **8**, 1–330.

Clouet, E. (2011*a*). Dislocation core field. I. Modeling in anisotropic linear elasticity theory. *Phys. Rev. B*, **84**, 224111(1–7).

Clouet, E. (2011*b*). Dislocation core field. II. Screw dislocation in iron. *Phys. Rev. B*, **84**, 224107(1–12).

Clouet, E., Ventelon, L., and Willaime, F. (2009). Dislocation core energies and core fields from first principles. *Phys. Rev. Lett.*, **102**, 055502(1–4).

Cordier, P., Amodeo, J., and Carrez, P. (2012). Modelling the rheology of MgO under earth's mantle pressure, temperature and strain rates. *Nature*, **481**, 177–180.

Cordier, P. and Rubie, D.C. (2000). Plastic deformation of minerals under extreme pressure using a multi-anvil apparatus. *Mater. Sci. Eng. A*, **309–310**, 38–43.

Cottrell, A.H. (1953). *Dislocations and Plastic Flow in Crystals*. Clarendon Press, Oxford.

Cottrell, A.H. and Stokes, R.J. (1955). Effects of temperature on the plastic properties of aluminium crystals. *Proc. R. Soc. Lond. A*, **233**, 17–34.

Couret, A., Caillard, D., Püschl, W., and Schoeck, G. (1991). Prismatic glide in divalent h.c.p. metals. *Phil. Mag. A*, **63**, 1045–1057.

Cousland, S. McK. (1970). The relation between Cottrell-Stokes ratios and stacking fault energies in pure F.C.C. crystals. *Phys. Stat. Sol. (b)*, **37**, 159–164.

Csikor, F.F., Motz, C., Weygand, D., Zaiser, M., and Zapperi, S. (2007). Dislocation avalanches, strain bursts, and the problem of plastic forming at the micrometer scale. *Science*, **318**, 251–254.

Dahmen, K.A., Ben-Zion, Y., and Uhl, J.T. (2009). Micromechanical model for deformation in solids with universal predictions for stress-strain curves and slip avalanches. *Phys. Rev. Lett.*, **102**, 175501(1–4).

de Crécy, A., Bourret, A., Naka, S., and Lasalmonie, A. (1983). High resolution determination of the core structure of $1/3\langle 11\bar{2}0\rangle\{10\bar{1}0\}$ edge dislocation in titanium. *Phil. Mag.*, **47**, 245–254.

de Sansal, C., Devincre, B., and Kubin, L; (2010). Grain-Size strengthening in microcrystalline copper: A three-dimensional dislocation dynamics simulation. *Key Eng. Materials*, **423**, 25–32.

de Wit, R. (1967). Some relations for straight dislocations. *Phys. Stat. Sol. (b)*, **20**, 575–580.

Dehm, G. (2009). Miniaturized single-crystalline fcc metals deformed in tension: New insights in size-dependent plasticity. *Progr. Mater. Sci.*, **54**, 664–688.

Deng, J. and El-Azab, A. (2009). Mathematical and computational modelling of correlations in dislocation dynamics. *Model. Simul. Mater. Sci. Eng.*, **17**, 075010(1–31).

Deng, J. and El-Azab, A. (2010). Temporal statistics and coarse graining of dislocation ensembles. *Phil. Mag.*, **90**, 3651–3678.

Denoual, C. (2007). Modeling dislocation by coupling Peierls–Nabarro and element-free Galerkin methods. *Comput. Methods Appl. Mech. Engrg.*, **196**, 1915–1923.

Deo, C.S., Srolovitz, D.J., Cai, W., and Bulatov, V.V. (2005*a*). Kinetic Monte Carlo method for dislocation migration in the presence of solute. *Phys. Rev. B*, **71**, 014106(1–12).

Deo, C.S., Srolovitz, D.J., Cai, W., and Bulatov, V.V. (2005*b*). Stochastic simulation of dislocation glide in tantalum and Ta-based alloys. *J. Mech. Phys. Solids*, **53**, 1223–1247.

Déprés, C., Robertson, C.F., and Fivel, M.C. (2004). Low-strain fatigue in AISI 316L steel surface grains: a three-dimensional discrete dislocation dynamics modelling of the early cycles I. Dislocation microstructures and mechanical behaviour. *Phil. Mag.*, **84**, 2257–2275.

Déprés, C., Robertson, C.F., and Fivel, M.C. (2006). Low-strain fatigue in 316L steel surface grains: a three dimension discrete dislocation dynamics modelling of the early cycles. Part 2: Persistent slip markings and micro-crack nucleation. *Phil. Mag.*, **86**, 79–97.

Devincre, B. (1995). Three dimensional stress field expressions for straight dislocation segments. *Solid State Comm.*, **93**, 875–878.

Devincre, B. (2012*a*). Dislocation Gallery. http://zig.onera.fr/DisGallery.

Devincre, B. (2012*b*). microMegas. http://zig.onera.fr/microMegas.

Devincre, B. (2013). Dislocation dynamics simulations of slip systems interactions and forest strengthening in ice crystals. *Phil. Mag.*, in press.

Devincre, B. and Condat, M. (1992). Model validation of a 3D simulation of dislocation dynamics: Discretization and line tension effects. *Acta Metall. Mater.*, **40**, 2629–2637.

Devincre, B., Hoc, T., and Kubin, L. (2008). Dislocation mean free paths and strain hardening of crystals. *Science*, **320**, 1745–1748.

Devincre, B. and Kubin, L. (1997). The modelling of dislocation dynamics: elastic behaviour versus core properties. *Phil. Trans. Roy. Soc. A*, **355**, 2003–2012.

Devincre, B. and Kubin, L. (2010). Scale transitions in crystal plasticity by dislocation dynamics simulations. *C.R. Physique*, **11**, 274–284.

Devincre, B., Kubin, L., and Hoc, T. (2006). Physical analyses of crystal plasticity by DD simulations. *Scripta Mater.*, **54**, 741–746.

Devincre, B., Kubin, L., and Hoc, T. (2007). Collinear superjogs and the low-stress response of fcc crystals. *Scripta Mater.*, **57**, 905–908.

Devincre, B., Madec, R., Monnet, G., Queyreau, S., Gatti, R., and Kubin, L. (2011). Modeling crystal plasticity with dislocation dynamics simulations: The 'microMegas' code. In *Mechanics of Nano-Objects* (ed. O. Thomas, A. Ponchet, and S. Forest), pp. 81–99. Presses des Mines, Paris.

deWit, G. (actually de Wit R.) and Koehler, J.S. (1959). Interactions of dislocations with an applied stress in anisotropic crystals. *Phys. Rev. B*, **116**, 1113–1120.

Diaz de la Rubia, T., Zbib, H.M., Khraishi, T.A., Wirth, B.D., Victoria, M., and Caturla, M.-J. (2000). Mutiscale modeling of plastic flow localization in irradiated materials. *Nature*, **406**, 871–874.

Diehl, J. (1956). Zugverformung von Kupfer-Einkristallen. I. *Z. Metallkde*, **47**, 331–343.

Dietiker, M., Buzzi, S., Pigozzi, G., Löffler, J.F., and Spolenak, R. (2011). Deformation behaviour of gold nano-pillars prepared by nanoimprinting and focused ion-beam milling. *Acta Mater.*, **59**, 2180–2192.

Dimiduk, D.M., Koslowski, M., and LeSar, R. (2006*a*). Preface to the viewpoint set on: Statistical mechanics and coarse graining of dislocation behavior for continuum plasticity. *Scripta Mater.*, **54**, 701–704.

Dimiduk, D.M., Nadgorny, E.M., Woodward, C., Uchic, M.D., and Shade, P.A. (2010). An experimental investigation of intermittent flow and strain burst scaling behavior in LiF crystals during microcompression testing. *Phil. Mag.*, **90**, 3621–3649.

Dimiduk, D.M., Uchic, M.D., and Parthasarathy, T.A. (2005). Size-affected single-slip behavior of pure nickel microcrystals. *Acta Mater.*, **53**, 4065–4077.

Dimiduk, D.M., Woodward, C., LeSar, R., and Uchic, M.D. (2006*b*). Scale-free intermittent flow in crystal plasticity. *Science*, **312**, 1188–1190.

Domain, C. (2002). *Simulations atomiques ab initio de l'effet de l'hydrogène et de l'iode dans le zirconium*. Ph.D. thesis, Université des Sciences et Technologies de Lille, France.

Domain, C. and Legris, A. (2004). Investigation of glide properties in hexagonal titanium and zirconium: An ab initio atomic scale study. In *IUTAM Symposium on Mesoscopic Dynamics of Fracture Processes and Materials Strength* (ed. H. Kitagawa and Y. Shibutani), Dordrecht (The Netherlands), pp. 411–420. Kluwer Academic Publishers.

Domain, C. and Monnet, G. (2005). Simulation of screw dislocation motion in iron by molecular dynamics simulations. *Phys. Rev. Lett.*, **95**, 215506(1–4).

Dou, R. and Derby, B. (2009). A universal scaling law for the strength of metal micropillars and nanowires. *Scripta Mater.*, **61**, 524–527.

Douin, J., Pettinary-Sturmel, F., and Coujou, A. (2007). Dissociated dislocations in confined plasticity. *Acta Mater.*, **55**, 6453–6458.

Dudarev, S.L., Bullough, R., and Derlet, P.M. (2008). Effect of the α–γ phase transition on the stability of dislocation loops in bcc iron. *Phys. Rev. Lett.*, **100**, 135503(1–4).

Dudarev, S.L. and Derlet, P.M. (2005). A 'magnetic' interatomic potential for molecular dynamics simulations. *J. Phys.: Condens. Matter*, **17**, 7097–7118.

Dudarev, S.L., Derlet, P.M., and Bullough, R. (2009). The magnetic origin of anomalous high-temperature stability of dislocation loops in iron and iron-based alloys. *J. Nucl. Mater.*, **386–388**, 45–48.

Duesbery, M.S. (1983*a*). On kinked screw dislocations in the b.c.c. lattice—I. The structure and Peierls stress of isolated kinks. *Acta Metall.*, **31**, 1747–1758.

Duesbery, M.S. (1983*b*). On kinked screw dislocations in the b.c.c. lattice—II. Kink energies and double kinks. *Acta Metall.*, **31**, 1759–1770.

Duesbery, M.S. (1989). The dislocation core and plasticity. In *Dislocations in Solids* (ed. F.R.N. Nabarro), Volume 8, pp. 67–173. Elsevier Science B.V., Amsterdam.

Duesbery, M.S. and Basinski, Z.S. (1993). The flow stress of potassium. *Acta Metall. Mater.*, **41**, 643–647.

Duesbery, M.S. and Joós, B. (1996). Dislocation motion in silicon: The shuffle-glide controversy. *Phil. Mag. Lett.*, **74**, 253–258.

Duesbery, M.S., Louat, N.P., and Sadananda, K. (1992*a*). The mechanics and energetics of cross-slip. *Acta Metall. Mater.*, **40**, 149–158.

Duesbery, M.S., Louat, N.P., and Sadananda, K. (1992*b*). The numerical simulation of continuum dislocations. *Phil. Mag. A*, **65**, 311–325.

Duesbery, M.S. and Richardson, G.Y. (1991). The dislocation core in crystalline materials. *Crit. Rev. Solid State Mater. Sci.*, **17**, 1–46.

Duesbery, M.S. and Vitek, V. (1998). Plastic anisotropy in b.c.c. transition metals. *Acta Mater.*, **46**, 1481–1492.

Duesbery, M.S. and Xu, W. (1998). The motion of edge dislocations in body-centered cubic metals. *Scripta Mater.*, **39**, 283–287.

Dunne, F.P.E. and Petrinic, N. (2005). *Introduction to Computational Plasticity*. Oxford University Press.

Dupuy, L. and Fivel, M.C. (2002). A study of dislocation junctions in fcc metals by an orientation dependent line tension model. *Acta Mater.*, **50**, 4873–4885.

Durinck, J., Devincre, B., Kubin, L., and Cordier, P. (2007). Modeling the plastic deformation of olivine by dislocation dynamics simulations. *American Mineralogist*, **92**, 1346–1357.

Edagawa, K., Suzuki, T., and Takeuchi, S. (1997). Motion of a screw dislocation in a two-dimensional Peierls potential. *Phys. Rev. B*, **55**, 6180–6187.

El-Awady, J.A., Biner, S.B., and Ghoniem, N.M. (2008). A self-consistent boundary element, parametric dislocation dynamics formulation of plastic flow in finite volumes. *J. Mech. Phys. Solids*, **56**, 2019–2035.

El-Awady, J.A., Ghoniem, N.M., and Mughrabi, H. (2007). Dislocation modelling of localized plasticity in persistent slip bands. In *Mechanics and Materials Modeling and Materials Design Methodologies* (ed. B.L. Adams and A. Garmestani), 136th Annual Meeting and Exhibition, pp. 23–35. TMS, Warrendale PA, USA.

El-Awady, J.A., Wen, M., and Ghoniem, N.M. (2009). The role of weakest-link mechanism in controlling the plasticity of micropillars. *J. Mech. Phys. Solids*, **57**, 32–50.

El-Azab, A. (2006). Statistical mechanics of dislocation systems. *Scripta Mater.*, **54**, 723–727.

Escaig, B. (1968*a*). Cross-slipping process in the f.c.c. structure. In *Dislocation Dynamics* (ed. A.R. Rosenfeld, G.T. Hahn, A.L. Bement, and R.I. Jaffee), New York, pp. 655–677. McGraw–Hill.

Escaig, B. (1968*b*). L'activation thermique des déviations sous faibles contraintes dans les structures h.c. et c.c. *Phys. Stat. Sol. (b)*, **283**, 463–474.

Escaig, B. (1968*c*). Sur le glissement dévié des dislocations dans la structure cubique à faces centrées. *J. de Physique*, **29**, 225–239.

Escaig, B. (1968*d*). Sur l'origine de la limite élastique des métaux cubiques centrés à basse température. *J. de Physique*, **27**, C3-205–208.

Espinosa, H.D., Berbenni, S., Panico, M., and Schwarz, K.W. (2005). An interpretation of size-scale plasticity in geometrically confined thin films. *PNAS*, **102**, 16933–16938.

Espinosa, H.D., Panico, M., Berbenni, S., and Schwarz, K.W. (2006). Discrete dislocation dynamics simulations to interpret plasticity size and surface effects in freestanding FCC films. *Int. J. Plasticity*, **22**, 2091–2117.

Essmann, U. and Mughrabi, H. (1979). Annihilation of dislocations during tensile and cyclic deformation and limits of dislocation densities. *Phil. Mag. A*, **40**, 731–756.

Farber, B.Y., Iunin, Y.L., and Nikitenko, V.I. (1986). Experimental study of the double-kink formation kinetics and kink mobility on the dislocation line in Si single crystals. *Phys. Stat. Sol. (a)*, **97**, 469–478.

Farenc, S., Caillard, D., and Couret, A. (1993). An *in situ* study of prismatic glide in α titanium at low temperatures. *Acta Metall. Mater.*, **41**, 2701–2709.

Farenc, S., Caillard, D., and Couret, A. (1995). A new model for the peak of activation area of α titanium. *Acta Metall. Mater.*, **43**, 3669–3678.

Ferré, D., Carrez, P., and Cordier, P. (2008). Modeling dislocation cores in $SiTiO_3$ using the Peierls–Nabarro model. *Phys. Rev. B*, **77**, 014106(1–7).

Fertig, R.S. and Baker, S.P. (2009). Simulations of dislocation and strength in thin films: A review. *Progr. Mater. Sci.*, **54**, 874–908.

Fertig, R.S. III and Baker, S.P. (2010). Dislocation dynamics simulations of dislocation interactions and stresses in thin films. *Acta Mater.*, **58**, 5206–5218.

Fisher, D.S. (1998). Collective transport in random media: From superconductors to earthquakes. *Phys. Rep.*, **301**, 113–150.

Fisher, J.C., Hart, E.W., and Pry, R.H. (1952). Theory of slip-band formation. *Phys. Rev.*, **87**, 958–961.

Fitzgerald, S.P. (2010). Frank–Read sources and the yield of anisotropic crystals. *Phil. Mag. Lett.*, **90**, 209–218.

Fitzgerald, S.P. and Aubry, S. (2010). Self force on dislocation segments in anisotropic crystals. *J. Phys.: Condens. Matter*, **22**, 29540(3–9).

Fivel, M.C. (2012). Numerical Modelling of Dislocations. http://www.numodis.com.

Fivel, M., Chang, H.-J., and Verdier, M. (2011). Multiscale modeling of indentation: from atomic to continuum. In *Plasticity of Crystalline Materials From Dislocation to Continuum* (ed. I.R. Ionescu, S. Bouvier, O. Cazacu, and P. Franciosi), Chapter 2, pp. 37–56. Wiley–ISTE, ISBN:978-1-84821-278-7.

Fleck, N.A., Muller, G.M., Ashby, M.F., and Hutchinson, J.W. (1994). Strain gradient plasticity: theory and experiment. *Acta Metall. Mater.*, **42**, 475–487.

Fleischer, R.L. (1959). Cross slip of extended dislocations. *Acta Metall.*, **7**, 134–135.

Foreman, A.J.E. (1967). The bowing of a dislocation segment. *Phil. Mag.*, **15**, 1011–1021.

Franciosi, P. (1984). *Plasticité à froid des monocristaux c.f.c.: Étude du durcissement latent*. Ph.D. thesis, Université Paris VII.

Franciosi, P. (1985). The concepts of latent hardening and strain hardening in meallic single crystals. *Acta Metall.*, **33**, 1601–1612.

Franciosi, P., Berveiller, M., and Zaoui, A. (1980). Latent hardening in copper and aluminium single crystals. *Acta Metall.*, **28**, 273–283.

Franciosi, P. and Zaoui, A. (1982). Multislip tests on copper crystals: A junctions hardening effect. *Acta Metall.*, **30**, 2141–2151.

Franěk, A., Kalus, R., and Kratochvíl, J. (1991). Model of early stage of dislocation structure formation in cyclically deformed metal crystals. *Phil. Mag.*, **64**, 497–511.

Frederiksen, S.L. and Jacobsen, K.W. (2003). Density functional studies of screw dislocation core structures in bcc metals. *Phil. Mag.*, **83**, 365–375.

Fressengeas, C., Beaudoin, A.J., Entemeyer, D., Lebedkina, T., Lebyodkin, M., and Taupin, V. (2009). Dislocation transport and intermittency in the plasticity of crystalline solids. *Phys. Rev. B*, **79**, 014108(1–9).

Freund, L.B. (1987). The stability of a dislocation threading a strained layer on a substrate. *J. Appl. Mech.*, **54**, 553–557.

Friedel, J. (1957). Regarding Seeger's paper on work hardening. In *Dislocations and Mechanical Properties of Crystals* (ed. J. Fisher, W. Johnston, R. Thomson, and T. Vreeland Jr), New York, pp. 330–332. J. Wiley and Sons.

Friedel, J. (1967). *Dislocations*. Pergamon Press, Oxford.

Friedman, N., Jennings, A.T., Tsekenis, G., Kim, J.-Y., Tao, M., and Uhl, J.T. *et al.* (2012). Statistics of dislocation slip avalanches in nanosized single crystals show tuned critical behavior predicted by a simple mean field model. *Phys. Rev. Lett.*, **109**, 095507(1–5).

Fusenig, K.-D. and Nembach, E. (1993). Dynamic dislocation effects in precipitation hardened materials. *Acta Metall. Mater.*, **41**, 3181–3189.

Gale, W.F. and Totemeier, T.C. (ed.) (2004). *Smithells Metals Reference Book (8th. edition)*, Oxford. Elsevier (Butterworth-Heinemann).

Gally, B.J. and Argon, A.S. (2001). Brittle to ductile transitions in the fracture of silicon single crystals by dynamic crack arrest. *Phil. Mag.*, **81**, 699–740.

Gao, Y., Liu, Z.L., You, X.C., and Zhuang, Z. (2010). A hybrid multiscale computational framework of crystal plasticity at submicron scales. *Comput. Mater. Sci.*, **49**, 672–681.

Gavazza, S.D. and Barnett, D.M. (1976). The self-force on a planar dislocation loop in an anisotropic linear-elastic medium. *J. Mech. Phys. Solids*, **24**, 171–185.

Gehlen, P.C., Hirth, J.P., Hoagland, R.G., and Kanninen, M.F. (1972). A new representation of the strain field associated with the cube-edge dislocation in a model of a α-iron. *J. Appl. Phys.*, **43**, 3921–3933.

George, A. (1979). Measurements of dislocation velocities in silicon. *J. de Physique*, **40**, C6(133–137).

George, A. (1997). Plastic deformation of semiconductors: Some recent advances and persistent challenges. *Mater. Sci. Eng. A*, **233**, 88–102.

George, A., Jacques, A., and Legros, M. (2007). Low-cycle fatigue in silicon: Comparison with fcc metals. *Fat. Fract. Eng. Mater. Struct.*, **30**, 41–56.

George, A. and Rabier, J. (1987). Dislocations and plasticity in semiconductors. I—Dislocation structures and dynamics. *Rev. Phys. Appl.*, **22**, 941–966.

George, A. and Yip, S. (2001). Preface to the viewpoint set on: Dislocation mobility in silicon. *Scripta Mater.*, **45**, 1233–1238.

Gerberich, W.W., Michler, J., Mook, W.M., Ghisleni, R., Östlund, F., and Stauffer, D.D. *et al.* (2009). Scale effects for strength, ductility, and toughness in "brittle" materials. *J. Mater. Res.*, **24**, 898–906.

Gerold, V. (1979). Precipitation hardening. In *Dislocations in Solids* (ed. F.R.N. Nabarro), Volume 4, pp. 219–260. North-Holland, Amsterdam.

Gerold, V. and Karnthaler, H.P. (1989). On the origin of planar slip in f.c.c. alloys. *Acta Metall.*, **37**, 2177–2183.

Ghazisaeidi, M. and Trinkle, D.R. (2012). Core structure of a screw dislocation in Ti from density functional theory and classical potentials. *Acta Mater.*, **60**, 1287–1292.

Ghoniem, N.M. (2012). MODEL. https://bitbucket.org/model/model/wiki/Home.

Ghoniem, N.M. and Amodeo, R. (1988). Computer simulation of dislocation pattern formation. *Solid State Phenom.*, **3–4**, 377–388.

Ghoniem, N.M. and Sun, L.Z. (1999). A fast sum method for the elastic field of 3-D dislocation ensembles. *Phys. Rev. B*, **60**, 128–140.

Ghoniem, N.M., Tong, S.-H., Singh, B.N., and Sun, L.Z. (2001). On dislocation interaction with radiation-induced defect clusters and plastic flow localization in fcc metals. *Phil. Mag. A*, **81**, 2743–2764.

Ghoniem, N.M., Tong, S.-H., and Sun, L.Z. (2000). Parametric dislocation dynamics: A thermodynamics-based approach to investigations of mesoscopic plastic deformation. *Phys. Rev. B*, **61**, 913–927.

Gil Sevillano, J. (1993). Flow stress and work hardening. In *Plastic Deformation and Fracture of Materials* (ed. H. Mughrabi), Volume 6, Materials Science and Technology, pp. 19–88. VCH, Weinheim (FRG).

Gil Sevillano, J., Bouchaud, E., and Kubin, L.P. (1991). The fractal nature of gliding dislocation lines. *Scripta Metall. Mater.*, **25**, 355–360.

Gil Sevillano, J., Ocaña Arizcorreta, I, and Kubin, L.P. (2001). Intrinsic size effects in plasticity by dislocation glide. *Mater. Sci. Eng. A*, **309–310**, 393–405.

Godet, J., Brochard, S., Pizzagalli, L., Beauchamp, P., and Soler, J.M. (2006). Dislocation formation from a surface step in semiconductors: An *ab initio* study. *Phys. Rev. B*, **73**, 092105(1–4).

Godet, J., Hirel, P., Brochard, S., and Pizzagalli, L. (2009). Evidence of two plastic regimes controlled by dislocation nucleation in silicon nanostructures. *J. Appl. Phys.*, **105**, 026104(1–3).

Godet, J., Pizzagalli, L., Brochard, S., and Beauchamp, P. (2004). Theoretical study of dislocation nucleation from simple surface defects in semiconductors. *Phys. Rev. B*, **70**, 054109(1–8).

Gómez-García, D., Devincre, B., and Kubin, L.P. (2006). Dislocation patterns and the similitude principle: 2.5D mesoscale simulations. *Phys. Rev. Lett.*, **96**, 125503(1–4).

Gordon, P.A., Neeraj, T., Li, Y., and Li, J. (2010). Screw dislocation core mobility in BCC metals: The role of the compact core on double-kink nucleation. *Model. Simul. Mater. Sci. Eng.*, **18**, 085008(1–13).

Gottschalk, H., Hiller, N., Sauerland, S., Specht, P., and Alexander, H. (1993). Constricted dislocations and their use for TEM experiments of the velocity of edge and 60° dislocations in silicon. *Phys. Stat. Sol. (a)*, **138**, 547–555.

Gouldstone, A., Chollacoop, N., Dao, M., Li, J., Minor, A.M., and Shen, Y.-L. (2007). Indentation across scales and disciplines: Recent developments in experimentation and modeling. *Acta Mater.*, **55**, 4015–4039.

Granato, A. and Lücke, K. (1956). Application of dislocation theory to internal friction phenomena at high frequency. *J. Appl. Phys.*, **27**, 789–805.

Greengard, L. (1987). *The rapid evaluation of potential fields in particle systems*. The M.I.T. Press, Cambridge, MA (USA).

Greengard, L. (1997). Fast algorithms for classical physics. *Science*, **265**, 909–914.

Greengard, L. and Rokhlin, V. (1987). A fast algorithm for particle simulations. *J. Comput. Phys.*, **73**, 325–348.

Greengard, L. and Rokhlin, V. (1997). A new version of the Fast Multipole Method for the Laplace equation in three dimensions. *Acta Numerica*, **6**, 229–269.

Greer, J.R. (2006). Bridging the gap between computational and experimental length scales: a review on nano-scale plasticity. *Rev. Adv. Mater. Sci.*, **13**, 59–70.

Greer, J.R. and De Hosson, J.T.M. (2011). Plasticity in small-sized metallic systems: Intrinsic versus extrinsic size effects. *Progr. Mater. Sci.*, **56**, 654–724.

Greer, J.R. and Nix, W.D. (2006). Nanoscale gold pillars strengthened by dislocation starvation. *Phys. Rev. B*, **73**, 245410(1–6).

Greer, J.R., Oliver, W.C., and Nix, W.D. (2005). Size dependence of mechanical properties of gold at the micron scale in the absence of strain gradients. *Acta Mater.*, **53**, 1821–1830.

Greer, J.R., Oliver, W.C., and Nix, W.D. (2006). Corrigendum to "size dependence of mechanical properties of gold at the micron scale in the absence of strain gradients". *Acta Mater.*, **54**, 1705.

Greer, J.R., Weinberger, C.R., and Cai, W. (2008). Comparing the strength of f.c.c. and b.c.c. sub-micrometer pillars: Compression experiments and dislocation dynamics simulations. *Mater. Sci. Eng. A*, **493**, 21–25.

Gröger, R., Bailey, A.G., and Vitek, V. (2008*a*). Multiscale modelling of plastic deformation of molybdenum and tungsten: I. Atomistic studies of the core structure and glide of $1/2\langle 111\rangle$ screw dislocations at 0 K. *Acta Mater.*, **56**, 5401–5411.

Gröger, R., Racherla, V., Bassani, J.L., and Vitek, V. (2008*b*). Multiscale modelling of plastic deformation of molybdenum and tungsten: II. Yield criterion for single crystals based on atomistic studies of glide of $1/2\langle 111\rangle$ screw dislocations. *Acta Mater.*, **56**, 5412–5425.

Gröger, R. and Vitek, V. (2007). Explanation of the discrepancy between the measured and atomistically calculated yield stress in body-centred cubic metals. *Phil. Mag. Lett.*, **87**, 113–120.

Gröger, R. and Vitek, V. (2008). Multiscale modelling of plastic deformation of molybdenum and tungsten: III. Effects of temperature and plastic strain rates. *Acta Mater.*, **56**, 5426–5439.

Gröger, R. and Vitek, V. (2012). Constrained nudged elastic band calculation of the Peierls barrier with atomic relaxations. *Model. Simul. Mater. Sci. Eng.*, **20**, 035019(1–13).

Grosskreutz, H. and Mughrabi, H. (1975). Description of the work-hardened structure at low temperature in cyclic deformation. In *Constitutive Equations in Plasticity* (ed. A.S. Argon), Cambridge, MA (USA), pp. 251–326. The M.I.T. Press.

Gruber, P.A., Böhm, J., Onuseit, F., Wanner, A., Spolenak, R., and Arzt, E. (2008). Size effects on yield strength and strain hardening for ultra-thin Cu films with and witout passivation: A study by synchrotron and bulge stress techniques. *Acta Mater.*, **56**, 2318–2335.

Grydlik, M., Boioli, F., Groiss, H., Gatti, R., Brehm, M., and Montalenti, F. *et al.* (2012). Misfit dislocation gettering by substrate pit-patterning in SiGe films on Si(001). *J. Appl. Phys.*, **101**, 013119(1–4).

Guiu, F. (1969). Slip asymetry in molybdenum single crystals deformed in direct shear. *Scripta Metall.*, **3**, 449–454.

Gumbsch, P. and Gao, H. (1999). Dislocations faster than the speed of sound. *Science*, **283**, 965–968.

Haasen, P. (1979). Solution hardening in f.c.c. metals. In *Dislocations in Solids* (ed. F.R.N. Nabarro), Volume 4, pp. 155–189. North-Holland, Amsterdam.

Haasen, P. (1996). *Physical Metallurgy*. Cambridge University Press, Cambridge.

Hähner, P. (1996*a*). Stochastic dislocation patterning during cyclic deformation. *Appl. Phys. A*, **63**, 43–55.

Hähner, P. (1996*b*). A theory of dislocation cell formation based on stochastic dislocation dynamics. *Acta Mater.*, **44**, 2345–2352.

Hähner, P., Bay, K., and Zaiser, M. (1998). Fractal dislocation patterning during plastic deformation. *Phys. Rev. Lett.*, **81**, 2470–2473.

Hähner, P. and Zaiser, M. (1997). From mesoscopic heterogeneity of slip to macroscopic fluctuations of stress and strain. *Acta Mater.*, **45**, 1067–1075.

Haken, H. (2004). *Synergetics. Introduction and Advanced Topics*. Springer, Berlin.

Han, X., Ghoniem, N.M., and Wang, Z. (2003). Parametric dislocation dynamics of anisotropic crystals. *Phil. Mag.*, **83**, 3705–3721.

Hansen, N., Huang, X., and Hughes, D.A. (2001). Microstructural evolution and hardening parameters. *Mater. Sci. Eng. A*, **317**, 3–11.

Hansen, N., Huang, X., and Winther, G. (2010). Effect of grain boundaries and grain orientation on structure and properties. *Metall. Mater. Trans. A*, **42**, 613–625.

Hazzledine, P.M., Karnthaler, H.P., and Wintner, E. (1975). Non-parallel dissociation of dislocations in thin films. *Phil. Mag. A*, **31**, 81–97.

Hemker, K.J. and Nix, W.D. (2008). Seeing is believing. *Nature Mater.*, **7**, 97–98.

Henager Jr., C.H. and Hoagland, R.G. (2004). Dislocation core fields and forces in FCC metals. *Scripta Mater.*, **50**, 1091–1095.

Henager Jr., C.H. and Hoagland, R.G. (2005). Dislocation and stacking fault core fields in fcc metals. *Phil. Mag.*, **85**, 4477–4508.

Henkelman, G. and Jónsson, H. (2000). Improved tangent estimate in the nudged elastic band method for finding energy minimum paths and saddle points. *J. Chem. Phys.*, **113**, 9978–9985.

Henkelman, G., Uberuaga, B.P., and Jónsson, H. (2000). A climbing image nudged elasic band method for finding energy minimum paths and saddle points. *J. Chem. Phys.*, **113**, 9901–9904.

Hesselbarth, H.W. and Steck, E. (1992). A simulation of dislocation patterning derived from cellular automata. *Solid State Phenom.*, **23&24**, 445–454.

Hillert, M. and Agren, J. (2006). Extremum principles for irreversible processes. *Acta Mater.*, **54**, 2063–2066.

Hiritani, H. and Zbib, H.M. (2003). On dislocation–defect interactions and patterning: stochastic discrete dislocation dynamics (SDD). *J. Nucl. Mater.*, **323**, 290–303.

Hirsch, P.B. (1960). Comment. In *Proc. of the 5th. Int. Conf. of Crystallography*, Cambridge, p. 139. Cambridge University Press.

Hirsch, P.B. (1975). *The Physics of Metals: 2. Defects*. Cambridge University Press, Cambridge.

Hirsch, P.B., Ourmazd, A., and Pirouz, P. (1981). Relaxation of dislocations in deformed silicon. In *Microscopy of Semiconducting Materials, 1981* (ed. A.C. Cullis and D.C. Joy), Conference Series Number 60, pp. 29–34. The Institute of Physics, Bristol, London.

Hirth, J.P. (1985). A brief history of dislocation theory. *Metall. Trans. A*, **16**, 2085–2090.

Hirth, J.P. and Kubin, L. (ed.) (2010). *Dislocations in Solids. The 30th Anniversary Volume*, Volume 16, Dislocations in Solids. Elsevier Science B.V., Amsterdam.

Hirth, J.P. and Lothe, J. (1992). *Theory of Dislocations*. Krieger, Malabar (Florida).

Hoc, T., Rey, C., and Raphanel, J.-L. (2001). Experimental and numerical analyses of localization during sequential test for an IF-Ti steel. *Acta Mater.*, **49**, 1835–1846.

Hollang, L. (2002). Mechanical properties. In *Purification process and chacracterization of ultra high purity metals: Application of basic science to metallurgical processing* (ed. Y. Waseda and M. Isshiki), Springer Series in Materials Processing, pp. 305–348. Springer, Berlin.

Hollang, L., Brunner, D., and Seeger, A. (2001). Work hardening and flow stress of ultrapure molybdenum single crystals. *Mater. Sci. Eng. A*, **319–321**, 2177–2187.

Holt, D.L. (1970). Dislocation cell formation in metals. *J. Appl. Phys.*, **41**, 3197–3201.

Holzwarth, U. and Essmann, U. (1993). Transformation of dislocation patterns in fatigued copper single crystals. *Mater. Sci. Eng. A*, **164**, 206–210.

Holzwarth, U. and Essmann, U. (1994). Temperature-induced rearrangement of the dislocation pattern of Persistent Slip Bands in copper single crystals. *Appl. Phys. A*, **58**, 197–210.

Honeycombe, R.W.K. (1968). *The Plastic Deformation of Metals*. Edward Arnold Ltd., London.

Hornstra, J. (1958). Dislocations in the diamond lattice. *J. Phys. Chem. Solids*, **5**, 129–141.

Hosford, W.F. (1993). *The Mechanics of Crystals and Textured Polycrystals*. Oxford Engineering Science Series. Oxford University Press, New York.

Hosford, W.F. (2005). *Mechanical Behavior of Materials*. Cambridge University Press, New-York.

Huang, J. and Ghoniem, N.M. (2002). The dynamics of dislocation interaction with sessile self-interstitial atom (SIA) defect cluster atmospheres. *Comput. Mater. Sci.*, **23**, 225–234.

Huang, M., Zhao, L., and Tong, J. (2012). Discrete dislocation dynamics modelling of mechanical deformation of nickel-based single crystal superalloys. *Int. J. Plasticity*, **28**, 141–158.

Huang, X. (1998). Grain orientation effect on microstructure in tensile strained copper. *Scripta Mater.*, **38**, 1697–1703.

Huang, X. and Winther, G. (2007). Dislocation structures. Part I. grain orientation dependence. *Phil. Mag.*, **86**, 5189–5214.

Hughes, D.A. (2001). Microstructure evolution, slip patterns and flow stress. *Mater. Sci. Eng. A*, **319–321**, 46–54.

Hughes, D.A., Chrzan, D.C., Liu, Q., and Hansen, N. (1998). Scaling of misorientation angle distribution. *Phys. Rev. Lett.*, **81**, 4664–4667.

Hughes, D.A. and Hansen, N. (2000). Microstructure and strength of nickel at large strains. *Acta Mater.*, **48**, 2985–3004.

Hughes, D.A., Liu, Q., Chrzan, D.C., and Hansen, N. (1997). Scaling of microstructural parameters: Misorientations of deformation induced boundaries. *Acta Mater.*, **45**, 105–112.

Hull, D. and Bacon, D.J. (2001). *Introduction to Dislocations* (4th. edn). Reprinted 2002 (5th. edn 2011). Butterworth-Heinemann, Oxford.

Imai, M. and Sumino, K. (1983). In situ X-ray topographic study of the dislocation mobility in high-purity and impurity-doped silicon crystals. *Phil. Mag. A*, **47**, 599–621.

Indenbom, V.L., Pethukov, B.V., and Lothe, J. (1992). Dislocation motion over the Peierls barrier. In *Elastic strain fields and dislocation mobility* (ed. V.I. Indenbom and J. Lothe), Volume 31, Modern Problems in Condensed Matter Science, pp. 489–516. Elsevier Science B.V., Amsterdam.

Irwin, G.J., Guiu, F., and Pratt, P.L. (1974). The influence of orientation on slip and strain hardening of molybdenum single crystals. *Phys. Stat. Sol. (a)*, **22**, 685–698.

Ismail-Beigi, S. and Arias, T.A. (2000). Ab initio study of screw dislocations in Mo and Ta: A new picture of plasticity in bcc transition metals. *Phys. Rev. Lett.*, **84**, 1499–1502.

Itakura, M., Kaburaki, H., and Yamaguchi, M. (2012). First-principles study on the mobility of screw dislocations in bcc iron. *Acta Mater.*, **60**, 3698–3710.

Ito, K. and Vitek, V. (2001). Atomistic study of non-Schmid effects in the plastic yielding of bcc metals. *Phil. Mag. A*, **81**, 1387–1407.

Iunin, Y.L. and Nikitenko, V.I. (2001). Modes of kink motion on dislocations in semiconductors. *Scripta Mater.*, **45**, 1239–1246.

Izumi, S., Miyake, T., Sakai, S., and Ohta, H. (2005). Application of three-dimensional dislocation dynamics simulation to the STI semiconductor structure. *Mater. Sci. Eng. A*, **395**, 62–69.

Izumi, S., Ohta, H., Takahashi, C., Suzuki, T., and Saka, H. (2010). Shuffle-set dislocation nucleation in semiconductor silicon device. *Phil. Mag. Lett.*, **90**, 707–714.

Izumi, S and Yip, S. (2008). Dislocation nucleation from a sharp corner in silicon. *J. Appl. Phys.*, **104**, 033513(1–4).

Jackson, P.J. (1968). Double coplanar slip in silver. *Scripta Metall.*, **2**, 411–418.

Jackson, P.J. (1985). Dislocation modelling of shear in fcc crystals. *Progr. Mater. Sci.*, **29**, 139–175.

Jacques, A. and George, A. (2001). Ideal and possible experiments for further progress on dislocation mobility in silicon. *Scripta Mater.*, **45**, 1279–1285.

Jenkins, M.L. (1972). Measurement of the stacking-fault energy of gold using the weak-beam technique of electron microscopy. *Phil. Mag.*, **26**, 747–751.

Jennings, A.T., Burek, M.J., and Greer, J.R. (2010). Microstructure versus size: Mechanical properties of electroplated single crystalline Cu nanopillars. *Phys. Rev. Lett.*, **104**, 135503(1–4).

Jennings, A.T., Li, Y., and Greer, J.R. (2011). Emergence of strain-rate sensitivity in Cu nanopillars: Transition from dislocation multiplication to dislocation nucleation. *Acta Mater.*, **59**, 5627–5637.

Jones, R. (2000). Do we really understand dislocations in semiconductors? *Mater. Sci. Eng. B*, **71**, 24–29.

Jones, R. and Blumenau, A.T. (2000). Interactions of dislocations in Si with intrinsic defects. *Scripta Mater.*, **45**, 1253–1258.

Jónsson, H., Mills, G., and Jacobsen, K.W. (1998). Nudged elastic band method for finding minimum energy paths of transitions. In *Classical and Quantum Dynamics in Phase Transitions* (ed. B.J. Berne, G. Ciccotti, and D.F. Coker), pp. 385–403. World Scientific, Singapore.

Joós, B. and Duesbery, M.S. (1997). The Peierls stress of dislocations: An analytic formula. *Phys. Rev. Lett.*, **78**, 266–269.

Kammler, M., Chidambarrao, D., Schwarz, K.W., Black, C.T., and Ross, F.M. (2005). Controlled nucleation of dislocations by a spatially localized stress field. *Appl. Phys. Lett.*, **87**, 133116(1–3).

Kawasaki, Y. and Takeuchi, T. (1980). Cell structures in copper single crystals deformed in the [001] and [111] axes. *Scripta Metall.*, **14**, 183–188.

Keh, A.S. and Weissmann, S. (1963). Deformation substructure in body-centered cubic metals. In *Electron Microscopy and Strength of Crystals* (ed. G. Thomas and J. Washburn), New-York, pp. 231–300. Interscience Publishers.

Kelly, A. and Nicholson, R.B. (1971). *Strengthening Methods in Crystals*. Applied Science Publishers, London.

Khachaturyan, A. G. (1983). *Theory of Structural Transitions in Solids*. John Wiley and Sons, New York.

Khater, H.A. and Bacon, D.J. (2010). Dislocation core structure and dynamics in two atomic models of α-zirconium. *Acta Mater.*, **58**, 2978–2987.

Khraishi, T.A., Yan, L., and Shen, Y.L. (2004). Dynamic simulations of the interaction between dislocations and dilute particle concentrations in metal–matrix composites. *Int. J. Plasticity*, **20**, 1039–1057.

Khraishi, T.A., Zbib, H.M., and Diaz de la Rubia, T. (2001). The treatment of dislocation-free boundary conditions in three-dimensional dislocation dynamics using generalized image stress analysis. *Mater. Sci. Eng. A*, **309–310**, 283–287.

Khraishi, T.A., Zbib, H.M., Diaz de la Rubia, T., and Victoria, M. (2002). Localized deformation and hardening in irradiated materials: 3-D dislocation dynamics simulations. *Metall. Mater. Trans. B*, **33**, 285–296.

Kibey, S., Liu, J.B., Johnson, D.D., and Sehitoglu, H. (2007). Predicting twinning stress in fcc metals: Linking twin-energy pathways to twin nucleation. *Acta Mater.*, **55**, 6843–6851.

Kiener, D., Grosinger, W., and Dehm, G. (2009*a*). On the importance of sample compliance in uniaxial microtesting. *Scripta Mater.*, **60**, 148–151.

Kiener, D., Grosinger, W., Dehm, G., and Pippan, R. (2008). A further step towards an understanding of size-dependent crystal plasticity: In situ tension experiments of miniaturized single-crystal copper samples. *Acta Mater.*, **56**, 580–592.

Kiener, D., Motz, C., and Dehm, G. (2009*b*). Micro-compression testing: A critical discussion of experimental constraints. *Mater. Sci. Eng. A*, **505**, 79–87.

Kiener, D., Motz, C., Rester, M., Jenko, M., and Dehm, G. (2007). FIB damage of Cu and possible consequences for miniaturized mechanical tests. *Mater. Sci. Eng. A*, **459**, 262–272.

Kim, J.-Y., Jang, D., and Greer, J.R. (2010). Tensile and compressive behavior of tungsten, molybdenum, tantalum and niobium at the nanoscale. *Acta Mater.*, **58**, 2355–2363.

Kim, J.-Y. and Greer, J.R. (2009). Tensile and compressive behavior of gold and molybdenum single crystals at the nano-scale. *Acta Mater.*, **57**, 5245–5253.

Knoesen, D. and Kritzinger, S. (1982). Dislocation cell boundary widths and dislocation cell sizes in deformed copper. *Acta Metall.*, **30**, 1219–1222.

Kocks, U.F. (1960). Polyslip in single crystals. *Acta Metall.*, **8**, 345–352.

Kocks, U.F. (1976). Laws for work-hardening and low-temperature creep. *J. Eng. Materials and Eng. Technol.*, **98**, 76–85.

Kocks, U.F. (1979). Superposition of alloy hardening, strain hardening and dynamic recovery. In *Strength of Metals and Alloys, Proc. ICSMA 5* (ed. P. Haasen, V. Gerold, and G. Kostorz), Volume 3, Oxford, pp. 1661–1680. Pergamon Press.

Kocks, U.F., Argon, A.S., and Ashby, M.F. (1975). Thermodynamics and kinetics of slip. *Progr. Mater. Sci.*, **19**, 1–291.

Kocks, U.F. and Mecking, H. (2003). Physics and phenomenology of strain hardening: The fcc case. *Progr. Mater. Sci.*, **48**, 171–273.

Koester, A., Ma, A., and Hartmaier, A. (2012). Atomistically informed crystal plasticity model for body-centered cubic iron. *Acta Mater.*, **60**, 3894–3901.

Kohler, C., Kizler, P., and Schmauder, S. (2005). Atomic simulation of precipitation hardening in α-iron: Influence of precipitate shape and chemical composition. *Model. Simul. Mater. Sci. Eng.*, **13**, 35–45.

Kolar, H.R., Spence, J.C.H., and Alexander, H. (1996). Observation of moving kinks and unpinning. *Phys. Rev. Lett.*, **77**, 4031–4034.

Korner, A. and Karnthaler, H.P. (1983). Weak-beam study of glide dislocations in h.c.p. cobalt. *Phil. Mag. A*, **48**, 469–477.

Korte, S., Barnard, J.S., Stearn, R.J., and Clegg, W.J. (2011). Deformation of silicon—insights from microcompression testing at 25–500 °C. *Int. J. Plasticity*, **27**, 1823–1866.

Koslowski, M., Cuitiño, A.M., and Ortiz, M. (2002). A phase-field theory of dislocation dynamics, strain hardening and hysteresis in ductile single crystals. *J. Mech. Phys. Solids*, **50**, 2597–2635.

Koslowski, M., LeSar, R., and Thomson, R. (2004). Avalanches and scaling in plastic deformation. *Phys. Rev. Lett.*, **93**, 125502(1–4).

Kraft, O., Gruber, P.A., Mönig, R., and Weygand, D. (2010). Plasticity in confined dimensions. *Annu. Rev. Mater. Res.*, **40**, 293–317.

Kratochvíl, J. and Libovický, S. (1986). Dipole drift mechanism of early stage of dislocation pattern formation in deformed metal single crystals. *Scripta Metall.*, **20**, 1625–1630.

Křišťán, J. and Kratochvíl, J. (2007). Interactions of glide dislocations in a channel of a persistent slip band. *Phil. Mag.*, **87**, 4593–4613.

Kröner, E. (2001). Benefits and shortcomings of the continuous theory of dislocations. *Int. J. Solid Structures*, **38**, 1115–1134.

Kroupa, F. (1961). The force between non-parallel dislocations. *Czech. J. Phys.*, **11**, 847–848.

Kubin, L.P. (1976). The low temperature plastic deformation of bcc metals. *Rev. Deform. Behav. Mater.*, **1**, 244–288.

Kubin, L.P. (1982). The low temperature mechanical properties of bcc metals and their alloys II. *Rev. Deform. Behav. Mater.*, **IV**, 181–275.

Kubin, L.P. (1993). Dislocation patterning. In *Plastic Deformation and Fracture of Materials* (ed. H. Mughrabi), pp. 137–190. VCH, Weinheim (FRG).

Kubin, L.P., Canova, G., Condat, M., Devincre, B., Pontikis, V., and Bréchet, Y. (1992). Dislocation microstructures and plastic flow: A 3 D simulation. *Solid State Phenom.*, **23–24**, 455–472.

Kubin, L., Devincre, B., and Hoc, T. (2008). Modeling dislocation storage rates and mean free paths in face-centered cubic crystals. *Acta Mater.*, **56**, 6040–6049.

Kubin, L.P., Fressengeas, C., and Ananthakrishna, G. (2002). Collective behaviour of dislocations in plasticity. In *Dislocations in Solids* (ed. F.R.N. Nabarro and M.S. Duesbery), Volume 11, pp. 101–192. Elsevier Science B.V., Amsterdam.

Kubin, L., Hoc, T., and Devincre, B. (2009). Dynamic recovery and its orientation dependence in face-centered cubic crystals. *Acta Mater.*, **57**, 2567–2575.

Kubin, L.P. and Louchet, F. (1979). Analysis of softening in the Fe-C system from in situ and conventional experiments-I. In situ experiments. *Acta Metall.*, **27**, 337–342.

Kubin, L.P., Madec, R., and Devincre, B. (2003). Dislocations intersections and reactions in fcc and bcc crystals. In *Multiscale Phenomena in Materials—Experiments and Modeling related to Mechanical Behavior* (ed. H. Zbib, D. Lassila, L. Levine, and K. Hemker), Volume 779, MRS Symposium Proceedings, pp. 25–36. Materials Research Society, Warrendale, PA (USA).

Kubin, L.P. and Martin, J.L. (1979). In situ deformation experiments in the H.V.E.M. In *Strength of Metals and Alloys, Proc. ICSMA 5* (ed. P. Haasen, V. Gerold, and G. Kostorz), Volume 3, Oxford, pp. 1639–1660. Pergamon Press.

Kuhlmann-Wilsdorf, D. (1962). A new theory of work hardening. *Trans. Met. Soc. AIME*, **224**, 1047–1061.

Kuhlmann-Wilsdorf, D. (2002). The LES theory of solid plasticity. In *Dislocations in Solids* (ed. F.R.N. Nabarro and M.S. Duesbery), Volume 11, pp. 211–342. Elsevier Science B.V., Amsterdam.

Kumagai, J., Takaki, S., Suzuki, S., and Kimura, H. (1990). Temperature dependence of work-hardening in high purity iron single crystals. *Materials Transactions, JIM*, **31**, 118–128.

Kuramoto, E., Aono, Y., and Kitajima, K. (1979). Thermally activated slip deformation of high purity iron single crystals between 4.2 K and 300 K. *Scripta Metall.*, **13**, 1039–1042.

Labusch, A. and Schröter, W. (1983). Electrical properties of dislocations in semiconductors. In *Dislocations in Solids* (ed. F.R.N. Nabarro), Volume 5, pp. 127–191. North-Holland, Amsterdam.

Lachenmann, R. and Schultz, H. (1970). Inherent lattice hardening and interstitial solution hardening in tantalum. *Scripta Metall.*, **4**, 709–713.

Laird, C. (1996). Fatigue. In *Physical Metallurgy* (fourth edn) (ed. R.W. Cahn and P. Haasen), Volume 3, pp. 2294–2397. Elsevier Science B.V., Amsterdam.

Laurson, L. and Alava, M.J. (2006). $1/f$ noise and avalanche scaling in plastic deformation. *Phys. Rev. E*, **74**, 066106(1–5).

Lavrentev, F.F. (1980). The type of junction interaction as the factor determining work hardening. *Mater. Sci. Eng.*, **46**, 191–208.

Lee, S.-W., Aubry, S., Nix, W.D., and Cai, W. (2011). Dislocation junctions and jogs in a free-standing FCC thin film. *Model. Simul. Mater. Sci. Eng.*, **19**, 025002(1–14).

Lee, S.-W., Han, S.M., and Nix, W.D. (2009). Uniaxial compression of fcc Au nanopillars on an MgO substrate: The effects of prestraining and annealing. *Acta Mater.*, **57**, 4404–4415.

Lee, S.-W. and Nix, W.D. (2011). Geometrical analysis of 3D dislocation dynamics simulations of FCC micro-pillar plasticity. *Mater. Sci. Eng. A*, **527**, 1903–1910.

Lefebvre, S., Devincre, B., and Hoc, T. (2007). Yield stress strengthening in ultrafine-grained metals: A two-dimensional simulation of dislocation dynamics. *J. Mech. Phys. Solids*, **55**, 788–802.

Legrand, B. (1984). Le glissement dans les métaux de structure hexagonale compacte: Structure de coeur et structure électronique. In *Dislocations 1984* (ed. P. Veyssière, L. Kubin, and J. Castaing), pp. 73–92. Editions du CNRS, Paris.

Legrand, B. (1985). Structure du coeur des dislocations vis (1/3)<11$\bar{2}$0> dans le titane. *Phil. Mag. A*, **52**, 83–97.

Legros, M., Gianola, D.S., and Motz, C. (2010). Quantitative *in situ* mechanical testing in electron microscopes. *MRS Bulletin*, **35**(5), 354–360.

Legros, M., Jacques, A., and George, A. (2004). Cyclic deformation of silicon single crystals: Mechanical behaviour and dislocation arrangements. *Mater. Sci. Eng. A*, **387–389**, 495–500.

Lehto, N. and Öberg, S. (1998). Effects of dislocations interactions: Application to the period-doubled core of the 90° partial dislocation in silicon. *Phys. Rev. Lett.*, **80**, 5568–5571.

Lemarchand, C., Devincre, B., and Kubin, L.P. (2001). Homogenization method for a discrete-continuum simulation of dislocation dynamics. *J. Mech. Phys. Solids*, **49**, 1969–1982.

Lépinoux, J. and Kubin, L.P. (1987). The dynamic organization of dislocation structures: A simulation. *Scripta Metall.*, **21**, 833–838.

LeSar, R. and Rickman, J.M. (2002). Multipole expansion of dislocation interactions. *Phys. Rev. B*, **65**, 144110(1–6).

Li, B., Cao, B.Y., Ramesh, K.T., and Ma, E. (2009). A nucleation mechanism of deformation twins in pure aluminum. *Acta Mater.*, **57**, 4500–4507.

Li, H., Wurster, S., Motz, C., Romaner, L., Ambrosch-Draxl, C., and Pippan, R. (2012). Dislocation-core symmetry in tungsten alloys: Ab initio calculations and microcantilever bending experiments. *Acta Mater.*, **60**, 748–758.

Li, J.C.M. (1964). Stress field of a dislocation segment. *Phil. Mag.*, **10**, 1097–1098.

Lilleodden, E. (2009). Novel techniques for nanomechanical characterization. *JOM*, **61**(3), 13.

Liu, Q. and Hansen, N. (1995). Geometrically necessary boundaries and incidental dislocation boundaries formed during cold deformation. *Scripta Metall. Mater.*, **32**, 1289–1295.

Liu, Q., Juul Jensen, D., and Hansen, N. (1998). Effect of grain orientation on deformation structure in cold-rolled polycrystalline aluminium. *Acta Mater.*, **46**, 5819–5838.

Liu, X., Golubov, S.I., Woo, C.H., and Huang, H. (2004). Glide of edge dislocations in tungsten and molybdenum. *Mater. Sci. Eng. A*, **365**, 96–100.

Liu, X.H. and Schwarz, K.W. (2005). Modelling of dislocations intersecting a free surface. *Model. Simul. Mater. Sci. Eng.*, **13**, 1233–1247.

Louchet, F. (1981). On the mobility of dislocations in silicon by in situ straining in a high-voltage electron microscope. *Phil. Mag. A*, **43**, 1289–1297.

Louchet, F. and George, A. (1983). Dislocation mobility measurements: An essential tool for understanding the atomic and electronic core structures of dislocations in semiconductors. *Journal de Physique*, **44**, C4–51–60.

Louchet, F. and Kubin, L.P. (1975). Dislocation substructures in the anomalous slip plane of single crystal niobium strained at 50 K. *Acta Metall.*, **23**, 17–21.

Louchet, F., Kubin, L.P., and Vesely, D. (1979). In situ deformation of b.c.c. crystals at low temperatures in a high voltage electron microscope. Dislocation mechanisms and strain-rate equation. *Phil. Mag. A*, **39**, 433–454.

Lu, G. (2005). The Peierls–Nabarro model of dislocations: A venerable theory and its current development. In *Handbook of Materials Modelling* (ed. S. Yip), Chapter 2.20, pp. 793–811. Springer, Berlin.

Lu, G., Bulatov, V.V., and Kioussis, N. (2002). Dislocation constriction and cross-slip: An *ab initio* study. *Phys. Rev. B*, **66**, 144103(1–5).

Lu, G. and Kaxiras, E. (2005). Overview of multiscale simulations of materials. In *Handbook of Theoretical and Computational Nanotechnology* (ed. M. Rieth and W. Schommers), Volume X, Chapter 22, pp. 1–33. American Scientific Publishers.

Lu, G., Kioussis, N., Bulatov, V., and Kaxiras, E. (2000). Generalized stacking fault energy surfaces and dislocation properties of aluminum. *Phys. Rev. B*, **62**, 3099(1–10).

Lubarda, V., Blume, J.A., and Needleman, A. (1993). An analysis of equilibrium dislocations distributions. *Acta Mater.*, **41**, 625–642.

Luft, A. (1991). Microstructural processes of plastic instabilities in strengthened metals. *Progr. Mater. Sci.*, **35**, 97–204.

Ma, A., Roters, F., and Raabe, D. (2006). A dislocation density based constitutive model for crystal plasticity FEM including geometrically necessary dislocations. *Acta Mater.*, **54**, 2169–2179.

Maass, R., Van Petegem, S., Ma, D., Zimmermann, J., Grolimund, D., and Roters, F. *et al.* (2009). Smaller is stronger: The effect of strain hardening. *Acta Mater.*, **57**, 5996–6005.

Madec, R., Devincre, B., and Kubin, L.P. (2001). New line model for optimized dislocation dynamics simulations. In *Multiscale Modeling of Materials—2000* (ed. L. Kubin, R. Selinger, J. Bassani, and K. Cho), Volume 653, MRS Symposium Proceedings, pp. Z1.8(1–6). Materials Research Society, Warrendale, PA (USA).

Madec, R., Devincre, B., and Kubin, L.P. (2002*a*). From dislocation junctions to forest hardening. *Phys. Rev. Lett.*, **89**, 255508(1–4).

Madec, R., Devincre, B., and Kubin, L.P. (2002*b*). Simulations of dislocation patterns in multislip. *Scripta Mater.*, **47**, 689–695.

Madec, R., Devincre, B., and Kubin, L. (2004). On the use of periodic boundary conditions in dislocation dynamics simulations. In *IUTAM Symposium on Mesoscopic Dynamics of Fracture Processes and Materials Strength* (ed. H. Kitagawa and Y. Shibutani), Dordrecht (The Netherlands), pp. 35–44. Kluwer Academic Publishers.

Madec, R., Devincre, B., Kubin, L.P., Hoc, T., and Rodney, D. (2003). The role of collinear interaction in dislocation-induced hardening. *Science*, **301**, 1879–1882.

Madec, R. and Kubin, L.P. (2004). Dislocation interactions and symmetries in bcc crystals. In *Mesoscopic Dynamics of Fracture Processes and Materials Strength* (ed. H. Kitagawa and Y. Shibutani), Dordrecht (The Netherlands), pp. 69–78. Kluwer Academic Publishers.

Madec, R. and Kubin, L.P. (2008). Second-order junctions and strain hardening in bcc and fcc crystals. *Scripta Mater.*, **58**, 767–770.

Mader, S. (1957). Elektronenmikroskopische Untersuchung der Gleitlinienbildung auf Kupfereinkristallen. *Z. f. Physik A*, **149**, 73–102.

Maeda, K. and Takeuchi, S. (1996). Enhancement of dislocation mobility in semiconducting crystals by electronic excitation. In *Dislocations in Solids* (ed. F.R.N. Nabarro and M.S. Duesbery), Volume 10, pp. 443–504. Elsevier Science B.V., Amsterdam.

Maeda, K. and Yamashita, Y. (1993). Dislocation motion in strained thin films. Are kinks colliding with each other? *Phys. Stat. Sol. (a)*, **138**, 523–532.

Magid, K.R., Florando, J.N., Lassila, D.H., Leblanc, M.M., Tamura, N., and Morris Jr., J.W. (2009). Mesoscale heterogeneity in the plastic deformation of a copper crystal. *Phil. Mag.*, **89**, 77–107.

Malygin, G.A. (1999). Dislocation self-organization processes and crystal plasticity. *Physics-Uspeckhi*, **42**, 887–916.

Malygin, G.A. (2010). Size effects under plastic deformation of microcrystals and nanocrystals. *Phys. Solid State*, **52**, 49–57.

Marian, J., Cai, W., and Bulatov, V.V. (2004). Dynamic transitions from smooth to rough to twinning in dislocation motion. *Nature Mater.*, **3**, 158–162.

Marian, J. and Caro, A. (2006). Moving dislocations in disordered alloys: Connecting continuum and discrete models with atomistic simulations. *Phys. Rev. B*, **74**, 024113(1–12).

Marian, J., Martinez, E., Lee, H.J., and Wirth, B.D. (2009). Micro/mesoscale computational study of dislocation-stacking-fault tetrahedron interactions in copper. *J. Mater. Res.*, **24**, 3628–3635.

Marklund, S. (1985). The entropy factor involved in the Hirth and Lothe theory of dislocation velocity in silicon. *Solid State Comm.*, **54**, 555–558.

Martin, J.W. (1998). *Precipitation Hardening*. Butterworth-Heinemann, Oxford.

Martin, R.M. (2004). *Electronic Structure. Basic Theory and Practice*. Cambridge University Press, Cambridge.

Martinez, E., Marian, J., Arsenlis, A., Victoria, M., and Perlado, J.M. (2008). Atomistically informed dislocation dynamics in fcc crystals. *J. Mech. Phys. Solids*, **56**, 869–895.

Matsui, H. and Kimura, H. (1975). Anomalous {110} slip and the role of co-planar double slip in bcc metals. *Scripta Metall.*, **9**, 971–978.

Matsui, H. and Kimura, H. (1976). Anomalous {110} slip in high-purity molybdenum single crystals and its comparison with that in V(a) metals. *Mater. Sci. Eng.*, **24**, 247–256.

Matthews, J.W., Mader, S., and Light, T.B. (1970). Accommodation of misfit across the interface between crystals of semiconducting elements or compounds. *J. Appl. Phys.*, **41**, 3800–3804.

Mecking, H. and Kocks, U.F. (1981). Kinetics of flow and strain-hardening. *Acta Metall.*, **29**, 1865–1875.

Medvedeva, N.I., Gornostyrev, Yu. N., and Freeman, A.J. (2007). Solid solution softening and hardening in the group-V and group-VI bcc transition metals alloys: First principles calculations and atomistic modeling. *Phys. Rev. B*, **76**, 212104(1–4).

Mendelev, M.I., Han, S., Srolovitz, D.J., Ackland, G.J., Sun, D.Y., and Asta, M. (2003). Development of new interatomic potentials for crystalline and liquid iron. *Phil. Mag.*, **83**, 3977–3994.

Messerschmidt, U. (2010*a*). *Dislocation Dynamics during Plastic Deformation*. Springer Series in Materials Science 129. Springer, Berlin.

Messerschmidt, U. (2010*b*). Video clips of in situ deformation. http://extras.springer.com/2010/978-3-642-03177-9/.

Metsue, A., Carrez, P., Denoual, C., Mainprice, D., and Cordier, P. (2010). Plastic deformation of wadsleyite: IV. Dislocation core modelling based on the Peierls–Nabarro–Galerkin model. *Acta Mater.*, **58**, 1467–1478.

Meyer, E. (1908). Untersuchungen über Härteprüfung und Härte. *Z. Ver. Dtsch. Ing.*, **52**, 645–654.

Meyers, M.A., Jarmakani, H., Bringa, E.M., and Remington, B.A. (2009). Dislocations in shock compression and release. In *Dislocations in Solids* (ed. J.P. Hirth and L. Kubin), Volume 15, pp. 91–197. Elsevier B.V.

Michot, G., Azzouzi, H., Maloufi, N., Loyola de Oliveira, M.A., Scandian, C., and George, A. (2000). Possible mechanisms for dislocation multiplication at, or close to, a crack tip. In *Multiscale Phenomena in Plasticity* (ed. J. Lépinoux, D. Mazière, V. Pontikis, and G. Saada), Dordrecht (The Netherlands), pp. 117–125. Kluwer Academic Publishers.

Miguel, M.C., Laurson, L., and Alava, M.J. (2008). Material yielding and irreversible deformation mediated by dislocation motion. *Eur. Phys. J. B*, **64**, 443–450.

Miguel, M.C., Vespignani, A., Zapperi, S., Weiss, J., and Grasso, J.-R. (2001). Intermittent dislocation flow in viscoplastic deformation. *Nature*, **410**, 667–671.

Miguel, M.C. and Zapperi, S. (2006). Fluctuations in plasticity at the microscale. *Science*, **312**, 1151–1152.

Mishin, Y. (2004). Atomistic modeling of the γ and γ'-phases of the Ni–Al system. *Acta Mater.*, **52**, 1451–1467.

Mishin, Y., Mehl, M.J., Papaconstantopoulos, D.A., Voter, A.F., and Kress, J.D. (2001). Structural stability and lattice defects in copper: *Ab initio*, tight-binding and embedded-atom calculations. *Phys. Rev.*, **63**, 224106(1–16).

Mitchell, T.E. (1964). Dislocations and plasticity in single crystals of face-centred cubic metals and alloys. *Progr. Appl. Mater. Res.*, **6**, 117–237.

Miura, S. and Hamashima, K. (1980). Effect of orientation on work-softening in aluminum single crystals. *J. Mater. Sci.*, **15**, 2550–2558.

Mohles, V. (2001). Simulations of dislocation glide in overaged precipitation-hardened crystals. *Phil. Mag. A*, **81**, 971–990.

Mohles, V. (2003). Superposition of dispersion strengthening and size-mismatch strengthening: Computer simulations. *Phil. Mag. Lett.*, **83**, 9–19.

Mohles, V. (2004*a*). The critical resolved stress of single crystals with long-range ordered precipitates calculated by dislocation dynamics simulations. *Mater. Sci. Eng. A*, **365**, 144–150.

Mohles, V. (2004*b*). Dislocation dynamics simulations of particle strengthening. In *Continuum Scale Simulation of Engineering Materials* (ed. D. Raabe, F. Roters, F. Barlat, and L. Chen), Chapter 17, pp. 375–395. Wiley-VCH, Weinheim (FRG).

Mohles, V. (2012). Simulations of Strengthening Mechanisms in Metals and Alloys. http://mohles.eu/disloc.

Möller, H.J. (1978). The movement of dissociated dislocations in the diamond-cubic structure. *Acta Metall.*, **26**, 963–973.

Mompiou, F., Legros, M., Sedlmayr, A., Gianola, D.S., Caillard, D., and Kraft, O. (2012). Source-based strengthening of sub-micrometer Al fibers. *Acta Mater.*, **60**, 977–983.

Monnet, G. and Devincre, B. (2006). Solute friction and forest interaction. *Phil. Mag.*, **86**, 1555–1565.

Monnet, G., Devincre, B., and Kubin, L.P. (2004). Dislocation study of prismatic slip systems and their interactions in hexagonal close packed metals: Application to zirconium. *Acta Mater.*, **52**, 4317–4328.

Monnet, G., Osetsky, Y.N., and Bacon, D.J. (2010). Mesoscale thermodynamic analysis of atomic-scale dislocation-obstacle interactions simulated by molecular dynamics. *Phil. Mag.*, **90**, 1001–1018.

Mordehai, D., Ashkenazy, Y., Kelson, I., and Makov, G. (2003). Dynamic properties of screw dislocations in Cu: A molecular dynamics study. *Phys. Rev. B*, **67**, 024112(1–9).

Mordehai, D., Clouet, E., Fivel, M., and Verdier, M. (2008). Introducing dislocation climb by bulk diffusion in dislocation dynamics. *Phil. Mag.*, **88**, 899–925.

Mordehai, D., Clouet, E., Fivel, M., and Verdier, M. (2009). Annealing of dislocation loops in dislocation dynamics simulations. *IOP Conf. Series: Mater. Sci. Eng.*, **3**, 012001(1–6).

Moriarty, J.A., Benedict, L.X., Glosli, J.N., Hood, R.Q., Orlikowski, D.A., and Patel, M.V. *et al.* (2006). Robust quantum-based interatomic potentials for multiscale modeling in transition metals. *J. Mater. Res.*, **21**, 563–573.

Moriarty, J.A., Belak, J.F., Rudd, R.E., Söderlind, P., Streitz, F.H., and Yang, L.H. (2002*a*). Quantum-based atomistic simulations of material properties in transition metals. *J. Phys.: Condens. Matter*, **14**, 2825–2857.

Moriarty, J.A., Vitek, V., Bulatov, V.V., and Yip, S. (2002*b*). Atomistic simulations of dislocations and defects. *J. Comput.-Aided Materials Design*, **9**, 99–132.

Motz, C., Schöberl, T., and Pippan, R. (2005). Mechanical properties of micro-sized copper bending beams machined by the focused ion beam technique. *Acta Mater.*, **53**, 4269–4279.

Motz, C., Weygand, D., Senger, J., and Gumbsch, P. (2008). Micro-bending tests: A comparison between three-dimensional discrete dislocation dynamics simulations and experiments. *Acta Mater.*, **56**, 1942–1955.

Motz, C., Weygand, D., Senger, J., and Gumbsch, P. (2009). Initial dislocation structures in 3-D discrete dislocation dynamics and their influence on microscale plasticty. *Acta Mater.*, **57**, 1744–1754.

Moulin, A., Condat, M., and Kubin, L.P. (1997). Simulation of Frank–Read sources in silicon. *Acta Mater.*, **45**, 2339–2348.

Moulin, A., Condat, M., and Kubin, L.P. (1999*a*). Mesoscale modelling of the yield point properties of silicon crystals. *Acta Mater.*, **47**, 2879–2888.

Moulin, A., Condat, M., and Kubin, L.P. (1999*b*). Perfect and partial Frank–Read sources in silicon: A simulation. *Phil. Mag. A*, **79**, 1995–2011.

Mrovec, M., Gröger, R., Bailey, A.G., Nguyen-Manh, D., Elsässer, C., and Vitek, V. (2007). Bond-order potential for simulations of extended defects in tungsten. *Phys. Rev. B*, **75**, 104119 (1–16).

Mrovec, M., Nguyen-Manh, D., Elsässer, C., and Gumbsch, P. (2011). Magnetic bond-order potential for iron. *Phys. Rev. Lett.*, **106**, 246402(1–4).

Mrovec, M., Nguyen-Manh, D., Pettifor, D.G., and Vitek, V. (2004). Bond-order potential for molybdenum: Application to dislocation behavior. *Phys. Rev. B*, **69**, 094115(1–16).

Mughrabi, H. (1975). Description of the dislocation structure after unidirectional deformation at low temperatures. In *Constitutive Equations in Plasticity* (ed. A.S. Argon), Cambridge, MA (USA), pp. 199–250. The M.I.T. Press.

Mughrabi, H. (ed.) (1993). *Plastic Deformation and Fracture of Materials*, Volume 6, Materials Science and Technology, Weinheim (FRG). VCH.

Mughrabi, H. (2009). Cyclic stress irreversibilities and the evolution of fatigue damage. *Metall. Mater. Trans. A*, **40A**, 1257–1279.

Mughrabi, H., Ackermann, F., and Herz, K. (1979). Persistent slip bands in fatigued face-centered and body-centered cubic metals. In *Fatigue Mechanisms, ASTM-STP 675* (ed. J. Fong), Philadelphia, PA, USA, pp. 69–105. American Society for Testing and Materials.

Mughrabi, H. and Ungár, T. (2002). Long-range internal stresses in deformed single-phase materials: The composite model and its consequences. In *Dislocations in Solids* (ed. F.R.N. Nabarro and M.S. Duesbery), Volume 11, pp. 343–411. Elsevier Science B.V., Amsterdam.

Mura, T. (1963). Continuous distributions of moving dislocations. *Phil. Mag.*, **8**, 843–857.

Mura, T. (1991). *Micromechanics of Defects in Solids* (2nd edn). Kluwer Academic Publishers, Dordrecht (The Netherlands).

Naamane, S., Monnet, G., and Devincre, B. (2010). Low temperature deformation in iron studied with dislocation dyanmics simulations. *Int. J. Plasticity*, **26**, 84–92.

Nabarro, F.R.N. (1947). Dislocations in a simple cubic lattice. *Proc. Phys. Soc.*, **59**, 256–272.

Nabarro, F.R.N. (1967). *The Theory of Crystal Dislocations*. Oxford University Press.

Nabarro, F.R.N. (1986). Work hardening of face-centered cubic single crystals. In *Strength of Metals and Alloys, Proc. ICSMA 7* (ed. H.J. McQueen, J.P. Bailon, J.I. Dickson, J.J. Jonas, and M.G. Akben), Volume 3, Oxford, pp. 1667–1700. Pergamon Press.

Nabarro, F.R.N. (1997*a*). Fifty-year study of the Peierls–Nabarro stress. *Mater. Sci. Eng. A*, **234–236**, 67–76.

Nabarro, F.R.N. (1997*b*). Theoretical and experimental estimates of the Peierls stress. *Phil. Mag. A*, **75**, 703–711.

Nabarro, F.R.N. and Duesbery, M.S. (ed.) (2002). *Dislocations in Solids*, Volume 11. Elsevier Science B.V., Amsterdam.

Naka, S., Lasalmonie, A., Costa, P., and Kubin, L.P. (1988). The low-temperature deformation of α-titanium and the core structure of a-type screw dislocations. *Phil. Mag.*, **57**, 717–740.

Needleman, A. and Gil Sevillano, J. (2003). Preface to the viewpoint set on: Geometrically necessary dislocations and size dependent plasticity. *Scripta Mater.*, **48**, 109–111.

Nembach, E. (1996). *Particle Strengthening of Metals and Alloys*. John Wiley and Sons, New York.

Neubert, M. and Rudolph, P. (2001). Growth of semi-insulating GaAs crystals in low temperature gradients by using the vapour pressure controlled Czochralski method (VCz). *Progr. Cryst. Growth Charact. Mater.*, **43**, 119–185.

Neuhaus, R. and Schwink, Ch. (1992). On the flow stress of [100]- and [111]-oriented Cu-Mn single crystals: A transmission electron microscopy study. *Phil. Mag. A*, **65**, 1463–1484.

Neuhäuser, H. (1983). Slip line formation and collective dislocation motion. In *Dislocation in Solids* (ed. F.R.N. Nabarro), Volume 6. North-Holland, Amsterdam.

Neuhäuser, H. and Schwink, Ch. (1993). Solid solution strengthening. In *Plastic Deformation and Fracture of Materials* (ed. H. Mughrabi), Volume 6, Materials Science and Technology, pp. 191–250. VCH, Weinheim (FRG).

Newman, M.E.J. (2005). Power laws, Pareto distributions and Zipf's law. *Contemp. Phys.*, **46**, 323–351.

Ng, K.S. and Ngan, A.H.W. (2008*a*). Breakdown of Schmid's law in micropillars. *Scripta Mater.*, **59**, 796–799.

Ng, K.S. and Ngan, A.H.W. (2008*b*). Stochastic nature of plasticity of aluminum micro-pillars. *Acta Mater.*, **56**, 1712–1720.

Nikitenko, V.I., Farber, B.Y., and Iunin, Y.L. (1985). Possibility of experimentally studying the formation kinetics and mobility of kinks on a dislocation line. *Sov. Phys. JETP Lett.*, **41**, 124–127.

Nikitenko, V.I., Farber, B.Y., Iunin, Y.L., and Orlov, V.I. (1993). Dislocation-point defect interaction in semiconductors and kink mobility. *Mater. Sci. Eng. A*, **164**, 346–349.

Nix, W.D. (1989). Mechanical properties of thin films. *Metall. Mater. Trans. A*, **20**, 2217–2245.

Nix, W.D. and Lee, S.-W. (2011). Micro-pillar plasticity controlled by dislocation nucleation at surfaces. *Phil. Mag.*, **91**, 1084–1096.

Nogaret, T., Rodney, D., Fivel, M., and Robertson, C. (2008). Clear band formation simulated by dislocation dynamics: Role of helical turns and pile-ups. *J. Nucl. Mater.*, **380**, 22–29.

Norfleet, D.M., Dimiduk, D.M., Polasik, S.J., Uchic, M.D., and Mills, M. (2008). Dislocation structures and their relationship to strength in deformed nickel microcrystals. *Acta Mater.*, **56**, 2988–3001.

Ogata, S., Li, J., and Yip, S. (2005). Energy landscape of deformation twinning in fcc and bcc metals. *Phys. Rev. B*, **71**, 224102(1–11).

Okazaki, K. (1996). Solid-solution hardening and softening in binary iron alloys. *J. Mater. Sci.*, **31**, 1087–1099.

Olfe, J. and Neuhäuser, H. (1988). Dislocation groups, multipoles, and friction stresses in α-CuZn alloys. *Phys. Stat. Sol. (a)*, **109**, 149–160.

Omri, M., Tete, C., Michel, J.P., and George, A. (1987). On the the yield point of floating-zone silicon single crystals I. Yield stresses and activation parameters. *Phil. Mag. A*, **55**, 601–616.

Pant, P., Schwarz, K.W., and Baker, S.P. (2003). Dislocation interactions in thin fcc metal films. *Acta Mater.*, **51**, 3243–3258.

Pantleon, W. (2004). Stage IV work-hardening related to disorientations in dislocation structures. *Mater. Sci. Eng. A*, **387–389**, 257–261.

Pantleon, W. and Hansen, N. (2001). Dislocation boundaries—The distribution function of disorientation angles. *Acta Mater.*, **49**, 1479–1493.

Parthasarathy, T.A., Rao, S.I., Dimiduk, D.M., Uchic, M.D., and Trinkle, D.R. (2007). Contribution to size effect of yield strength from the stochastics of dislocation source lengths in finite samples. *Scripta Mater.*, **56**, 313–316.

Patinet, S. and Proville, L. (2011). Dislocation pinning by substitutional impurities in an atomic-scale model for Al(Mg) solid solutions. *Phil. Mag.*, **91**, 1581–1606.

Peach, M.O. and Koehler, U.F. (1950). The forces exerted on dislocations and the stress fields produced by them. *Phys. Rev.*, **80**, 436–439.

Peeters, B., Bacroix, B., Teodosiu, C., Van Houtte, P., and Aernoudt, E. (2001*a*). Work-hardening/softening behaviour of b.c.c. polycrystals during changing strain paths: II. *Acta Mater.*, **49**, 1621–1632.

Peeters, B., Seefeldt, M., Teodosiu, C., Kalidindi, S.R., Van Houtte, P., and Aernoudt, E. (2001*b*). Work-hardening/softening behaviour of b.c.c. polycrystals during changing strain paths: I. *Acta Mater.*, **49**, 1607–1619.

Peierls, R. (1940). The size of a dislocation. *Proc. Phys. Soc.*, **52**, 34–37.

Phillips, R. (2001). *Crystals, Defects and Microstructures*. Cambridge University Press, Cambridge (UK).

Picu, R.C. and Soare, M.A. (2010). Asymmetric dislocation junctions exhibit a broad range of strengths. *Scripta Mater.*, **62**, 508–511.

Pink, E. and Arsenault, R.J. (1980). Low-temperature softening in body-centered cubic alloys. *Progr. Mater. Sci.*, **24**, 1–50.

Pizzagalli, L. and Beauchamp, P. (2006). First principles determination of the Peierls stress of the shuffle screw dislocation in silicon. *Phil. Mag. Lett.*, **84**, 729–736.

Pizzagalli, L., Demenet, J.L., and Rabier, J. (2009*a*). Theoretical study of pressure effect on the dislocation core properties of semiconductors. *Phys. Rev. B*, **79**, 045203(1–7).

Pizzagalli, L., Godet, J., and Brochard, S. (2009*b*). Glissile dislocations with transient cores in silicium. *Phys. Rev. Lett.*, **103**, 065505(1–4).

Pizzagalli, L., Godet, J., Guénolé, J., and Brochard, S. (2011). Dislocation cores in silicon: New aspects from numerical simulations. *J. Phys.: Conf. Series*, **281**, 012002(1–8).

Pizzagalli, L., Pedersen, A., Arnaldsson, A., Jónsson, H., and Beauchamp, P. (2008). Theoretical study of kinks on screw dislocations in silicon. *Phys. Rev. B*, **77**, 064106(1–14).

Pretorius, T. and Nembach, E. (2001). Strengthening of an $L1_2$-ordered γ'-intermetallic by disordered γ-particles. Part I: computer simulations. *Acta Mater.*, **49**, 1971–1980.

Pretorius, T. and Nembach, E. (2004). The effect of the degree of randomness of arrays of shearable particles on the critical resolved shear stress of particle strengthened materials. *Acta Mater.*, **52**, 3807–3812.

Prinz, F. and Argon, A.S. (1980). Dislocation cell formation during plastic deformation of copper single crystals. *Phys. Stat. Sol. (a)*, **57**, 741–753.

Proville, L. (2009). Depinning of a discrete elastic string from a random array of weak pinning points with finite dimensions. *J. Stat. Phys.*, **137**, 717–727.

Proville, L., Rodney, D., and Marinica, M.-C. (2012). Quantum effect on thermally activated glide of dislocations. *Nature Mater.*, **11**, 845–849.

Püschl, W. (1985). Reactions between glide dislocations and forest dislocations in anisotropic bcc metals. *Phys. Stat. Sol. (a)*, **90**, 181–189.

Püschl, W. (2002). Models for dislocation cross-slip in close-packed crystal structures: A critical review. *Progr. Mater. Sci.*, **47**, 415–461.

Püschl, W., Frydman, R., and Schoeck, G. (1982). The strength of the dislocation forest for 30° and 60° dislocations. *Phys. Stat. Sol. (a)*, **74**, 211–216.

Püschl, W. and Schoeck, G. (1984). Forest interactions of dislocations in hcp structures. *Crystal Res. Technol.*, **19**, 303–314.

Püschl, W. and Schoeck, G. (1994). Dissociation and constriction energies in extended dislocations. In *Strength of Materials, Proc. ICSMA 10* (ed. H. Oikawa, K. Maruyama, S. Takeuchi, and M. Yamaguchi), pp. 97–100. The Japan Institute of Materials.

Qin, Q. and Bassani, J.L. (1992*a*). Non-associated plastic flow in single crystals. *J. Mech. Phys. Solids*, **40**, 835–862.

Qin, Q. and Bassani, J.L. (1992*b*). Non-Schmid yield behavior in single crystals. *J. Mech. Phys. Solids*, **40**, 813–833.

Quek, S.S., Xiang, Y., Zhang, Y.W., Srolovitz, D.J., and Lu, C. (2006). Level set simulation of dislocation dynamics in thin films. *Acta Mater.*, **54**, 2371–2381.

Queyreau, S., Monnet, G., and Devincre, B. (2009). Slip system interactions in α-iron determined by dislocation dynamics simulations. *Int. J. Plasticity*, **25**, 361–377.

Raabe, D., Roters, F., Barlat, F., and Chen, L.Q. (ed.) (2004). *Continuum Scale Simulation of Engineering Materials*, Weinheim (FRG). Wiley-VCH.

Rabier, J. (2007). High-stress plasticity and the core structures of dislocations in silicon. *Phys. Stat. Sol. (a)*, **204**, 2248–2255.

Rabier, J., Cordier, P., Demenet, J.L., and Garem, H. (2001). Plastic deformation of Si at low temperature under high confining pressure. *Mater. Sci. Eng. A*, **309–310**, 74–77.

Rabier, J., Cordier, P., Tondellier, T., Demenet, J.L., and Garem, H. (2000). Dislocation microstructures in Si deformed at RT. *J. Phys.: Condens. Matter*, **12**, 10059–10064.

Rabier, J. and Demenet, J.L. (2000). Low temperature, high stress plastic deformation of semiconductors: The silicon case. *Phys. Stat. Sol. (b)*, **63**, 63–74.

Rabier, J. and Demenet, J.L. (2005). On the nucleation of shuffle dislocations in Si. *Phys. Stat. Sol. (a)*, **202**, 944–948.

Rabier, J., Denanot, M.F., Demenet, J.L., and Cordier, P. (2004). Plastic deformation by shuffle dislocations in silicon. *Mater. Sci. Eng. A*, **387–389**, 124–128.

Rabier, J. and George, A. (1987). Dislocations and plasticity in semiconductors. II. The relation between dislocation dynamics and plastic deformation. *Rev. Phys. Appl.*, **22**, 1327–1351.

Rabier, J., Pizzagalli, L., and Demenet, J.L. (2010). Dislocations in silicon at high stress. In *Dislocations in Solids. The 30th Anniversary Volume* (ed. J.P. Hirth and L. Kubin), Volume 16, pp. 47–108. Elsevier B.V., Amsterdam.

Rabier, J., Renault, P.O., Eyidi, D., Demenet, J.L., Chen, J., and Couvy, H. *et al.* (2007). Plastic deformation of silicon between 20 °C and 425 °C. *Phys. Stat. Sol. (c)*, **4**, 3110–3114.

Raj, S.V. and Pharr, G.M. (1986). A compilation and analysis of data for the stress dependence of the subgrain size. *Mater. Sci. Eng. A*, **81**, 217–237.

Rao, S.I., Dimiduk, D.M., El-Awady, J.A., Parthasarathy, T.A., Uchic, M.D., and Woodward, C. (2009). Atomistic simulations of cross-slip nucleation at screw dislocation intersections in face-centered cubic nickel. *Phil. Mag.*, **89**, 3351–3369.

Rao, S.I., Dimiduk, D.M., El-Awady, J.A., Parthasarathy, T.A., Uchic, M.D., and Woodward, C. (2010). Activated states for cross-slip at screw dislocation intersections in face-centered cubic nickel and copper via atomistic simulation. *Acta Mater.*, **58**, 5547–5557.

Rao, S.I., Dimiduk, D.M., Parthasarathy, T.A., El-Awady, J.A., Woodward, C., and Uchic, M.D. (2011). Calculations of intersection coss-slip activation energies in fcc metals using nudged elastic band method. *Acta Mater.*, **59**, 7135–7144.

Rao, S.I., Dimiduk, D.M., Parthasarathy, T.A., Uchic, M.D., Tang, M., and Woodward, C. (2008). Athermal mechanisms of size-dependent crystal flow gleaned from three-dimensional discrete simulations. *Acta Mater.*, **56**, 3245–3259.

Rao, S.I., Dimiduk, D.M., Tang, M., Parthasarathy, T.A., Uchic, M.D., and Woodward, C. (2007). Estimating the strength of single-ended dislocation sources in micron-sized crystals. *Phil. Mag.*, **87**, 4777–4794.

Rao, S.I., Parthasarathy, T.A., Dimiduk, D.M., and Hazzledine, P.M. (2004). Discrete dislocation simulation of precipitation in superalloys. *Phil. Mag.*, **84**, 3195–3215.

Rao, S., Parthasarathy, T.A., and Woodward, C. (1999). Atomistic simulation of cross-slip processes in model fcc structures. *Phil. Mag. A*, **79**, 1167–1192.

Rasmussen, T., Jacobsen, K.W., Leffers, T., and Pedersen, O.B. (1997*a*). Simulations of the atomic structure, energetics and cross-slip of screw dislocations in copper. *Phys. Rev. B*, **56**, 2977–2990.

Rasmussen, T., Jacobsen, K.W., Leffers, T., Pedersen, O.B., Srinivasan, S.G., and Jónsson, H. (1997*b*). Atomistic determination of cross-slip pathway and energetics. *Phys. Rev. Lett.*, **79**, 3676–3679.

Rasmussen, T., Vegge, T., Leffers, T., Pedersen, O.B., and Jacobsen, K.W. (2000). Simulation of structure and annihilation of screw dislocation dipoles. *Phil. Mag. A*, **80**, 1273–1290.

Rauch, E.F. (1994). The relation between forest dislocations and stress in bcc metals. *Key Eng. Materials*, **97–98**, 371–376.

Read, W.T. (1953). *Dislocations in Crystals*. McGraw-Hill, New York.

Rees, D. (2006). *Basic Engineering Plasticity*. Elsevier (Butterworth-Heinemann), Amsterdam.

Reid, C.N. (1973). *Deformation Geometry for Materials Scientists*. Pergamon Press, Oxford.

Rhee, M., Stolken, J.S., Bulatov, V.V., Diaz de la Rubia, T., Zbib, H.M., and Hirth, J.P. (2001). Dislocation stress fields for dynamic codes using anisotropic elasticity: methodology and analysis. *Mater. Sci. Eng. A*, **309–310**, 288–293.

Rhee, M., Zbib, H.M., Hirth, J.P., Huang, H., and de la Rubia, T. (1998). Models for long-/short-range interactions and cross-slip in 3D dislocation simulation of bcc single crystals. *Model. Simul. Mater. Sci. Eng.*, **6**, 467–492.

Rice, J.R. (1999). Foundations of solid mechanics. In *Mechanics of Materials* (ed. M.A. Meyers, R.W. Armstrong, and H.O.K. Kirchner), pp. 33–69. John Wiley and Sons, New York.

Rice, J.R. and Beltz, G.E. (1994). The activation energy for dislocation nucleation at a crack. *J. Mech. Phys. Solids*, **42**, 333–360.

Richeton, T., Dobron, P., Chmelik, F., Weiss, J., and Louchet, F. (2006*a*). On the critical character of plasticity in metallic single crystals. *Mater. Sci. Eng. A*, **424**, 190–195.

Richeton, T., Weiss, J., and Louchet, F. (2005*a*). Breakdown of avalanche critical behaviour in polycrystalline plasticity. *Nature Mater.*, **4**, 465–469.

Richeton, T., Weiss, J., and Louchet, F. (2005*b*). Dislocation avalanches: Role of temperature, grain size and strain hardening. *Acta Mater.*, **53**, 4463–4471.

Richeton, T., Weiss, J., and Louchet, F. (2006*b*). Hall–Petch law revisited in terms of collective dislocation dynamics. *Phys. Rev. Lett.*, **97**, 075504(1–4).

Richeton, T., Weiss, J., Louchet, F., Dobron, P., and Chmelik, F. (2007). Critical character of plasticity from AE experiments in hcp and fcc metals. *Kovove Mater.*, **45**, 149–152.

Richter, G., Hillerich, K., Gianola, D.S., Mönig, R., Kraft, O., and Volkert, C.A. (2009). Ultra-high strength single crystalline nanowhiskers grown by physical vapor deposition. *Nano Lett.*, **9**, 3048–3052.

Richter, J. (1970*a*). The influence of temperature on slip behaviour of molybdenum single crystals deformed in tension in the range from 293 to 573 K. I. Stress-strain characteristics. *Phys. Stat. Sol. (b)*, **40**, 565–572.

Richter, J. (1970*b*). The influence of temperature on slip behaviour of molybdenum single crystals deformed in tension in the range from 293 to 573 K. II. Slip geometry and structure of slip bands. *Phys. Stat. Sol. (b)*, **46**, 203–215.

Roberts, S.G., Booth, A.S., and Hirsch, P.B. (1994). Dislocation activity and brittle–ductile transitions in single crystals. *Mater. Sci. Eng. A*, **176**, 91–98.

Robertson, I.M., Schuh, C.A., Vetrano, J.S., Browning, N.D., Field, D.P., and Juul Jensen, D. *et al.* (2011). Towards and integrated materials characterization toolbox. *J. Mater. Res.*, **26**, 1344–1383.

Rodary, E., Rodney, D., Proville, L., Bréchet, Y., and Martin, G. (2004). Dislocation glide in model Ni(Al) solid solutions by molecular dynamics. *Phys. Rev. B*, **70**, 054111(1–11).

Rodney, D. (2007). Activation energy for kink-pair nucleation on dislocations: Comparison between static and dynamic atomic-scale simulations. *Phys. Rev. B*, **76**, 144108(1–9).

Rodney, D. and Phillips, R. (1999). Structure and strength of dislocation junctions: An atomic level analysis. *Phys. Rev. Lett.*, **82**, 1704–1707.

Rollet, A.D. and Kocks, U.F. (1994). A review of the stages of work hardening. *Solid State Phenom.*, **35–36**, 1–18.

Rong, Z., Mohles, V., Bacon, D.J., and Osetsky, Y.N. (2005). Dislocation dynamics modelling of dislocation-loop interactions in irradiated materials. *Phil. Mag.*, **85**, 171–188.

Roters, F., Eisenlohr, P., Hantcherli, L., Tjahjanto, D.D., Bieler, T.R., and Raabe, D. (2010). Overview of constitutive laws, kinematics, homogenization and multiscale methods in crystal plasticity finite-element modeling: Theory, experiments, applications. *Acta Mater.*, **58**, 1152–1211.

Rouby, D., Fleischmann, P., and Duvergier, C. (1983). Un modèle de sources d'émission acoustique pour l'analyse de l'émission continue et de l'émission par salves. I: Analyse théorique. *Phil. Mag. A*, **47**, 671–687.

Rudolph, P. (2005). Dislocation cell structures in melt-grown semiconductor compound crystals. *Cryst. Res. Technol.*, **40**, 7–20.

Rybin, V.V. and Orlov, A.N. (1970). Theory of dislocation mobility in the low velocity range. *Sov. Phys. Solid State*, **11**, 2635–2641.

Saada, G. (1960). Sur le durcissement dû à la recombinaison des dislocations. *Acta Metall.*, **8**, 841–847.

Saada, G. (1991). Cross-slip and work hardening of f.c.c. crystals. *Mater. Sci. Eng. A*, **137**, 177–183.

Saka, H., Doi, M., and Imura, T. (1970). Direct observation of multiplication of dislocations in iron single crystal by high voltage electron microscopy. *J. Phys. Soc. Japan*, **29**, 803–804.

Saka, H., Noda, K., and Imura, T. (1973). Tensile test of foil specimens of iron single crystals at room and low temperature under observation in high voltage electron microscope. *Crystal Lattice Defects*, **4**, 45–56.

Saka, H., Yamamoto, K., Arai, S., and Kuroda, K. (2006). In-situ TEM observation of transformation from shuffle to glide sets in Si under supersaturation of interstitials. *Phil. Mag.*, **86**, 4841–4850.

Samaras, M. and Victoria, M. (2008). Modelling in nuclear energy environments. *Materials Today*, **11**, 54–62.

Sato, A. and Meshii, M. (1973). Solid solution softening and solid solution hardening. *Acta Metall.*, **21**, 753–768.

Sauzay, M. and Kubin, L. (2011). Scaling laws for dislocation microstructures in monotonic and cyclic deformation of fcc metals. *Progr. Mater. Sci.*, **56**, 725–784.

Scattergood, R.O. and Bacon, D.J. (1975). The Orowan mechanism in anisotropic crystals. *Phil. Mag.*, **31**, 179–198.

Scattergood, R.O. and Bacon, D.J. (1982). The strength effect of voids. *Acta Metall.*, **30**, 1665–1677.

Schmid, E. (1924). Neuere Untersuchungen an Metallkristallen. In *Proc. 1st Int. Congr. Appl. Mech., Delft, 1924*, Delft, pp. 342–353. Waltman Jr.

Schmid, E. and Boas, W. (1935). *Kristallplastizität*. Julius Springer, Berlin.

Schneider, A.S., Kaufmann, D., Clark, B.G., Frick, C.P., P.A., Gruber, and Mönig, R. *et al.* (2009). Correlation between critical temperature and strength of small-scale bcc pillars. *Phys. Rev. Lett.*, **103**, 105501(1–4).

Schoeck, G. (1965). The activation energy of dislocation movement. *Phys. Stat. Sol. (b)*, **8**, 499–507.

Schoeck, G. (1980). Thermodynamics and thermal activation of dislocations. In *Moving Dislocations* (ed. F.R.N. Nabarro), Volume 3, Dislocation in Solids, pp. 65–163. North-Holland, Amsterdam.

Schoeck, G. (2000). Thermal activation in plastic deformation. In *Multiscale Phenomena in Plasticity* (ed. J. Lépinoux, D. Mazière, V. Pontikis, and G. Saada), Dordrecht (The Netherlands), pp. 33–55. Kluwer Academic Publishers.

Schoeck, G. (2001). The core structure of dislocations. Peierls model vs. atomic simulations in Pd. *Comput. Mater. Sci.*, **21**, 124–134.

Schoeck, G. (2002). The core structure of dislocations in Al: A critical assessment. *Mater. Sci. Eng. A*, **333**, 390–396.

Schoeck, G. (2005). The Peierls model: Progress and limitations. *Mater. Sci. Eng. A*, **400–401**, 7–17.

Schoeck, G. (2006). The core structure of dislocations: Peierls model vs. atomic simulation. *Acta Mater.*, **54**, 4865–4870.

Schoeck, G. and Frydman, R. (1972). The contribution of the dislocation forest to the flow stress. *Phys. Stat. Sol. (b)*, **53**, 661–673.

Schoeck, G. and Krystian, M. (2005). The Peierls energy and kink energy in fcc metals. *Phil. Mag. A*, **85**, 949–966.

Schoeck, G. and Seeger, A. (1955). Activation energy problems associated with extended dislocations. In *Proceedings of the Conference on Defects in Crystalline Solids, Bristol 1964*, London, pp. 340–346. The Physical Society.

Schwarz, K.W. (1988). Three-dimensional vortex dynamics in superfluid ^{4}He: Homogeneous superfluid turbulence. *Phys. Rev. B*, **38**, 2398–2417.

Schwarz, K.W. (1999). Simulation of dislocations on the mesoscopic scale. 1. Methods and examples. *J. Appl. Phys.*, **85**, 108–119.

Schwarz, K.W. (2003*a*). Discrete dislocation dynamics study of strained-layer relaxation. *Phys. Rev. Lett.*, **91**, 145503(1–4).

Schwarz, K.W. (2003*b*). Local rules for approximating strong dislocation interactions in discrete dislocation dynamics. *Model. Simul. Mater. Sci. Eng.*, **11**, 609–625.

Schwarz, K.W. (2007). Modelling of dislocation behaviour at the continuum level. In *Multiscale Materials Modelling* (ed. Z.X. Guo), Cambridge (UK), pp. 25–61. Woodhead Publishing.

Schwarz, K.W., Cai, J., and Mooney, P.M. (2004). Comparison of large-scale layer-relaxation simulations with experiment. *Appl. Phys. Lett.*, **85**, 2238–2240.

Schwarz, K.W., Liu, X.H., and Chidambarrao, D. (2000). Dislocation modeling for the silicon world. *Mater. Sci. Eng. A*, **309–310**, 229–232.

Schwarz, K.W. and Mughrabi, H. (2006). Interaction and passing stress of two threading dislocations of opposite sign in a confined channel. *Phil. Mag. Lett.*, **86**, 773–785.

Schwarz, K.W. and Tu, Y. (2009). Dislocation-interaction-based model of strained-layer relaxation. *J. Appl. Phys.*, **106**, 083510(1–9).

Schwerdtfeger, J., Nadgorny, E., Koutsos, V., Blackford, J.R., and Zaiser, M. (2010). Statistiscal heterogeneity of plastic deformation: An investigation based on surface profilometry. *Acta Mater.*, **58**, 4859–4870.

Schwerdtfeger, J., Nadgorny, E., Madani-Grasset, F., Koutsos, V., and Zaiser, M. (2007). Scale-free statistics of plasticity-induced surface steps on KCl single crystals. *J. Stat. Mech.*, **2007**, L04001(1–8).

Seeger, A. (1984). Structure and diffusion of kinks in monoatomic crystals. In *Dislocations 1984* (ed. P. Veyssière, L. Kubin, and J. Castaing), Paris, pp. 141–178. Editions du CNRS.

Seeger, A. and Hollang, L. (2000). The flow stress asymmetry of ultra-pure molybdenum single crystals. *Mater. Trans. JIM*, **41**, 141–151.

Seeger, A., Mader, S., and Kronmüller, H. (1963). Theory of work-hardening of fcc and hcp single crystals. In *Electron Microscopy and Strength of Crystals* (ed. G. Thomas and J. Washburn), New-York. Interscience Publishers.

Seeger, A. and Schiller, P. (1962). The formation and diffusion of kinks as the fundamental process of dislocation movement in internal friction measurements. *Acta Metall.*, **10**, 348–357.

Segall, D.E., Strachan, A., W. A. Goddard III, Ismail-Beigi, S., and Arias, T.A. (2003). Ab initio and finite-temperature molecular dynamics studies of lattice resistance in tantalum. *Phys. Rev. B*, **68**, 014104(1–11).

Senger, J., Weygand, D., Gumbsch, P., and Kraft, O. (2008). Discrete dislocation simulations of the plasticity of micro-pillars under uniaxial loading. *Scripta Mater.*, **58**, 587–590.

Senger, J., Weygand, D., Motz, C., Gumbsch, P., and Kraft, O. (2011). Aspect ratio and stochastic effects in plasticity of uniformly loaded micrometer-sized specimens. *Acta Mater.*, **59**, 2937–2947.

Sěsták, B. (1979). Plasticity and crystal structure. In *Strength of Metals and Alloys, Proc. ICSMA 5* (ed. P. Haasen, V. Gerold, and G. Kostorz), Volume 3, Oxford, pp. 1461–1482. Pergamon Press.

Sethna, J.P., Dahmen, K.A., and Myers, C.R. (2001). Crackling noise. *Nature*, **410**, 242–250.

Shade, P.A., Wheeler, R., Choi, Y.S., Uchic, M.D., Dimiduk, D.M., and Fraser, H.L. (2009). A combined experimental and simulation study to examine lateral constraint effects on microcompression of single-slip oriented crystals. *Acta Mater.*, **57**, 4580–4587.

Shan, Z.W., Mishra, R.K., Syed Asif, S.A., Warren, O.L., and Minor, A.M. (2008). Mechanical annealing and source-limited deformation in submicrometre-diameter Ni crystals. *Nature Mater.*, **7**, 115–119.

Shehadeh, M.A. (2012). Multiscale dislocation dynamics simulations of shock-induced plasticity in small volumes. *Phil. Mag.*, **92**, 1173–1197.

Shehadeh, M.A., Bringa, E.M., Zbib, H.M., McNaney, J.M., and Remington, B.A. (2006). Simulation of shock-induced plasticity including homogeneous and heterogeneous dislocation nucleations. *Appl. Phys. Lett.*, **89**, 171918(1–3).

Shehadeh, M.A., Zbib, H.M., and Diaz de la Rubia, T. (2005*a*). Modelling the dynamic deformation and patterning in fcc single crystals at high strain rates: dislocation dynamics plasticity analysis. *Phil. Mag.*, **85**, 1667–1685.

Shehadeh, M.A., Zbib, H.M., and Diaz de la Rubia, T. (2005*b*). Multiscale dynamics simulations of shock compression in copper single crystal. *Int. J. Plasticity*, **21**, 2639–2390.

Shenoy, V.B., Kukta, R.V., and Phillips, R. (2000). Mesoscopic analysis of structure and strength of dislocation junctions in fcc metals. *Phys. Rev. Lett.*, **84**, 1491–1494.

Shields, J.A., Goods, S.H., Gibala, R., and Mitchell, T.E. (1975). Deformation of high purity tantalum single crystals at 4.2 K. *Mater. Sci. Eng.*, **20**, 71–81.

Shim, S., Bei, H., Miller, M.K., Pharr, G.M., and George, E.P. (2009). Effects of focussed ion beam milling on the compressive behavior of directionally solidified micropillars and the nanoindentation response of an electropolished surface. *Acta Mater.*, **57**, 503–510.

Shima, K., Izumi, S., and Sakai, S. (2010). Reaction analysis for dislocation nucleation from a sharp corner in silicon: Glide set versus shuffle set. *J. Appl. Phys.*, **108**, 063505(1–3).

Shin, C., Fivel, M., and Kim, W.W. (2005). Three-dimensional computation of the interaction between a straight dislocation line and a particle. *Model. Simul. Mater. Sci. Eng.*, **13**, 1163–1173.

Shin, C.S., Fivel, M.C., Rodney, D., Phillips, R., Shenoy, V.B., and Dupuy, L. (2001). Formation and strength of dislocation junctions in fcc metals: A study by dislocation dynamics and atomistic simulations. *J. de Physique IV*, **11**, Pr5-19–26.

Shin, C.S., Fivel, M.C., Verdier, M., and Kwon, S.C. (2006). Numerical methods to improve computing efficiency of discrete dislocation dynamics simulations. *J. Comput. Phys.*, **215**, 417–429.

Shin, C.S., Fivel, M.C., Verdier, M., and Oh, K.H. (2003). Dislocation-impenetrable precipitate interaction: A three-dimensional discrete dislocation dynamics analysis. *Phil. Mag.*, **1**, 3691–3704.

Shin, C.S., Robertson, C.F., and Fivel, M.C. (2007). Fatigue in precipitation hardenend materials: a three-dimensional discrete dislocation dynamics modelling of the early cycles. *Phil. Mag.*, **87**, 3657–3669.

Singh, B.N., Foreman, A.J.E., and Trinkaus, H. (1997). Radiation hardening revisited: Role of intracascade clustering. *J. Nucl. Mater.*, **249**, 103–115.

Singh, B.N., Ghoniem, N.M., and Trinkaus, H. (2002). Experiment-based modelling of hardening and localized plasticity in metals irradiated under cascade damage conditions. *J. Nucl. Mater.*, **307–311**, 159–170.

Singh, B.N., Huang, X., Tähtinen, S., Moilanen, P., Jacquet, P., and Dekeyser, J. (2007). Final report on in-reactor uniaxial tensile deformation of pure iron and Fe-Cr alloy. Technical Report Risø-R161(EN), Risø National Laboratory, Technical University of Denmark.

Smialek, R.L., Webb, G.L., and Mitchell, T.E. (1970). Solid solution softening in bcc metal alloys. *Scripta Metall.*, **4**, 33–38.

Soo, P. and Higgins, G.T. (1968). The deformation of zirconium–oxygen single crystals. *Acta Metall.*, **16**, 177–186.

Spence, J.C.H. (1999). Atomistics and mechanical properties of silicon. *Acta Mater.*, **47**, 4153–4159.

Spence, J.C.H. (2007). Experimental studies of dislocation core defects. In *Dislocations in Solids* (ed. F.R.N. Nabarro and J.P. Hirth), Volume 13, pp. 419–452. Elsevier Science B.V., Amsterdam.

Spence, J. and Koch, C. (2001). Experimental evidence for dislocation core structures in silicon. *Scripta Mater.*, **45**, 1273–1278.

Spitzig, W.A. and Keh, A.S. (1970). The effect of orientation and temperature on the plastic flow properties of iron single crystals. *Acta Metall.*, **18**, 611–622.

Stach, E.A., Hull, R., Tromp, R.M., Ross, F.M., Reuter, M.C., and Bean, J.C. (2000). In-situ transmission electron microscopy studies of the interactions between dislocations in strained SiGe/Si(001) heterostructures. *Phil. Mag. A*, **80**, 2159–2200.

Staker, M.R. and Holt, D.L. (1972). The dislocation cell size and dislocation density in copper deformed at temperatures between 25 and 700 °C. *Acta Metall.*, **20**, 569–579.

Steeds, J.W. (1966). Dislocation arrangements in copper single crystals as a function of strain. *Proc. Phys. Soc. A*, **292**, 343–373.

Steinbach, I. (2009). Phase-field models in materials science. *Model. Simul. Mater. Sci. Eng.*, **17**, 073001(1–31).

Stephens, J.R. and Witzke, W.R. (1976). Alloys softening in binary iron solid solutions. *J. Less-Common Met.*, **48**, 285–308.

Stobbs, W.M. and Sworn, C.H. (1971). The weak-beam technique as applied to the determination of the stacking-fault energy in copper. *Phil. Mag. A*, **24**, 1365–1381.

Stoller, R.R. and Mansur, L.K. (2005). An assessment of radiation damage models and methods. Report ONRL/TM-2005/506, Oak Ridge National Laboratory.

Stroh, A.N. (1954). Constrictions and jogs in extended dislocations. *Proc. Phys. Soc. A*, **3**, 427–436.

Sudharshan Phani, P., Johanns, K.E., Duscher, G., Gali, A., George, E.P., and Pharr, G.M. (2011). Scanning transmission electron microscope observations of defects in as-grown and pre-strained Mo alloy fibers. *Acta Mater.*, **59**, 2172–2179.

Sumino, K. and Imai, A. (1983). Interactions of dislocations with impurities in silicon crystals studied by in situ X-ray topography. *Phil. Mag. A*, **47**, 753–766.

Sutton, A.P. and Balluffi, R.W. (1996). *Interfaces in Crystalline Materials*. Monographs on the Physics and Chemistry of Materials. Clarendon Press, Oxford.

Suzuki, H. (1979). Solution hardening in body-centred cubic alloys. In *Dislocations in Solids* (ed. F.R.N. Nabarro), Volume 4, pp. 191–217. North-Holland, Amsterdam.

Suzuki, T., Kamimura, Y., and Kirchner, H.O.K. (1999). Plastic homology of bcc metals. *Phil. Mag. A*, **79**, 1629–1642.

Suzuki, T., Takeuchi, T., and Yoshinaga, Y. (1985). *Dislocation Dynamics and Plasticity*. Volume 12, Springer Series in Materials Science. Springer-Verlag, Berlin.

Takahashi, A. and Ghoniem, N.M. (2008). A computational method for dislocation-precipitate interaction. *J. Mech. Phys. Solids*, **56**, 1534–1553.

Takeuchi, S. (1969). Solid-solution strengthening in single crystals of iron alloys. *J. Phys. Soc. Japan*, **27**, 929–940.

Takeuchi, S., Hashimoto, T, and Maeda, K. (1982). Plastic deformation of bcc metal single crystals at very low temperatures. *Trans. JIM*, **23**, 60–69.

Takeuchi, S., Kuramoto, E., and Suzuki, T. (1972). Orientation dependence of slip in tantalum single crystals. *Acta Metall.*, **20**, 909–915.

Takeuchi, S. and Maeda, K. (1977). Slip in high purity tantalum between 0.7 and 40 K. *Acta Metall.*, **25**, 1485–1490.

Takeuchi, T. (1975). Work hardening of copper single crystals with multiple glide orientations. *Trans. JIM*, **16**, 629–640.

Takeuchi, T. (1976). Orientation dependence of work hardening of copper single crystals near the [001] axis. *J. Phys. Soc. Japan*, **40**, 741–746.

Takeuchi, T. and Suzuki, T. (1988). Deformation of crystals controlled by the Peierls mechanism. In *Strength of Metals and Alloys, Proc. ICSMA 8* (ed. P. Kettunen, T. Lepistö, and M. Lehtonen), Volume 1, Oxford, pp. 161–178. Pergamon Press.

Tang, H., Schwarz, K.W., and Espinosa, H.D. (2007). Dislocation escape-related size effects in single-crystal micropillars under uniaxial compression. *Acta Mater.*, **55**, 1607–1616.

Tang, H., Schwarz, K.W., and Espinosa, H.D. (2008). Dislocation-source shutdown and the plastic behavior of single-crystal micropillars. *Phys. Rev. Lett.*, **100**, 185503(1–4).

Tang, M., Cai, W., Xu, G., and Bulatov, V.V. (2006). A hybrid method for computing forces on curved dislocations intersecting free surfaces in three-dimensional dislocation dyanmics. *Model. Simul. Mater. Sci. Eng.*, **14**, 1139–1151.

Tang, M., Devincre, B., and Kubin, L.P. (1999). Simulation and modelling of forest hardening in body centred cubic crystals at low temperature. *Model. Simul. Mater. Sci. Eng.*, **7**, 893–908.

Tang, M., Kubin, L.P., and Canova, G. (1998). Dislocation mobility and the mechanical response of b.c.c. single crystals: A mesoscopic approach. *Acta Mater.*, **46**, 3221–3235.

Taylor, G.I. (1928). The deformation of crystals of β-brass. *Proc. R. Soc. Lond. A*, **118**, 1–24.

Taylor, G.I. (1934). The mechanism of plastic deformation of crystals. part I. - Theoretical. *Proc. R. Soc. Lond.*, **145**, 362–387.

Taylor, G. (1992). Thermally-activated deformation of bcc metals and alloys. *Progr. Mater. Sci.*, **36**, 29–61.

Taylor, G.I. and Elam, C.F. (1926). The distortion of iron crystals. *Proc. R. Soc. Lond. A*, **112**, 337–361.

Taylor, G.I. and Farren, W.S. (1926). The distortion of crystals of aluminium under compression. Part I. *Proc. R. Soc. Lond. A*, **CXI**, 529–551.

Teodosiu, C., Raphanel, J.-L., and Tabourot, C. (1993). Finite-element simulation of the large elastoplastic deformation of multicrystals. In *Large Plastic Deformations* (ed. C. Teodosiu, J.-L. Raphanel, and F. Sidoroff), Rotterdam, pp. 153–168. A.A. Balkema.

Thornton, P.R., Mitchell, T.E., and Hirsch, P.B. (1962). The strain-rate dependence of the flow stress of copper single crystals. *Phil. Mag.*, **7**, 337–362.

Trinkaus, H., Singh, B.N., and Foreman, A.J.E. (1997*a*). Mechanisms for decoration of dislocations by small dislocation loops under cascade damage conditions. *J. Nucl. Mater.*, **249**, 91–102.

Trinkaus, H., Singh, B.N., and Foreman, A.J.E. (1997*b*). Segregation of cascade induced interstitial loops at dislocations: Possible effect on initiation of plastic deformation. *J. Nucl. Mater.*, **251**, 172–187.

Trinkle, D.R. and Woodward, C. (2005). The chemistry of deformation: How solutes soften pure metals. *Science*, **310**, 1665–1667.

Uchic, M.D., Dimiduk, D.M., Florando, J.N., and Nix, W.D. (2004). Sample dimensions influence strength and crystal plasticity. *Science*, **305**, 986–989.

Uchic, M.D., Shade, P.A., and Dimiduk, D.M. (2009). Plasticity of micrometer-scale single crystals in compression. *Annu. Rev. Mater. Res.*, **39**, 361–386.

Valladares, A. and Sutton, A.P. (2007). First principles simulations of kink defects on the SP 90° partial dislocation in silicon. *Progr. Mater. Sci.*, **52**, 421–463.

Valladares, A., White, J.A., and Sutton, A.P. (1998). First principles simulations of the structure, formation, and migration energies of kinks on the 90° partial dislocation in silicon. *Phys. Rev. Lett.*, **81**, 4903–4906.

Vallino, F., Château, J.P., Jacques, A., and George, A. (2001). Dislocation multiplication during the very first stages of plastic deformation in silicon observed by X-ray topography. *Mater. Sci. Eng. A*, **319–321**, 152–155.

Vallino, F., Jacques, A., and George, A. (2000). Non-planar slip and cross-slip at the onset of plastic deformation in silicon. *Phys. Stat. Sol. (b)*, **222**, 51–62.

Van der Giessen, E. and Needleman, A. (1995). Discrete dislocation plasticity: a simple planar model. *Model. Simul. Mater. Sci. Eng.*, **3**, 689–735.

Van Swygenhoven, H., Derlet, P.M., and Frøseth, A.G. (2004). Stacking fault energies and slip in nanocrystalline metals. *Nature Mater.*, **3**, 399–403.

Vanderschaeve, G. and Caillard, D. (2007). On the mobility of dislocations in semiconductor crystals. *Mater. Sci. Eng. A*, **462**, 418–421.

Vattré, A., Devincre, B., Feyel, F., Gatti, R., Groh, S., and Jamond, O. *et al.* (2012). Modelling crystal plasticity by 3D dislocation dynamics and the finite element method: the Discrete-Continuous Model revisited. Manuscript submitted for publication.

Vattré, A., Devincre, B., and Roos, A. (2009). Dislocation dynamics simulations of precipitation hardening in Ni-based superalloys with high γ' volume fraction. *Intermetallics*, **17**, 988–994.

Vattré, A., Devincre, B., and Roos, A. (2010). Orientation dependence of plastic deformation in nickel-based single crystal superalloys: Discrete–continuous model simulations. *Acta Mater.*, **58**, 1938–1951.

Vattré, A. and Fedelich, B. (2011). On the relationship between anisotropic yield strength and internal stresses in single crystal superalloys. *Mech. Mater.*, **43**, 930–951.

Vegge, T. and Jacobsen, K.W. (2002). Atomistic simulations of dislocation processes in copper. *J. Phys.: Condens. Matter*, **14**, 2929–2956.

Vegge, T., Rasmussen, T., Leffers, T., Pedersen, O.B., and Jacobsen, K.W. (2001). Atomistic simulations of cross-slip of jogged screw dislocations in copper. *Phil. Mag. Lett.*, **81**, 137–144.

Ventelon, L. and Willaime, F. (2007). Core structure and Peierls potential of screw dislocations in α-Fe from first principles: Cluster versus dipole approaches. *J. Comput.-Aided Materials Design*, **14**, 85–94.

Ventelon, L. and Willaime, F. (2010). Generalized stacking-faults and screw-dislocation core-structure in bcc iron: A comparison between ab initio calculations and empirical potentials. *Phil. Mag.*, **90**, 1063–1074.

Ventelon, L., Willaime, F., and Leyronnas (2009). Atomistic simulation of single kinks of screw dislocations in α-Fe. *J. Nucl. Mater.*, **386–388**, 26–29.

Verdier, M., Fivel, M., and Groma, I. (1998). Mesoscopic scale simulations of dislocation dynamics in fcc metals: Principles and applications. *Model. Simul. Mater. Sci. Eng.*, **6**, 755–770.

Veyssière, P. and Saada, G. (1996). Microscopy and Plasticity of the $L1_2$ Ordered Alloys. In *$L1_2$ Ordered Alloys* (ed. F.R.N. Nabarro and M.S. Duesbery), Volume 10, pp. 254–441. Elsevier Science B.V., Amsterdam.

Veyssière, P., Wang, H., Xu, D.S., and Chiu, Y.L. (2009). Local dislocation reactions, self-organization and hardening in single slip. *IOP Conf. Series: Mater. Sci. Eng.*, **3**, 012018(1–10).

Victoria, M., Baluc, N., Bailat, C., Dai, Y., Luppo, M.I., and Schäublin, R. *et al.* (2000). The microstructure and associated tensile properties of irradiated fcc and bcc materials. *J. Nucl. Mater.*, **276**, 114–122.

Vitek, V. (1966). Thermally activated motion of screw dislocations in b.c.c. metals. *Phys. Stat. Sol. (b)*, **18**, 687–702.

Vitek, V. (1968). Intrinsic stacking faults in body-centred cubic crystals. *Phil. Mag.*, **18**, 773–786.

Vitek, V. and Paidar, V. (2008). Non-planar dislocation cores: A ubiquitous phenomenon affecting mechanical properties of crystalline materials. In *Dislocations in Solids* (ed. J.P. Hirth), Volume 14, pp. 439–514. Elsevier Science B.V., Amsterdam.

Volterra, V. (1907). Sur l'équilibre des corps élastiques mutiplement connexes. *Ann. Sci. École Norm. Sup.*, **24**, 401–517.

von Blanckenhagen, B., Arzt, E., and Gumbsch, P. (2004). Discrete dislocation dynamics simulations of plastic deformation in metal thin films. *Acta Mater.*, **52**, 773–784.

von Blanckenhagen, B., Gumbsch, P., and Arzt, E. (2003). Dislocation sources and the flow stress of polycrystalline metal films. *Phil. Mag. Lett.*, **83**, 1–8.

Vorbrugg, W., Goetting, H.C., and Schwink, C. (1971). Work-hardening and surface investigations on Cu single crystals oriented for multiple glide. *Phys. Stat. Sol. (b)*, **46**, 257–264.

Walgraef, D. and Aifantis, E.C. (1985). Dislocation patterning in fatigued metals as a result of dynamical instabilities. *J. Appl. Phys.*, **58**, 688–691.

Wang, C.-Z., Li, J., Ho, K.-M., and Yip, S. (2006*a*). Undissociated screw dislocation in Si: Glide or shuffle set? *Appl. Phys. Lett.*, **89**, 051910(1–3).

Wang, G., Strachan, A., Çağin, T., and W. A. Goddard III (2003*a*). Role of core polarization curvature of screw dislocations in determining the Peierls stress in bcc Ta: A criterion for designing high-performance materials. *Phys. Rev. B*, **67**, 140101(1–4).

Wang, J.N. (1996). Prediction of Peierls stress for different crystals. *Mater. Sci. Eng. A*, **206**, 259–269.

Wang, Y., Durham, W.B., Getting, I.C., and Weidner, D.J. (2003*b*). The deformation–DIA: A new apparatus for high temperature triaxial deformation to pressures up to 15 GPa. *Rev. Sci. Instrum.*, **74**, 3002–3011.

Wang, Y.U., Jin, Y.M., and Khachaturyan, A. G. (2005). Dislocation dynamics—Phase field. In *Handbook of Materials Modelling* (ed. S. Yip), Chapter 7.12, pp. 2287–2305. Springer, Berlin.

Wang, Z.Q. and Beyerlein, I.J. (2011). An atomistically-informed dislocation dynamics model for the plastic anisotropy and tension-compression asymmetry of BCC metals. *Int. J. Plasticity*, **27**, 1471–1484.

Wang, Z.Q., Beyerlein, I.J., and LeSar, R. (2007*a*). Dislocation motion in high-strain rate deformation. *Phil. Mag.*, **87**, 2263–2279.

Wang, Z.Q., Beyerlein, I.J., and LeSar, R. (2007*b*). The importance of cross-slip in high-rate deformation. *Model. Simul. Mater. Sci. Eng.*, **15**, 675–690.

Wang, Z.Q., Beyerlein, I.J., and LeSar, R. (2008). Slip band formation and dislocation density generation in high rate deformation of fcc crystals. *Phil. Mag.*, **88**, 1321–1343.

Wang, Z.Q., Beyerlein, I.J., and LeSar, R. (2009). Plastic anisotropy in *fcc* single crystals in high rate deformation. *Int. J. Plasticity*, **25**, 26–48.

Wang, Z., Ghoniem, N., and LeSar, R. (2004*a*). Multipole representation of the elastic field of dislocation ensembles. *Phys. Rev. B*, **69**, 174102(1–7).

Wang, Z., Ghoniem, N., Swaminarayan, S., and LeSar, R. (2006*b*). A parallel algorithm for 3D dislocation dynamics. *J. Comput. Phys.*, **219**, 608–621.

Wang, Z. and Li, J. (2010). Phase field modeling of defects and deformation. *Acta Mater.*, **58**, 1212–1235.

Wang, Z., McCabe, R.J., Ghoniem, N.M., LeSar, R., Misra, A., and Mitchell, T.E. (2004*b*). Dislocation motion in thin Cu foils: a comparison between computer simulations and experiment. *Acta Mater.*, **52**, 1535–1542.

Washburn, J. (1965). Intersection cross-slip. *Appl. Phys. Lett.*, **7**, 183–185.

Wasserbäch, W. (1995). Anomalous slip in high-purity niobium and tantalum crystals. *Phys. Stat. Sol. (a)*, **147**, 417–446.

Weertman, J. and Weertman, J.R. (1980). Moving dislocations. In *Moving Dislocations* (ed. F.R.N. Nabarro), Volume 3, Dislocation in Solids, pp. 1–59. North-Holland, Amsterdam.

Weertman, J. and Weertman, J.R. (1992). *Elementary Dislocation Theory*. Oxford University Press.

Weinberger, C.R., Aubry, S., Lee, S.-W., Nix, W.D., and Cai, W. (2009). Modelling dislocations in a free-standing film. *Model. Simul. Mater. Sci. Eng.*, **17**, 075007(1–26).

Weinberger, C.R. and Cai, W. (2011). The stability of Lomer-Cottrell jogs in nanopillars. *Scripta Mater.*, **64**, 529–532.

Weinberger, C.R. and Cai, W. (2012). Plasticity of metal nanowires. *J. Mater. Chem.*, **22**, 3277–3292.

Weinberger, C.R., Jennings, A.T., Kang, K., and Greer, J.R. (2012). Atomistic simulations and continuum modeling of dislocation nucleation and strength in nanowires. *J. Mech. Phys. Solids*, **60**, 84–103.

Weiss, J., Grasso, J.-R., Miguel, M.C., Vespignani, A., and Zapperi, S. (2002). Complexity in dislocation dynamics experiments. *Mater. Sci. Eng. A*, **309–31**, 360–364.

Weiss, J. and Marsan, D. (2003). Three-dimensional mapping of dislocation avalanches: Clustering and space/time coupling. *Science*, **299**, 89–92.

Weiss, J. and Miguel, M.C. (2004). Dislocation avalanche correlations. *Mater. Sci. Eng. A*, **387–389**, 292–296.

Weiss, J. and Montagnat, M. (2007). Long-range spatial correlations and scaling in dislocation slip patterns. *Phil. Mag.*, **87**, 1161–1174.

Weiss, J., Richeton, T., Louchet, F., Chmelik, F., Dobron, P., and Entemeyer, D. *et al.* (2007). Evidence for universal intermittent crystal plasticity from acoustic emission and high-resolution extensometry experiments. *Phys. Rev. B*, **76**, 224110(1–8).

Weisstein, E.W. (2002). *CCR Concise Encyclopedia of Mathematics*. Chapman and Hall, New-York.

Wen, M. and Ngan, A.H.W. (2000). Atomistic simulations of kink-pairs of screw dislocations in bcc Fe. *Acta Mater.*, **48**, 4255–4265.

Werner, M. (1987). Temperature and strain-rate dependence of the flow stress of ultrapure tantalum crystals. *Phys. Stat. Sol. (a)*, **104**, 63–78.

Wessel, K. and Alexander, H. (1977). On the mobility of partial dislocations in silicon. *Phil. Mag.*, **35**, 1523–1536.

Weygand, D., Friedman, L. H., Van der Giessen, E., and Needleman, A. (2002). Aspects of boundary-value problem solutions with three-dimensional dislocation dynamics. *Model. Simul. Mater. Sci. Eng.*, **10**, 437–468.

Weygand, D. and Gumbsch, P. (2005). Study of dislocation reactions and rearrangements under different loading conditions. *Mater. Sci. Eng. A*, **400–401**, 158–161.

Weygand, D., Poignant, M., Gumbsch, P., and Kraft, O. (2008). Three-dimensional dislocation dynamics simulations of the influence of sample size on the stress−strain relation of fcc single-crystalline pillars. *Mater. Sci. Eng. A*, **483–484**, 188–190.

Weygand, D., Senger, J., Motz, C., Augustin, W., Heuveline, V., and Gumbsch, P. (2009). High performance computing and discrete dislocation dynamics: Plasticity of micrometer-sized specimens. In *High Performance Computing in Science and Engineering '08* (ed. W.E. Nagel, D.B. Kröner, and M.M. Rescho), Berlin, pp. 507–523. Springer.

Wickham, L.K., Schwarz, K.W., and Stölken, J.S. (1999). Rules for forest interactions between dislocations. *Phys. Rev. Lett.*, **83**, 4574–4577.

Wolf, H. (1960). Die Aktivierungsenergie für die Quergleitung aufgespaltener Schraubenversetzungen. *Z. Naturforschg.*, **15A**, 180–193.

Woodward, C. (2005). First principles simulations of dislocation cores. *Mater. Sci. Eng. A*, **400–401**, 59–67.

Woodward, C. and Rao, S.I. (2001). Ab initio simulation of isolated screw dislocations in bcc Mo and Ta. *Phil. Mag. A*, **81**, 1305–1316.

Woodward, C. and Rao, S.I. (2002). Flexible ab initio conditions: Simulating isolated dislocations in bcc Mo and Ta. *Phys. Rev. Lett.*, **88**, 216402(1–4).

Woodward, C., Trinkle, D.R., Hector Jr., L.G., and Olmsted, D.L. (2008). Prediction of dislocation cores in aluminum from density functional theory. *Phys. Rev. Lett.*, **100**, 045507(1–4).

Wu, T.-Y., Bassani, J.L., and Laird, C. (1991). Latent hardening in single crystals I. Theory and experiments. *Proc. R. Soc. Lond. A*, **435**, 1–19.

Wu, X., Wang, R., and Wang, S. (2010). Generalized-stacking-fault energy and surface properties for HCP metals: A first-principles study. *Appl. Surf. Science*, **256**, 6345–6349.

Xiang, Y., Cheng, L.-T., Srolovitz, D.J., and Weinan, E. (2003). A level set method for dislocation dynamics. *Acta Mater.*, **51**, 5499–5518.

Xiang, Y. and Vlassak, J.J. (2006). Bauschinger and size effects in thin film plasticity. *Acta Mater.*, **54**, 5449–5460.

Yamashita, Y., Maeda, K., Fujita, K., Usami, N., Suzuki, K., and Fukatsu, S. *et al.* (1993). Dislocation glide motion in heteroepitaxial thin films of $Si_{1-x}Ge/Si(100)$. *Phil. Mag. Lett.*, **67**, 165–171.

Yang, L.H., Söderlind, P., and Moriarty, J. (2001). Accurate atomistic simulations of (a/2)<111> screw dislocations and other defects in bcc tantalum. *Phil. Mag. A*, **81**, 1355–1385.

Yang, L.H., Tang, M., and Moriarty, J.A. (2010). Dislocations and plasticity in bcc transition metals at high pressure. In *Dislocations in Solids* (ed. J.P. Hirth and L. Kubin), Volume 16, pp. 1–47. Elsevier Science B.V., Amsterdam.

Yang, X., He, W., Pingbing, M., and E., Weinan (2008). A generalized Peierls–Nabarro model for curved dislocations and core structures of dislocation loops in Al and Cu. *Acta Mater.*, **56**, 1447–1460.

Yasi, J.A., Hector Jr., L.G., and Trinkle, D.R. (2011). Prediction of thermal cross-slip stress in magnesium alloys from direct first-principles data. *Acta Mater.*, **59**, 5652–5660.

Yasi, J.A., Hector Jr., L.G., and Trinkle, D.R. (2012). Prediction of thermal cross-slip stress in magnesium alloys from a geometric interaction model. *Acta Mater.*, **60**, 2350–2358.

Yasi, J.A., Nogaret, T., Trinkle, D.R., Qi, Y., Hector, L.G., and Curtin, W.A. (2009). Basal and prism dislocation cores in magnesium: Comparison of first-principles and embedded-atom-potential methods predictions. *Model. Simul. Mater. Sci. Eng.*, **17**, 055012(1–13).

Yasin, H., Zbib, H.M., and Khaleel, M.A. (2001). Size and boundary effects in discrete dislocation dynamics. *Mater. Sci. Eng. A*, **309–310**, 294–299.

Yin, J., Barnett, D.M., and Cai, W. (2010). Efficient computation of forces on dislocation segments in anisotropic elasticity. *Model. Simul. Mater. Sci. Eng.*, **18**, 045013(1–13).

Yin, J., Barnett, D.M., Fitzgerald, S.P., and Cai, W. (2012). Computing dislocation stress fields in anisotropic elastic media using fast multipole expansions. *Model. Simul. Mater. Sci. Eng.*, **20**, 045015(1–18).

Zaiser, M. (2006). Scale invariance in plastic flow of crystalline solids. *Adv. Phys.*, **55**, 185–245.

Zaiser, M., Bay, K., and Hähner, P. (1999). Fractal analysis of deformation-induced dislocation patterns. *Acta Mater.*, **47**, 2463–2476.

Zaiser, M. and Hähner, P. (1999). The flow stress of fractal dislocation arrangements. *Mater. Sci. Eng. A*, **270**, 299–307.

Zaiser, M. and Moretti, P. (2005). Fluctuation phenomena in crystal plasticity. *J. Stat. Mech.*, **2005**, P08004(1–17).

Zaiser, M. and Nikitas, N. (2007). Slip avalanches in crystal plasticity: Scaling of the avalanche cut-off. *J. Stat. Mech.*, **2007**, P04013(1–10).

Zaiser, M., Schwerdtfeger, J., Schneider, A.S., Frick, C.P., Clark, B.G., and Gruber, P.A. *et al.* (2008). Strain bursts in plastically deforming molybdenum micro- and nanopillars. *Phil. Mag.*, **88**, 3861–3874.

Zaiser, M. and Seeger, A. (2002). Long-range internal stresses, dislocation patterning and work-hardening in crystal plasticity. In *Dislocations in Solids* (ed. F.R.N. Nabarro and M.S. Duesbery), Volume 11, pp. 1–100. Elsevier Science B.V., Amsterdam.

Zapperi, S., Cizeau, P., Durin, G., and Stanley, H.E. (1998). Dynamics of a ferromagnetic domain wall: Avalanches, depinning transition, and the Barkhausen effect. *Phys. Rev. B*, **58**, 6353–6366.

Zbib, H.M. (2012). CMMS: Downloads, Research (video clips). http://www.cmms.wsu.edu.

Zbib, H.M. and Diaz de la Rubia, T. (2002). A multiscale model of plasticity. *Int. J. Plasticity*, **18**, 1133–1163.

Zbib, H.M., Diaz de la Rubia, T., Rhee, M., and Hirth, J.P. (2000). 3D dislocation dynamics: stress–strain behavior and hardening mechanisms in fcc and bcc metals. *J. Nucl. Mater.*, **276**, 154–165.

Zbib, H.M. and Khraischi, T.A. (2008). Size effects and dislocation–wave interaction in dislocation dynamics. In *Dislocations in Solids.* (ed. J.P. Hirth), Volume 14, pp. 209–249. Elsevier Science B.V., Amsterdam.

Zbib, H.M., Overman, C.T., Akasheh, F., and Bahr, D. (2011). Analysis of plastic deformation in nanoscale metallic multilayers with coherent and incoherent interfaces. *Int. J. Plasticity*, **27**, 1618–1639.

Zbib, H.M., Rhee, M., and Hirth, J.P. (1998). On plastic deformation and the dynamics of 3D dislocations. *Int. J. Mech. Sci.*, **40**, 113–127.

Zhao, D., Huang, J., and Xiang, Y. (2010). A new version Fast Multipole Method for evaluating the stress fields of dislocation ensembles. *Model. Simul. Mater. Sci. Eng.*, **18**, 045006(1–25).

Zheng, H., Cao, A., Weinberger, C.R., Huang, J.Y., Du, K., and Wang, J. *et al.* (2010). Discrete plasticity in sub-10-nm-sized gold crystals. *Nat. Commun.*, **1**, 144(1–8).

Zhou, C., Beyerlein, I.J., and LeSar, R. (2011). Plastic deformation mechanisms of fcc single crystals at small scales. *Acta Mater.*, **59**, 7673–7682.

Zhou, C., Biner, S.B., and LeSar, R. (2010). Discrete dislocation dynamics simulations of plasticity at small scales. *Acta Mater.*, **58**, 1565–1577.

Zhou, C. and LeSar, R. (2012). Dislocation dynamics simulations of plasticity in polycrystalline thin films. *Int. J. Plasticity*, **30–31**, 185–201.

Zhu, T., Li, J., Samanta, A., Leach, A., and Gall, K. (2008). Temperature and strain-rate dependence of surface dislocation nucleation. *Phys. Rev. Lett.*, **100**, 025502(1–4).

Zhu, T. and Li, Y. (2010). Ultra-strength materials. *Progr. Mater. Sci.*, **55**, 710–757.

Zhu, T., Samanta, A., Kim, H.G., and Suresh, S. (2007). Interfacial plasticity governs strain rate sensitivity and ductility in nanostructured materials. *PNAS*, **104**, 3031–3036.

Zikry, M.A. and Kao, M. (1996). Inelastic microstructural failure mechanisms in crystalline materials with high angle grain boundaries. *J. Mech. Phys. Solids*, **44**, 1765–1798.

INDEX